Basic Mathematics for Welding

Len Mrachek
Hennepin Technical Centers

with Frank Arendt
Hennepin Technical Centers

Breton Publishers
a division of Wadsworth, Inc.
North Scituate, Massachusetts

Cover photograph courtesy of the Lincoln Electronic
Company.

BRETON PUBLISHERS
A Division of Wadsworth, Inc.

Library of Congress Cataloging in Publication Data

Mrachek, Leonard, 1936-
   Basic mathematics for welding.

   Includes index.
   1. Welding--Mathematics. I. Title.
TS227.2.M73   512'.12   81-18058
ISBN 0-534-01110-1   AACR2

Printed in the United States of America
1 2 3 4 5 6 7 8 9 - 86 85 84 83 82

Contents

Preface

Basic Mathematics for Welding provides the material
necessary for the welding student to develop the
mathematical skills needed for successful job performance. The text is written both for students who are
first encountering the material and for students who
have covered the material earlier but need to refresh
their knowledge. Both groups of students may suffer
from "math anxiety" and wish to avoid anything that
involves mathematics.

Each chapter is organized into two parts. The
first part is a general, but practical, presentation
of mathematics. The second part contains a series of
applied problems that directly involve welding. These
applied problems give the student an immediate application of the concepts given in the general mathematics section.

Basic Mathematics for Welding may be used in a
traditional classroom setting, in a program of individualized, self-paced instruction, or in a combination of both. The easy-to-read explanations, numerous examples, and actual on-the-job applications
provide an interesting approach for a traditional
classroom setting, while the test program allows students to progress according to individualized levels
of ability. Preview and self-tests are provided at
the beginning of each chapter. Applied welding
survey tests are given at the end of each chapter.
Additional chapter tests are found in the Instructor's
Manual.

Electronic calculators are a part of our life and
a valuable tool for workers in all vocational areas.
While the use of a calculator will vary according to
the individual concerned and the job situation, it is
suggested that a calculator be employed once the basic
manipulative math skills have been mastered.

TO THE STUDENT:

Many students need to learn or review the basic math
skills that are essential for their vocations. When
math fundamentals are learned, the practical skills
should immediately be developed by studying applications in the field. Thus, this text is developed with

two sections: the math section, where the math skills
are learned, and the applied section, where the math
skills are applied to the field of welding.

For the best learning results, the following procedure is recommended for studying <u>Basic Mathematics
for Welding</u>:

- Read the objectives and skim the contents of the
  chapter,
- Complete the Self-Test and self-check (the
  answers for the Self-Test appear in the back of
  the text),
- Study the chapter by reading the chapter, studying the examples, and completing the exercises,
- Complete the Chapter Test (which the instructor
  will provide),
- Study the applied section for the chapter by
  studying the examples and completing the applied
  exercises,
- Complete the Applied Welding Problems Survey Test
- Complete the same procedure for each of the following chapters.

For some students, it will not be necessary to complete
all the exercises. Once the skills have been developed,
the next chapter may be started.

ACKNOWLEDGMENTS

The authors acknowledge and thank the many people who
have contributed to the development of this text. A
special thanks to Merri Lundquist, who typed the original manuscript and made helpful corrections. The
greatest thanks goes to the students who reported what
they need to know and helped develop the whole concept.
The authors have worked with numbers for many years
and realize that unfortunately there will be errors.
For these errors the authors assume all responsibility
and would appreciate and solicit comments and sugges-
tions for the improvement of future editions.

Chapter 1: Whole Numbers
Preview and Self-Test

Name ____

Course/Sec. _________________ Date __________________

Chapter 1 covers whole numbers and the fundamental
operations on whole numbers. Before you turn to the
chapter, complete the Self-Test to determine which
sections in the chapter you need to study carefully.

Self-Test

READING AND WRITING WHOLE NUMBERS

1-1. Write ninety-nine in digits.   1-1. ________________

1-2. Write 66 in words.   1-2. ________________

1-3. Write forty-two million, forty-two in digits.   1-3. ________________

     Reading and writing whole numbers score   ________________

ADDITION OF WHOLE NUMBERS

```
1-4.    424      1-5.      29      1-6.    5280          1-4. ________________
       + 67               392              280
                        + 6740           +  28          1-5. ________________

                                                        1-6. ________________
```

     Addition of whole numbers score   ________________

SUBTRACTION OF WHOLE NUMBERS

```
1-7.   74569     1-8.    146931                         1-7. ________________
      - 21989           - 64391
                                                        1-8. ________________
```

     Subtraction of whole numbers score   ________________

MULTIPLICATION OF WHOLE NUMBERS

1-9.   679   1-10.   5280   1-9. ___________
       x 9   x 17   1-10. ___________

                    Multiplication of whole numbers score ___________

DIVISION OF WHOLE NUMBERS

1-11.   7⟌2156   1-12.   4304⟌327104   1-11. ___________

                                                          1-12. ___________

                    Division of whole numbers score   ___________

FINDING AVERAGES OF WHOLE NUMBERS

1-13.   Find the average of the following math test   1-13. ___________
        scores: 91, 86, 79, 98, 85 and 95.

1-14.   A 52-acre field produced 1924 bushels of wheat.   1-14. ___________
        Find the average number of bushels per acre.

                    Finding averages of whole numbers score ___________

PRACTICAL APPLICATIONS OF WHOLE NUMBERS

1-15.   If a tank holds 500 gal, how much is left after   1-15. ___________
        379 gal have been used?

1-16.   There are 5280' in a mile. How many feet are   1-16. ___________
        there in 7 mi?

1-17.   If you use 24 gal of gasoline to travel 768 mi,   1-17. ___________
        what is your rate in miles per gallon?

1-18.   A packaging machine will fill 5220 bottles in an   1-18. ___________
        hour. How many bottles can the machine fill per
        minute?

1-19.   How much wire is left on a 500' roll if lengths   1-19. ___________
        of 12', 27', 92', and 311' have been used?

1-20.   If you earned $597 for a landscaping job that   1-20. ___________
        took 34 hours, and you spent $223 for plants and
        materials, how much profit did you earn per hour?

                    Practical applications of whole numbers score ___________

After you have checked your answers on the Self-Test,
transfer your scores to the Chapter 1 Objectives.

Chapter 1

••

Whole Numbers

Upon successful completion of this chapter, you will be
able to read and write whole numbers; perform fundamental operations on whole numbers, including addition,
subtraction, multiplication, and division; find averages;
and make practical applications of whole numbers.

Objectives

	Self-Test Scores	If you did poorly on the Self-Test, turn to:
• Reading and writing whole numbers	________	Section 1-2
• Addition of whole numbers	________	Section 1-3
• Subtraction of whole numbers	________	Section 1-4
• Multiplication of whole numbers	________	Section 1-5
• Division of whole numbers	________	Section 1-6
• Finding averages of whole numbers	________	Section 1-7
• Practical applications of whole numbers	________	Section 1-8

1–1   Introduction to Whole Numbers

Numbers play an important part in our everyday lives.
Most occupations require the development of technical
skills based on the ability to read, write, and work
with whole numbers.

You must also know how to read and write numbers.
When writing a check, for example, you must be able to
spell out numbers. You need to know the procedures of
addition, subtraction, multiplication, and division to
be able to perform the fundamental operations on whole
numbers. Although we are living in an age of computers
and electronic calculators, much of the simple arithmetic used in business and industry is done by hand. As
a matter of fact, many trade and technical jobs require
that you "show" you can do arithmetic calculations
before you get the job. Thus, learning the fundamental
operations on whole numbers is vitally important.

1–2   Reading and Writing Whole Numbers

Every number can be written by using one or a combination of ten digits: 0, 1, 2, 3, 4, 5, 6, 7, 8, and 9.
These digits are symbols that name the whole numbers
from zero through nine.

Our number system is called the "decimal" (meaning
ten) number system because it uses ten digits. This
system is the Hindu-Arabic number system that was
developed in India about 500 B.C. It is used in most
countries of the world, including the United States.
The ten digits and their words names are:

0	zero	5	five
1	one	6	six
2	two	7	seven
3	three	8	eight
4	four	9	nine

Many whole numbers contain two or more digits
written in definite places. When numbers of more than
four digits are written, the numbers may be separated by
commas in groups of three, starting at the right, to
make reading easier. An example is 5,837,261.

Some two-digit numbers have a one-word name. These
numbers and their word names are:

10	ten	20	twenty
11	eleven	30	thirty
12	twelve	40	forty
13	thirteen	50	fifty
14	fourteen	60	sixty
15	fifteen	70	seventy
16	sixteen	80	eighty
17	seventeen	90	ninety
18	eighteen		
19	nineteen		

The word names for all other two-digit numbers contain a hyphen (-). For example, the number 44 is written "forty-four," and the number 69 is written "sixty-nine."

The names of the first ten places in whole numbers, starting at the right, are: ones or units, tens, hundreds, thousands, ten thousands, hundred thousands, millions, ten millions, hundred millions, and billions:

<pre>
billions
  hundred millions
    ten millions
      millions
        hundred thousands
          ten thousands
            thousands
              hundreds
                tens
                  ones (units)

6,   5   3   9,   8   2   9,   4   7   5
</pre>

Thus, the number 6,539,829,475 is read as "six billion, five hundred thirty-nine million, eight hundred twenty-nine thousand, four hundred seventy-five."

Example 1–1

PROBLEM

Write the following numbers in words:

474, 5280, and 7,142,572

ANSWER

```
      474 = four hundred seventy-four
     5280 = five thousand, two hundred eighty
7,142,572 = seven million, one hundred forty-two
            thousand, five hundred seventy-two
```

Example 1–2

PROBLEM

Write the following numbers in digits:

    eighty-nine
    forty-two thousand, three hundred fifty-six
    seven billion, forty-six thousand, twenty-nine

ANSWER

```
eighty-nine = 89
forty-two thousand, three hundred fifty-six = 42,356
seven billion, forty-six thousand, twenty-nine =
    7,000,046,029
```

Exercises: Writing Numbers in Words and Digits

Write the following numbers in words:

1-1. 62   1-2. 733   1-3. 59,999

1-4. 218,469,572   1-5. 92,000,539

Write the following numbers in digits:

1-6.   three thousand, nine hundred thirty-three

1-7.   nineteen hundred eighty-four

1-8.   five hundred sixty-eight thousand, six

1-9.   seventy-seven thousand, fifteen

1-10. eight million, four hundred sixteen thousand,
       five hundred twenty-two

1–3   Addition of Whole Numbers

Addition is the process of combining, or "putting
together," two or more numbers to get a total amount.
The numbers we add are called addends. The result of
addition is called the sum. The sign for addition is
the plus sign: +.
       Addition may be written horizontally (on the line)
as follows:

       88 + 9 + 56

Addition may also be written vertically (up and down):

```
   88
    9
 +56
```

       To find the sum of these numbers, first add the
ones (or units) column:

       8 + 9 + 6 = 23

Write 3 in the ones column and carry the 2--that is,
transfer the 2 to the tens column. Then add the tens
column:

       2 + 8 + 5 = 15

Write the 5 in the tens column and the 1 in the hundreds
column. The answer is 153:

```
    2
   88
    9
 +56
  153
```

The sum of an addition problem can be checked by adding each column upward and following the same procedure.

Procedure for Adding Whole Numbers

• •

1. Write the numbers one under another with each digit in its proper column.
2. Add all the numbers in the ones column.
3. Write the ones digit in the ones column.
4. Carry the tens digit(s) to the tens column and add it with the rest of the numbers in the column.
5. Continue the same procedure with the remaining columns.

• •

Example 1–3

PROBLEM

Add 9667, 829, 916, and 608.

SOLUTION

1. Arrange the addends, or numbers to be added, in their correct columns:

```
  9667
   829
   916
+  608
```

2. Add the digits in the units column:

    $7 + 9 + 6 + 8 = 30$

    Write 0 in the ones column and carry the 3.

3. Add the digits in the tens column:

    $3 + 6 + 2 + 1 + 0 = 12$

    Write 2 in the tens column and carry the 1.

4. Add the digits in the hundreds column:

    $1 + 6 + 8 + 9 + 6 = 30$

    Write 0 in the hundreds column and carry the 3.

5. Add the digits in the thousands column:

    $3 + 9 = 12$

    Write 2 in the thousands column and 1 in the ten

thousands column.   Your work should look like the
following:

```
   313
  9667
   829
   916
 + 608
 12020
```

ANSWER

    9667 + 829 + 916 + 608 = 12,020

The sum is read as "twelve thousand, twenty."

Example 1–4

PROBLEM

Add 5280, 1798, 444, and 12 and check the result.

SOLUTION

1.   Arrange the addends in their correct columns:

```
   5280
   1798
    444
 +   12
```

2.   Add the ones:

    0 + 8 + 4 + 2 = 14

    Write 4, carry 1.

3.   Add the tens:

    1 + 8 + 9 + 4 + 1 = 23

    Write 3, carry 2.

4.   Add the hundreds:

    2 + 2 + 7 + 4 = 15

    Write 5, carry 1.

5.   Add the thousands:

    1 + 5 + 1 = 7

    Write 7.   Your work should look like the following:

```
   121
  5280
  1798
   444
 +  12
  7534
```

ANSWER

    5280 + 1798 + 444 + 12 = 7534

CHECK

Add the columns upward.

1. Add the ones upward:

    2 + 4 + 8 + 0 = 14

    Write 4, carry 1.

2. Add the tens upward:

    1 + 1 + 4 + 9 + 8 = 23

    Write 3, carry 2.

3. Add the hundreds upward:

    2 + 4 + 7 + 2 = 15

    Write 5, carry 1.

4. Add the thousands upward:

    1 + 1 + 5 = 7

    Write 7. Your work should look like the following:

```
  ₁ ₂ ₁
  5280
  1798
   444
+   12
  7534
```

The results are the same, so the answer is correct.

Exercises: Adding Whole Numbers

Add the following:

1-11.   913 + 843   1-12.   203 + 515 + 892

1-13.   5793 + 429 + 792   1-14.   7502 + 95 + 991

1-15.   917 + 659 + 39   1-16.   5018 + 23 + 700
                                      + 69 + 321

Add the following and check:

1-17.   5015 + 727 + 39   1-18.   32 + 169 + 7765

1-19.   108127   1-20.   16927
          16527   5911
           7744   117
        +   329   +   36

1-21.   The owner of Phinque's Fashions worked the fol-
        lowing number of hours per week: 43, 47, 38, and
        44. How many hours did he work in all four weeks?

1-22.   The weights of the starting backfield of
        Alexander High School are 202 lb, 167 lb,
        171 lb, and 189 lb. What is the total weight
        of the starting backfield?

1-23.   A carpenter used the following lengths of 2 x 4s:
        16', 24', 8', 10', 14', and 12'. How many feet
        did he use?

1-24.   A tire salesperson drove 142 kilometers (km) on
        Monday, 82km on Tuesday, 121km on Wednesday,
        92km on Thursday, and 207km on Friday. Find the
        total number of kilometers driven during the week.

1-25.   The attendance on January 10, 1979, at five
        National Hockey League games was as follows:

        Colorado-Rangers game   5,923
        Pittsburgh-Canadians game   10,065
        Detroit-Islanders game   14,475
        Chicago-Kings game   8,221
        Toronto-North Stars game   16,361

        What was the total attendance at all five games?

1–4   Subtraction of Whole Numbers

Subtraction is the process of finding the difference
between two numbers. This process is the reverse of
addition. The smaller number placed on the bottom is
called the subtrahend. It is subtracted from the larger
number on top, which is called the minuend. The result
of subtraction is called the difference, or remainder.
The sign for subtraction is the minus sign: -. These
elements are shown in the following example:

        48   (minuend)
       -26   (subtrahend)
        22   (difference, or remainder)

     Occasionally, the digit in the bottom number
(subtrahend) is larger than the digit in the top number
(minuend). In such a case, we borrow from the digit
to the left. Borrowing is shown in the examples that
follow.

Procedure for Subtracting Whole Numbers
••

1. Write the larger number (minuend) on top.
2. Write the number to be subtracted (subtrahend) below
    it and place the digits in their proper columns.

3. Start with the ones column and subtract the number
    in the subtrahend from the number in the minuend.
    Write the result in the ones column below the
    problem rule.
4. Continue the same procedure with the tens, hundreds,
    thousands, and so on. Borrow when necessary to
    obtain the remainder, or the difference.
5. Check the result by adding the difference and the
    subtrahend to obtain the minuend.

●●

Example 1–5

PROBLEM

Subtract 446 from 739.

SOLUTION

1. Arrange the numbers by placing the digits in their
    proper columns, the larger number (minuend) above the
    smaller (subtrahend):

         739
        -446

2. Starting with the ones column, subtract 6 from 9:

    9 - 6 = 3

    Write 3 in the ones column.

3. In the tens column, attempt to subtract 4 from 3.
    Because 4 is larger than 3, we must borrow from the
    7 in the hundreds column. To borrow, draw a diago-
    nal line across the 7, write 6 above it, and place
    the borrowed 1 to the right of the 3. The 3 thus
    becomes 13. Now subtract 4 from 13:

    13 - 4 = 9

    Write 9 in the tens column.

4. Finally, in the hundreds column, subtract 4 from 6:

    6 - 4 = 2

    Write 2 in the hundreds column:

          6 1
         739
        -446
         293   (difference, or remainder)

ANSWER

    739 - 446 = 293

CHECK

Add the difference and the subtrahend:

```
  293  (difference)
+446  (subtrahend)
 739  (minuend)
```

Because the result, 739, equals the minuend, the answer
is correct.

Example 1–6

PROBLEM

Subtract 597 from 8008.

SOLUTION

1. Arrange the numbers by placing the digits in their
    proper columns, the larger number (minuend) above
    the smaller (subtrahend):

    ```
     8008  (minuend)
    - 597  (subtrahend)
    ```

2. Starting in the ones column, subtract 7 from 8:

    $8 - 7 = 1$

    Write 1 in the ones column.

3. Borrowing is necessary to subtract 9 from 0 in the
    tens column. However, we cannot borrow from the 0 in
    the hundreds column, so we borrow from the thousands
    column instead.

4. Complete the subtraction for each column:

    ```
     7 9 9
     8008
    - 597
     7411  (remainder, or difference)
    ```

ANSWER

$8008 - 597 = 7411$

CHECK

Add the difference and the subtrahend:

```
 7411  (difference)
+ 597  (subtrahend)
 8008  (minuend)
```

Because the sum, 8008, equals the minuend, the answer is
correct.

Exercises: Subtracting Whole Numbers

Subtract the following and check:

1-26.   87 - 43   1-27.   90 - 58

1-28.   812 - 97   1-29.   7070 - 582

1-30.   8002 - 193   1-31.   8148 - 579

1-32.   8764 - 4125   1-33.   6000 - 4687

1-34.   Find the difference between 8612 and 1981.

1-35.   What is the difference between 30,004 and 28,647?

1-36.   An electrician used 87' of a 100' roll of wire.
        How much wire was left on the roll?

1-37.   A loaded grain truck weighs 28,674 lb. The
        truck, when empty, weighs 7854 lb. How many
        pounds of grain are on the truck?

1-38.   The odometer in Medora's car reads 13,984 mi.
        If the warranty is good for 25,000 mi, how many
        miles does she have left on warranty?

1-39.   The height of Mount McKinley in Alaska is 20,320'
        above sea level. The height of Mount Washburn in
        Yellowstone National Park is 10,243'. What is
        the difference in their heights?

1–5   Multiplication of Whole Numbers

Multiplication is a rapid or short-cut method of
repeated addition. When two numbers are multiplied, the
top number is called the multiplicand and the bottom
number is called the multiplier. The answer obtained by
multiplication is called the product. The sign for
multiplication is the times sign: x.
    The product of two numbers is the same even if
their positions are switched:

        8   (multiplicand)   7   (multiplicand)
       x7   (multiplier)   x8   (multiplier)
       ──   ──
       56   (product)   56   (product)

Knowing the multiplication table through 12 x 12 is
necessary for easy multiplication of all numbers.

Procedure for Multiplying Whole Numbers

•••

1. Write the larger number (multiplicand) on top.
2. Write the smaller number (multiplier) below. Place
    the digits in their proper columns.

3. Multiply the digit in the ones, or units, column of the multiplicand by the digit in the units column of the multiplier. Write the ones result in the units column and carry the tens.
4. Multiply the digit in the tens column of the multiplicand by the digit in the units column of the multiplier. Add the tens remainder to this product.
5. Write the first digit on the right in the result in the tens column. Carry any digits.
6. Continue to multiply every digit in the multiplicand by every digit in the multiplier. Be sure to move one column to the left for each digit in the multiplier.
7. Add the results of each column.
8. Check the results by reversing the multiplicand and the multiplier and multiplying.

••

Example 1–7

PROBLEM

Multiply 39 by 7.

SOLUTION

1. Write the 39 on top (multiplicand) and the 7 on the bottom (multiplier):

    ```
     39   (multiplicand)
    x 7   (multiplier)
    ```

2. Multiply 7 by 9:

    7 x 9 = 63

    Write the 3 in the ones, or units, column and carry the 6.

3. Multiply 7 by 3 and add 6:

    7 x 3 + 6 = 27

    Write 7 in the tens column and 2 in the hundreds column:

    ```
      6
     39
    x 7
    273   (product)
    ```

ANSWER

    39 x 7 = 273

Example 1–8

PROBLEM

Multiply 96 by 48.

SOLUTION

1. Write the 96 on top (multiplicand) and the 48 on the
    bottom (multiplier):

    ```
     96  (multiplicand)
    x48  (multiplier)
    ```

2. Multiply 8 by 6:

    8 x 6 = 48

    Write the 8 in the ones, or units, column and carry
    the 4.

3. Multiply 8 by 9 and add 4:

    8 x 9 + 4 = 76

    Write 6 in the tens column and 7 in the hundreds
    column.

4. Multiply 4 by 6:

    4 x 6 = 24

    Write the 4 under the 6 and carry the 2.

5. Multiply 4 by 9 and add 2:

    4 x 9 + 2 = 38

    Write 8 under the 7 and write 3 to the left of the 7.

6. Add the two numbers to obtain the product:

    ```
       96
    x  48
    ─────
      768
     384
    ─────
     4608   (product)
    ```

ANSWER

    96 x 48 = 4608

Exercises: Multiplying Whole Numbers

Multiply the following and check:

1-40. 49 x 8   1-41. 76 x 7   1-42. 39 x 28

1-43. 97 x 83   1-44. 496 x 27   1-45. 4500 x 42

1-46. 5697 x 274   1-47. 4080 x 386

1-48. Multiply 888 by 87.

1-49. Find the product of 9090 and 606.

1-50.   How many trees are in an apple orchard if there
        are 36 trees in a row and there are 25 rows?

1-51.   A machine shop spends $9 a day for cutting oil.
        How much will the shop spend in a year (365 days)?

1-52.   How many miles would you drive to and from work
        in 240 workdays if a round trip is 19 mi?

1-53.   A new car weighs 4050 lb. What is the total
        weight of 8 new cars?

1-54.   How many people attend the 6 home games of the
        Fighting Irish of the University of Notre Dame if
        stadium capacity is 59,075 people, and all the
        games are sellouts?

1–6   Division of Whole Numbers

Division is a rapid form of subtraction or a reverse
multiplication process. Division is a mathematical
procedure for finding out how many times one number
goes into another. The signs for division are used in
the following examples:

$$8 \div 4, \quad \frac{8}{4}, \quad 8/4, \quad \text{and} \quad 4\overline{)8}$$

All indicate that 8 is to be divided by 4.
   The mathematical phrase "15 ÷ 5" is read as "fifteen
divided by 5." The phrase asks how many 5s there are
in 15.
   The result of finding how many times one number,
the dividend, contains another number, the divisor, is
called the quotient. So 15 ÷ 5 can be written as follows:

$$\frac{15 \ \text{(dividend)}}{5 \ \text{(divisor)}} = 3 \ \text{(quotient)} \quad \text{or}$$

$$5 \ \text{(divisor)} \ \overline{)15} \ \text{(dividend)} \quad \overset{3 \ \text{(quotient)}}{}$$

   Short division is used when the divisor is a single
digit and the division can be carried out mentally without writing down intermediate steps. Long division is
performed when the divisor is a larger number and the
successive steps of division cannot be done mentally and
must be written down. Short and long division are shown
in the examples.

Procedure for Dividing Whole Numbers

• •

1.   Write the number to be divided as the dividend
     within the division sign as shown:

     $$\text{divisor}\overline{)\text{dividend}}$$

2. Determine how many times the numerals in the first few digits of the dividend can be divided by the divisor. This guess is called the <u>trial quotient</u>.
3. Multiply the trial quotient by the <u>divisor</u>. If the product is larger than the first digits of the dividend, then use a smaller number as a divisor. Write the quotient above the last digit of the number being divided.
4. Subtract this product from the dividend and bring down the next digit. Find a new trial quotient and repeat the process until all the digits in the dividend are used.
5. If the divisor does not divide into the dividend evenly, the resulting number is called the <u>remainder</u>.
6. <u>Check</u> the result by multiplying the quotient by the <u>divisor</u>. Then add the remainder to obtain the dividend.

• •

Example 1–9

PROBLEM

Divide 252 by 6.

SOLUTION

<u>Note</u>: Because 6 is a simple divisor, we can use short division.

1. Write $252 \div 6$ by using a division rule:

    (divisor) $6\overline{)252}$ (dividend)

2. Determine how many times 6 will go into 25. Recall that $6 \times 4 = 24$, so 4 is the first digit of the quotient. Write 4 over the 5 in the dividend.

3. Mentally subtract 24 from 25 and place the 1 above and to the right of the last 2 in the dividend.

4. Determine how many times 6 will go into 12. Recall that $6 \times 2 = 12$, so 2 is the next digit of the quotient. Write 2 over the 2 in the dividend:

    $$\begin{array}{r} 42 \text{ (quotient)} \\ 6\overline{)252} \end{array}$$

ANSWER

    $252 \div 6 = 42$

CHECK

Multiply 42 (quotient) by 6 (divisor) to obtain 252 (dividend).

Example 1–10

PROBLEM

Divide 24,486 by 42.

SOLUTION

<u>Note</u>: Because the numbers are not simple single digits, we must use long division.

1. Write 24,486 ÷ 42 by using a division rule:

    42)‾2‾4‾4‾8‾6‾

2. Determine how many times 244 can be divided by 42. The trial quotient appears to be 6.

3. Multiply 42 by 6:

    42 x 6 = 252

    Since 252 is greater than 244, we must reduce the trial quotient to 5.

4. Multiply 42 by 5:

    42 x 5 = 210

    Write 5 as the first digit of the quotient above the last 4 in 244 and subtract 210 from 244.

5. Bring down the 8 and determine how many times 348 can be divided by 42. The trial quotient appears to be 8.

6. Multiply 42 by 8:

    42 x 8 = 336

    Write 8 as the second digit of the quotient (over the 8) and subtract 336 from 348. So far, the work is as follows:

    ```
          58
    42)24486
        210
        ‾‾‾
        348
        336
        ‾‾‾
         12
    ```

7. Bring down the 6 and determine how many times 42 divides 126. The trial quotient appears to be 3.

8. Multiply 42 by 3:

    42 x 3 = 126

    Write 3 as the third digit of the quotient (over the 6) and subtract 126 from 126:

```
          583
    42)24486
       210
        348
        336
         126
         126
           0
```

ANSWER

    24,486 ÷ 42 = 583

CHECK

Multiply 583 (quotient) by 42 (divisor) to obtain 24,486 (dividend).

Example 1–11

PROBLEM

What is the quotient of 55,681 divided by 752?

SOLUTION

1. Write the problem using a division rule:

    ```
    752)55681
    ```

2. Determine how many times 5568 can be divided by 752. The trial quotient appears to be 7.

3. Multiply 752 by 7:

    752 x 7 = 5264

    Write 7 as the first digit of the quotient (over the 8 in the dividend) and subtract 5264 from 5568.

4. Bring down the 1 and determine how many times 752 will divide 3041. The trial quotient appears to be 4.

5. Multiply 752 by 4:

    752 x 4 = 3008

    Write 4 as the second digit of the quotient and subtract 3008 from 3041. The remainder is 33.

6. In the quotient, write r = 33:

    ```
           74   r = 33
    752)55681
       5264
        3041
        3008
          33
    ```

ANSWER

    55,681 ÷ 752 = 74   r = 33

CHECK

Multiply 752 by 74 and add 33 to obtain the dividend,
55,681:

```
      752
    x  74
    ------
     3008
    5264
    -----
    55648
    +  33
    -----
    55681
```

Exercises: Dividing Whole Numbers

Divide the following and check:

1-55.   8)576   1-56.   8)1728   1-57.   2919 ÷ 3

1-58.   1576 ÷ 8   1-59.   44,154 ÷ 9 1-60.   12)1728

1-61.   35)2625   1-62.   23)19872

1-63.   80,032 ÷ 976   1-64.   148,610 ÷ 385

1-65.   If you earn $210 in a 5-day work week, how much
        do you earn per day?

1-66.   How many cars would a salesperson need to sell to
        make $5000 if he or she makes $625 on each car?

1-67.   Kathy Quick can type 82 words per minute. How
        long will it take her to type a 738-word letter?

1-68.   If you get 690 mi on 46 gal of gas, how many
        miles per gallon are you getting?

1-69.   Rod Carew's contract with the California Angels
        calls for $4 million for 5 years. If he plays
        162 games per year, how much is he paid per game?

1–7   Finding Averages of Whole Numbers

The _average_ of a group of numbers is the number that
best represents the group of numbers. The average of a
group of whole numbers is found by adding all the whole
numbers, or addends, and then dividing the sum by the
number of addends:

$$\text{average} = \frac{\text{sum of the addends}}{\text{number of addends}}$$

The average is sometimes called the _arithmetic average_,
the _arithmetic mean_, or the _mean_.

Procedure for Finding the Average of Whole Numbers

1. Make sure each quantity is of the same unit of measure.
2. Add all quantities, or addends.
3. Divide the sum by the number of quantities added.
4. If there is a remainder and it is greater than half the divisor, round the quotient to the next unit.

Example 1–12

PROBLEM

Find the average temperature in Honolulu for 5 days if the temperature on Monday is 78°; Tuesday, 84°; Wednesday, 80°; Thursday, 86°; and Friday, 87°.

SOLUTION

Add the temperatures and divide by 5:

```
     78
     84
     80
     86
    +87
 5)‾415‾
     83
```

ANSWER

The average temperature is 83°.

Example 1–13

PROBLEM

Find the average attendance at the following National Basketball Association games on January 17, 1979:

Houston-Golden State	11,393
San Diego-Los Angeles	13,703
Atlanta-Portland	5,111
New Jersey-Detroit	2,836
Seattle-Indiana	9,280
Philadelphia-Cleveland	7,184
Washington-New Orleans	10,412
Kansas City-Boston	6,583
San Antonio-Denver	15,219
Total	81,721

SOLUTION

Divide the total attendance figures by 9:

```
       9080   r = 1
   9) 81721
      81
      ‾‾‾
      072
       72
      ‾‾‾
       01
```

ANSWER

The average attendance at the 9 NBA games was 9080. We can disregard the remainder because it is less than half the divisor.

Exercises: Finding the Average of Whole Numbers

Find the average of the following:

1-70.   8 and 14   1-71.   2, 4, and 6

1-72.   86, 94, and 96   1-73.   212, 240, and 262

1-74.   8 hours, 12 hours, 10 hours, 11 hours, and 9
        hours

1-75.   A student earned the following grades in math:
        85, 95, 96, 84, 86, 98, and 100. Determine the
        student's average grade.

1-76.   Find the average yield per acre if a 42-acre
        field produces 1554 bushels of wheat.

1-77.   Pete Maravich scored the following points during
        a road trip: 33, 23, 23, and 25. Find his
        average score per game.

1-78.   Find the average age of a family if the father
        is 42, the mother is 39, and the children are 15,
        13, 11, 9, and 4.

1-79.   The five largest countries in the world are the
        Soviet Union, 8,599,300 sq mi; Canada, 3,851,809
        sq mi; China, 3,691,500 sq mi; the United States,
        3,675,633 sq mi; and Brazil, 3,286,478 sq mi.
        Find the average size of these countries.

1–8   Practical Applications of Whole Numbers

Solving many everyday practical problems requires the use of whole numbers. The examples that follow illustrate some common applications.

Example 1–14

PROBLEM

A small painting contractor bid a job at $1500. His expenses were labor, $660; paint, $456; and various supplies and equipment, $127. What was left for profit?

SOLUTION

1. Add all the expenses:

```
     660  labor
     456  paint
    +127  supplies
    1243  expenses
```

2. Subtract the sum from the $1500 bid to find the profit:

```
     1500  bid
    -1243  expenses
      257  profit
```

ANSWER

His profit was $257.

Example 1–15

PROBLEM

The taxes on a 6-unit apartment building increased from $2056 to $5080 in one year. Find the increase in taxes per unit per month.

SOLUTION

1. First determine the amount of increase by sub-
    tracting:

```
     5080
    -2056
     3024
```

2. Then divide 6 into the increase to find the increase
    per unit per year:

```
      504   (per unit per year)
    6 )3024
```

3. Finally, divide $504 by 12 to determine the increase
    per unit per month:

```
       42
    12)504
```

ANSWER

The increase in taxes per unit per month was $42.

Example 1–16

PROBLEM

A mid-size car costs $8000 and gets 20 miles per gallon (mpg) of gasoline. A smaller foreign compact car costs $9600 and gets 40 mpg. If you drive 25,000 mi, what is the total cost for each car when gasoline costs $2 per gallon?

SOLUTION

1. Find the number of gallons of gasoline used for each car by dividing the miles per gallon into 25,000:

    Mid-size car   Compact car

    $$20 \overline{)25000} = 1250 \qquad 40 \overline{)25000} = 625$$

2. Multiply the quotients by $2 to find the gasoline cost for each car. Then add the car cost to the gasoline cost for each car to find the total cost:

Mid-size Car		Compact car	
1250		625	
x 2		x 2	
2500	gasoline cost	1250	gasoline cost
8000	car cost	9600	car cost
10500	total cost	10850	total cost

ANSWER

The total cost for the mid-size car is $10,500, and the total cost for the compact car is $10,850.

Exercises: Practical Applications of Whole Numbers

Solve the following:

1-80.   Taxes on a home increased from $1247 to $1631 per year. By how much will this increase raise the monthly payment?

1-81.   If each step is 7" high, what is the rise in 15 steps?

1-82.   Find the total cost of a business trip if the following charges are made: airline ticket, $259; hotel room for 5 days at $51 per day; rental car, $97; and food, $110.

1-83.   How long will it take to drive 624 mi if you average 52 miles per hour (mph) and allow 3 hours for refueling and meals?

1-84.   A master electrician receives $17 per hour. How
        much does he earn in a 50-week year if he works
        40 hours per week?

1-85.   Referring to Example 1-16, determine the costs of
        driving 40,000 mi for the mid-size and the compact
        cars.

1-86.   Kay Cobb, an interior decorator, hired an elec-
        trical installer to install new sewing machines.
        He worked for 7 hours. If the total bill was
        $671 and the cost of equipment $587, find his
        hourly rate.

1-87.   Find the total cost of a home entertainment
        center if you pay $75 down and make 24 payments
        of $55.

1-88.   Sunburst Chemical Company sells a special drain
        cleaner by the drum. How many 231-cubic inch
        (cu in) containers can be filled from a drum
        that contains 12,705 cu in?

1-89.   How long would a 1-acre lot need to be if it were
        165' wide? (1 acre = 43,560 sq ft.)

1-90.   Jeff Maki delivered newspapers according to the
        following times (in minutes): Monday, 51;
        Tuesday, 45; Wednesday, 56; Thursday, 62; Friday,
        37; Saturday, 56; and Sunday, 64. What was his
        average delivery time?

1-91.   How many boards will be needed for the floor of
        a 10' deck if the decking is 5-1/2" wide and a
        space of 1/2" is allowed for each board?

1-92.   Find the average number of bushels per load if
        the weights (in pounds) of the truck loads are as
        follows: 12,720, 13,200, 13,320, 13,500, and
        12,600. (Wheat weighs 60 lb per bushel.)

1-93.   You can buy a full-size car for $9900 or a sporty
        compact for $11,950. If the cost of gasoline is
        $2 per gallon, how many gallons of gas could you
        buy for the difference in price of the cars?

1-94.   Refer to Problem 1-91 and determine the total
        length of decking boards needed for the deck.

1-95.   Find the monthly premium per employee, if the
        company pays $7752 annually for 17 employees.

Chapter 1

Whole Numbers

•••

Solving Applied Welding Problems

Now that you have mastered the fundamental operations
on whole numbers, you are ready to apply these math
skills to typical problems that welders encounter on the
job. After studying the examples that follow, solve the
applied welding problems.

Example 1–17

PROBLEM

A metal rack contains 357 lb of angle iron, 1015 lb of
bar stock, and 869 lb of round stock. What is the total
weight of the metal in the rack?

SOLUTION

Arrange the quantities to be added in their proper
columns and find the sum:

```
    357  lb of angle iron
   1015  lb of bar iron
 +  869  lb of round iron
   2241  lb of metal
```

ANSWER

The total weight of the metal in the rack is 2241 lb:

    357 + 1015 + 869 = 2241

Example 1–18

PROBLEM

What is the inside dimension of the frame constructed
of 2" square tubing shown in Figure 1-1?

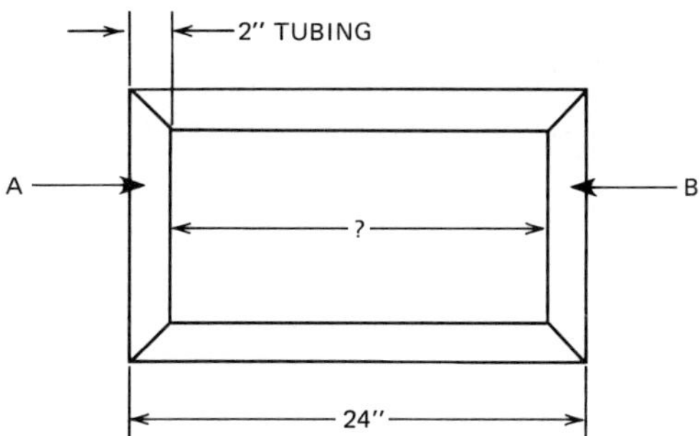

Figure 1—1

SOLUTION

Subtract the total width of the two sides of tubing from
the outside dimension of 24":

```
  2"  (side A)            24"  (outside dimension)
 +2"  (side B)            -4"  (total width of tubing)
  4"  (width of tubing)   20"
```

ANSWER

The inside dimension of the frame is 20":

    24 - 4 = 20

Example 1–19

PROBLEM

Gussets are triangular-shaped pieces of metal used for
support. If a barge requires 1865 gussets, how many
gussets are needed for 59 barges?

SOLUTION

Multiply the number of gussets needed for one barge
(1865) by the number of barges (59):

```
     1865
     x 59
    16785
     9325
   110035
```

ANSWER

The total number of gussets needed for 59 barges is
110,035:

    1865 x 59 = 110,035

Example 1–20

PROBLEM

1260 hand trucks are welded by 20 welders. How many
trucks does each welder weld?

SOLUTION

Divide the number of hand trucks by the number of welders:

```
        63
20) 1260
    120
     60
     60
      0
```

ANSWER

Each welder welds 63 trucks:

    1260 ÷ 20 = 63

Example 1–21

PROBLEM

A 450' length of pipe is divided into 25 equal pieces. How long is each piece?

SOLUTION

Divide the total length of pipe by the number of equal pieces:

```
      18
25) 450
    25
   200
   200
     0
```

ANSWER

Each piece is 18' long:

    450 ÷ 25 = 18

Applied Welding Problems

1-96. The electrode oven in a shop has a storage capacity of 2173 low-hydrogen electrodes and 1989 stainless steel electrodes. What is the total capacity of the oven?

1-97. A construction project required one thousand, nine hundred seventy-nine lengths of channel iron; five hundred forty-three lengths of I beam; and seven hundred sixty-three lengths of angle iron. How many lengths are needed to complete the job?

1-98.   The following hours were spent to repair a water tower: six thousand, two hundred fifty-one hours on cutting out and fitting in new sections; three thousand, nine hundred forty-eight hours on welding; and eight hundred thirty-two hours on grinding and cleanup. How many hours did the repair job take?

1-99.   A job shop had an order for 41,685 brackets. One thousand, six hundred ninety-nine brackets were added to the order on Tuesday. On Friday, 2730 brackets were added to the order. What was the total number of brackets on order?

1-100. The tubing frame in Figure 1-2 is constructed of 2" tubing.
   a.   What are the outside dimensions of the tubing frame?
   b.   How many inches of 2" angle iron are needed to build the frame?

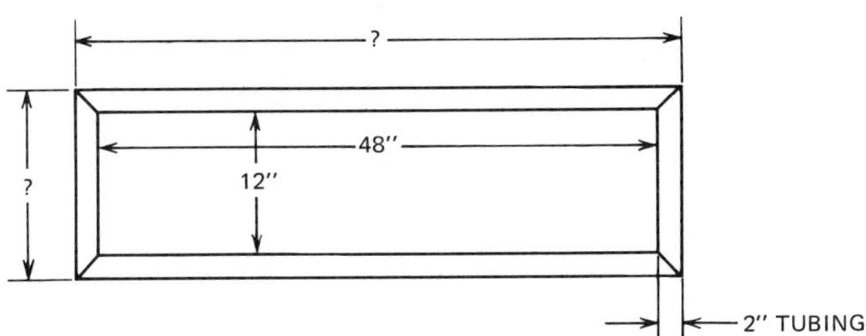

Figure 1—2

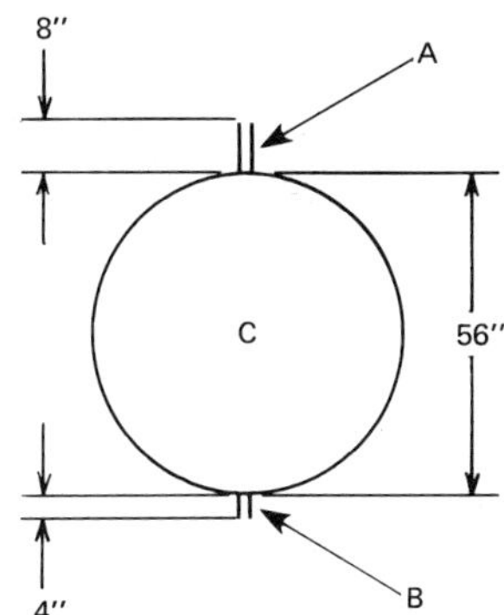

Figure 1—3

1-101. The tank in Figure 1-3 has an outside diameter of 56". An 8" pipe is welded on the top side, and a 4" pipe is welded on the bottom side.
   a.   What is the total height of pipes A and B and tank C?
   b.   What is the height of tank C and pipe B?

1-102. Prepare a list of materials to build 28 stands like the one in Figure 1-4.
   a.   How many inches of pipe W are required?
   b.   How many inches of pipe X are required?
   c.   How many inches of plate Y are required?
   d.   How many inches of plate Z are required?

1-103. A welding shop ordered forty thousand pounds of steel plates. Twenty-seven thousand, three hundred sixty-nine pounds were received on Monday, and twelve thousand, six hundred thirty-one pounds were received on Thursday. Three thousand pounds of the order were cancelled. How many pounds of steel remain on order?

1-104. A frame is shown in Figure 1-5.
   a.   Find dimension A.
   b.   The frame is cut from a 40' length of material. How many feet are left over?

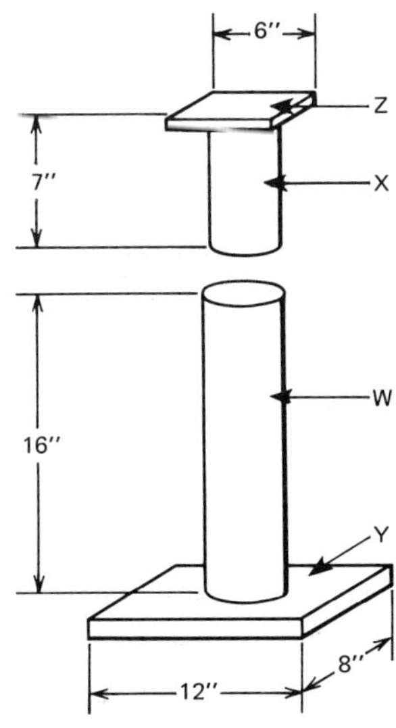

Figure 1—4

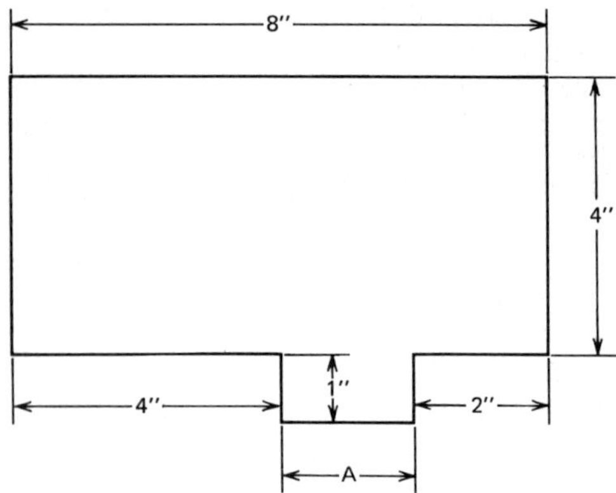

Figure 1—5

1-105.   A welder is to cut a centered hole 2" in
diameter between the two small holes in the steel
plate in Figure 1-6. What is dimension X?

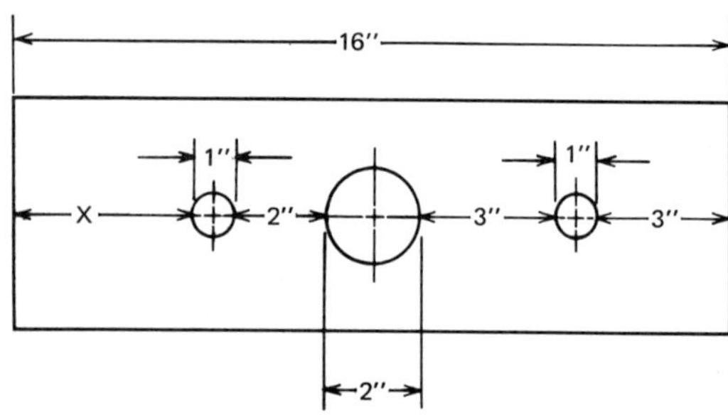

Figure 1—6

1-106.   Refer to Figure 1-7.
a.   What is the length of A?
b.   What is the length of B?

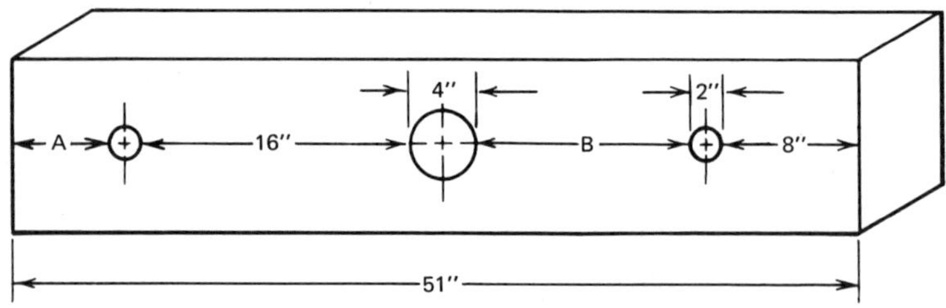

Figure 1—7

1-107.   Figure 1-8 shows a lug plate.
a.   How many inches of 2" x 9" material are
needed for 64 lug plates?
b.   How many inches of 11" x 16" material are
needed for 75 lug plates?

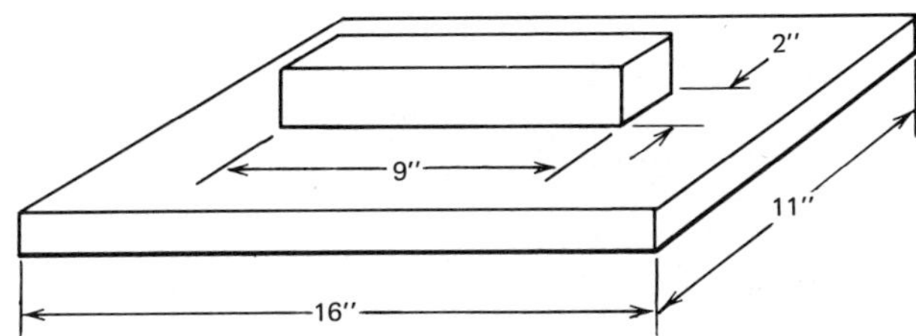

Figure 1—8

1-108. Sixteen utility tower base plates weigh a total of ten thousand, four hundred thirty-two pounds. How much does each plate weigh?

1-109. There are 43 support gussets on a quench tank, with a total of 602" of welding. How many inches of weld does each gusset require?

1-110. A welder is to lay out 21 evenly spaced holes for cutting in the slide rail shown in Figure 1-9. What is the center-to-center distance between each of the holes?

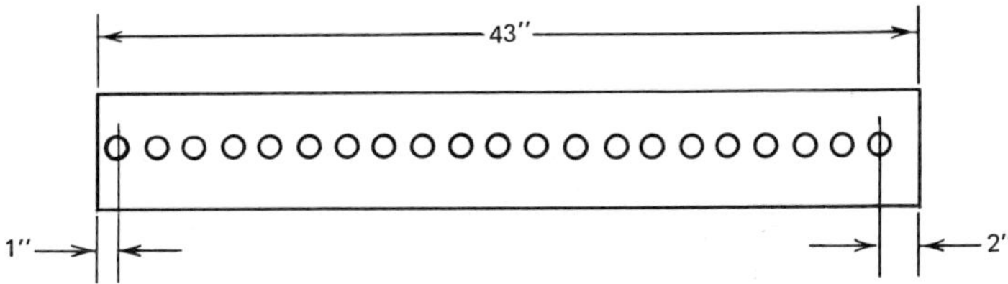

Figure 1—9

1-111. There are 3297 lb of welding electrodes stored in a shop warehouse. At 12 electrodes per pound, how many electrodes are in stock?

1-112. You are estimating the amount of metal needed to build 19 backhoe buckets. Each bucket will require 287 lb of metal. You have 1250 lb of metal in stock. How many additional pounds are needed?

1-113. The 3" angle iron frame shown in Figure 1-10 is 73" long and 11" wide. The frame has 4 evenly spaced cross braces of 3" angle iron. What is the distance between the braces?

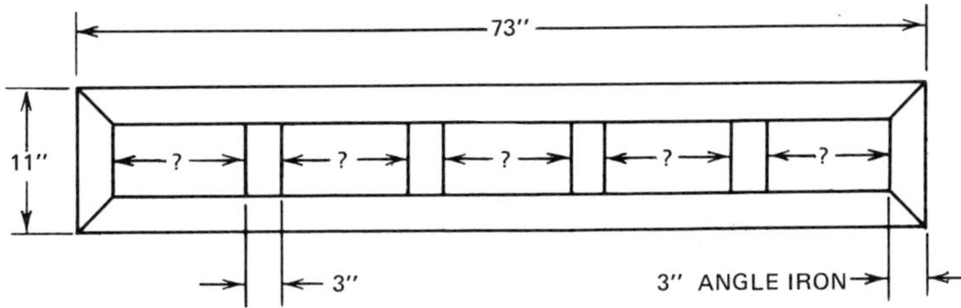

Figure 1—10

1-114.   An automatic welding machine welds 182" per
         minute. How long will it take to weld 2912"?

1-115.   The material needed for 6 large storage tanks is
         672 sheets of 10-gage steel, 888 sheets of
         12-gage steel, and 396 sheets of 14-gage steel.
         How many sheets of each gage are needed for each
         tank?

1-116.   Two thousand, nine hundred eighty-four parts
         have a 3/8" fillet weld in 563 places and a 1/4"
         fillet weld in 719 places. How many 1/4" and
         3/8" fillet welds are there?

1-117.   There were 53 welders at work on Monday, 50 on
         Tuesday, 46 on Wednesday, 55 on Thursday, and
         60 on Friday. What was the average number of
         welders at work this week?

1-118.   Ten thousand, five hundred fifty-six pounds of
         steel were used in the welding shop on Monday.
         Twelve thousand, nine hundred eighty-nine pounds
         were used on Tuesday; 15,046 on Wednesday; and
         8673 on Thursday. The shop was closed on Friday.
         What was the average amount of steel used per
         working day during the week?

1-119.   A welding shop received a shipment of steel.
         Random samples of the tensile strength of the
         steel in pounds per square inch (psi) were as
         follows: 40,128 psi, 60,000 psi, 38,900 psi,
         49,000 psi, 9998 psi, 58,000 psi, and 42,671 psi.
         What was the average tensile strength of the
         steel?

1-120.   What is the average length of the following
         welds completed by 4 welders in 6 hours:
         welder A, 46'; welder B, 43'; welder C, 45'; and
         welder D, 42'?

1-121.   An automatic machine welds various thicknesses
         of sheet metal at the following rate: 10-gage
         steel at 96" per minute, 12-gage steel at 128"
         per minute, and 18-gage steel at 180" per
         minute. What is the average welding speed per
         minute for this job?

1-122.   A welder earned nineteen thousand, one hundred
         eighty-eight dollars per year. How much did she
         average per month?

Chapter 1: Whole Numbers

Applied Welding Problems
Survey Test

Name ____

Course/Sec. ______________   Date _____________________

1-1.   A welder worked 48 hours, 14 hours, 7 hours, 9 hours, and 11 hours during a week. Find the total number of hours worked.

1-1. _______________

1-2.   A welder used the following lengths of pipe to reinforce a water tank: 36", 38", 8", 24", 7", and 35". How many inches of pipe were used to do the job?

1-2. _______________

1-3.   A welding shop ordered the following amounts of cold rolled steel: 714 lb, 1597 lb, 1043 lb, and 1767 lb. How many pounds of steel were ordered?

1-3. _______________

1-4.   During a 5-year period, a workman contributed the following amounts to his pension fund: $1327, $1305, $1403, and $1408. How much did he contribute to his pension fund during the 5-year period?

1-4. _______________

1-5.   Find the total weight of the following loads of scrap iron: 31,105 lb, 28,817 lb, 27,682 lb, 28,840 lb, and 28,795 lb.

1-5. _______________

1-6.   An inventory list stated that 2497 lb of welding rod were on hand January 7. On March 5, the list stated that 1362 lb of rod were on hand. How many pounds of rod were used in that period?

1-6. _______________

1-7.   A welder cut a 696' length cable from a 1250' roll. How much was left of the roll?

1-7. _______________

1-8.   A metal worker ordered 850 thread-cutting screws for a job. At the end of the job, he had 74 screws left. How many screws did he use?

1-8. _______________

1-9.   For the construction of a building, 81,397 lb of I beam were ordered. Due to a change, 86,456 lb were used. How many more pounds of I beam were needed than were originally ordered?

1-9. _______________

1-10.   A welding repair company has 1248 lb of stainless
steel. The company uses 929 lb of the steel to
build an acid tank. How many pounds are left?

1-10. _______________

1-11.   A spot welder welds 132 spot welds per minute.
How many spot welds are welded in an hour?

1-11. _______________

1-12.   There are 5280' of welding wire on a spool. How
many feet of wire are on 258 spools?

1-12. _______________

1-13.   The shipping weight of an electric arc welder is
258 lb. What is the shipping weight of 12 electric welders?

1-13. _______________

1-14.   How many inches of steel tubing are required to
cut 158 pieces, each 18" long? Disregard the
amount lost in cutting.

1-14. _______________

1-15.   A multiple stationary engine mounting frame
weighs 986 lb. What is the total weight of the
frame when the following hardware is attached:
21 mounting brackets at 4 lb each, 19 centering
clamps at 3 lb each, 14 rubber mounts at 2 lb
each, and 14 locking bolts at 1 lb each?

1-15. _______________

1-16.   An oxygen delivery truck traveled 64,560 mi over
a period of one year. If the truck averaged 12
mpg of diesel fuel, how many gallons of diesel
fuel were used that year?

1-16. _______________

1-17.   A subcontractor employed 57 persons during one
year and paid salaries totaling $1,068,750. If
each employee received the same salary, how much
was each paid?

1-17. _______________

1-18.   A shear cuts 3888 2" strips in 36 hours. How
many strips does it cut per hour?

1-18. _______________

1-19.   A welding shop makes a profit of $59 each time it
sells a welded utility trailer. How many trailers
must be sold to earn $708?

1-19. _______________

1-20.   A welder completes 432 tungsten inert gas welds
in an 8-hour day. What is her average per hour?

1-20. _______________

Chapter 2: Fractions
Preview and Self-Test

Name ____

Course/Sec. _________________ Date _______________

Chapter 2 covers fractions and the fundamental operations
on fractions. Before you turn to the chapter, complete
the Self-Test to determine which sections in the chapter
you need to study carefully.

Self-Test

PROPERTIES AND DEFINITIONS OF FRACTIONS

2-1. Change 3/8 to 64ths.   2-1. _______________

2-2. Reduce 144/204 to lowest terms.   2-2. _______________

       Properties and definitions of fractions score   _______________

MULTIPLICATION OF FRACTIONS

Multiply and reduce to lowest terms: 2-3. _______________

2-3. 1/6 x 12/2   2-4. 8/9 x 3/16   2-5. 8 x 5/16   2-4. _______________

2-6. 12-5/16 x 3-2/5   2-5. _______________

                                        2-6. _______________

           Multiplication of fractions score   _______________

DIVISION OF FRACTIONS

Divide and reduce:

2-7. 1/2 ÷ 7/8   2-8. 24 ÷ 3/8   2-9. 9 ÷ 3-1/2   2-7. _______________

2-10. 26-1/3 ÷ 79   2-8. _______________

                                        2-9. _______________

                                        2-10. _______________

          Division of fractions score   _______________

ADDITION OF FRACTIONS

Add and reduce:

2-11. 3/4 + 7/16   2-12. 13-1/5 + 11-2/3   2-11. ___________

2-13. 1-3/8 + 7-5/6 + 27-13/24   2-12. ___________

                                                          2-13. ___________

                    Addition of fractions score   ___________

SUBTRACTION OF FRACTIONS

Subtract and reduce: 2-14. ___________

2-14. 13/16   2-15.   4-7/8   2-16.   39   2-15. ___________
       - 3/4   - 3-9/10   - 14-31/64
                                                          2-16. ___________

                 Subtraction of fractions score   ___________

SIMPLIFYING COMPLEX FRACTIONS

Simplify the following:

2-17. $\dfrac{\frac{3}{4}}{\frac{5}{6}+\frac{7}{8}}$ 2-18. $\dfrac{1}{\frac{1}{2}+\frac{2}{3}-\frac{4}{5}}$ 2-19. $\dfrac{12\frac{3}{4}\times 10\frac{1}{2}}{3}\div 18$

2-17. ___________

2-18. ___________

2-19. ___________

              Simplifying complex fractions score   ___________

PRACTICAL APPLICATIONS OF FRACTIONS

2-20. Find A in Figure 2-1.   2-20. ___________

2-21. Find B in Figure 2-1.   2-21. ___________

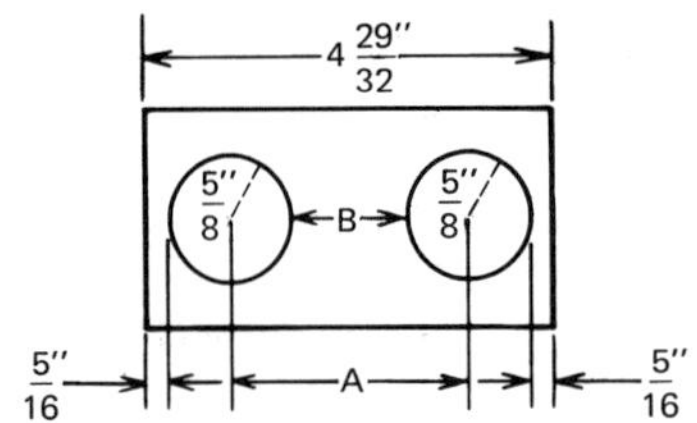

Figure 2—1

2-22. A pump raises 2-1/8 gal of water per stroke.   2-22. ___________
       How many gallons will it raise in 24 strokes?

2-23. How many shelves 18-3/4" long can be cut from an   2-23. ___________
       8' oak board? (Note: 1' = 12")

2-24. The width of a brick including the joint is 3-3/4". 2-24. ___________
       What is the width of 18 bricks and joints?

2-25. How much material is needed for the 4 legs of a   2-25. ___________
       table if each leg is 29-5/8" long?

              Practical applications of fractions score   ___________

After you have checked your answers on the Self-Test,
transfer your scores to the Chapter 2 Objectives.

Chapter 2

....

Fractions

Upon successful completion of this chapter, you will know
the definitions and properties of fractions. You will be
able to perform fundamental operations on fractions,
including multiplication, division, addition, and subtraction; simplify complex fractions; and make practical
applications of fractions.

Objectives

	Self-Test Scores	If you did poorly on the Self-Test, turn to:
• Properties and defini-tions of fractions	_________	Section 2-2
• Multiplication of fractions	_________	Section 2-3
• Division of fractions	_________	Section 2-4
• Addition of fractions	_________	Section 2-5
• Subtraction of fractions	_________	Section 2-6
• Simplifying complex fractions	_________	Section 2-7
• Practical applications of fractions	_________	Section 2-8

2–1   Introduction to Fractions

For some people, working with fractions can be more difficult than any other part of basic mathematics. The use
of calculators and the metric system would seem to make
working with fractions unnecessary. However, fractions
are vital in measurement and are used frequently on the
job and in everyday living.
    Fractions are essential for many functions in our
modern world. You can perform fundamental operations
with fractions by learning some basic procedures.

2–2   Properties and Definitions
of Fractions

A _fraction_ is a number indicating that a whole number or
quantity has been divided into equal parts. For example,
the fraction 3/4 indicates that a whole has been divided
into 4 equal parts and that 3 equal parts are represented.
The top number is called the _numerator_, and the bottom
number is called the _denominator_. These properties of
fractions can be shown as follows:

$$\text{a fraction} = \frac{3}{4} \frac{\text{(numerator)}}{\text{(denominator)}} = \frac{\text{number of parts}}{\text{number of (equal)}} \atop \text{parts of the whole}$$

Some students remember that the denominator is the bottom
number because it starts with letter d like the word _down_.
    The fraction 3/4 is read as "three-fourths." Always
read the top number first and then the bottom number.
Thus, 15/32 is read as "fifteen thirty-seconds."
    When the numerator is smaller than the denominator,
the fraction is called a _proper_ _fraction_. A fraction
whose numerator is larger than its denominator is called
an _improper_ _fraction_. A _mixed_ _number_ is a number that
consists of a whole number and a proper fraction. A mixed
number can be changed to an improper fraction, as the
examples show.

Procedure for Changing Fractions
by Multiplying

●●

The numerator and the denominator of a fraction can both
be multiplied by the same number, except zero, without
changing the value of the fraction.

●●

Example 2–1

PROBLEM

Change 7 to 4ths.

SOLUTION

1. Write 7 as 7/1.

Note: A whole number can always be written over a deno-minator of 1 without changing its value.

2. To obtain a denominator of 4, 1 must be multiplied by 4 (1 x 4 = 4).

3. Multiply both the numerator and the denominator by 4:

$$\frac{7 \times 4}{1 \times 4} = \frac{28}{4}$$

ANSWER

$$7 = \frac{28}{4}$$

Example 2–2

PROBLEM

Change 3/5 to 15ths.

SOLUTION

1. To obtain a denominator of 15, 5 must be multiplied by 3 (5 x 3 = 15).

2. Multiply both the numerator and the denominator by 3:

$$\frac{3 \times 3}{5 \times 3} = \frac{9}{15}$$

ANSWER

$$\frac{3}{5} = \frac{9}{15}$$

Exercises: Changing Fractions by Multiplying

Solve the following and show necessary work:

2-1. Change 3 to 5ths.   2-2. Change 9 to 36ths.

2-3. Change 6 to 16ths.   2-4. Change 8 to 4ths.

2-5. Change 3/4 to 16ths.   2-6. Change 1/2 to 24ths.

2-7. Change 7/8 to 48ths.   2-8. Change 11/12 to 36ths.

2-9. Change 29/32 to 128ths. 2-10. Change 144 to 8ths.

Procedure for Changing Fractions by Dividing

Both the numerator and the denominator of a fraction can be divided by the same number, except zero, without changing the value of the fraction. This process is called <u>reducing</u> the fraction.

Example 2–3

PROBLEM

Change, or reduce, 24/6 to a whole number.

SOLUTION

Divide 6 into 24 and 6:

$$\frac{24 \div 6}{6 \div 6} = \frac{4}{1} = 4$$

ANSWER

$$\frac{24}{6} = 4$$

Example 2–4

PROBLEM

Reduce the fraction 12/24 to lowest terms.

SOLUTION

1. To determine what number will evenly divide into both the numerator and the denominator, recall the multiplication facts:

   12 can be divided by 2, 3, 4, 6, and 12

   24 can be divided by 2, 3, 4, 6, 8, 12, and 24

2. Select the largest number that will evenly divide into both the numerator and the denominator. Selecting the largest number comes only with practice. Such a number is called the <u>highest</u> <u>common</u> <u>divisor</u>.

3. In this case, 12 is the highest common divisor. Divide 12 into both the numerator and the denominator:

   $$\frac{12 \div 12}{24 \div 12} = \frac{1}{2}$$

ANSWER

$$\frac{12}{24} = \frac{1}{2}$$

Example 2–5

PROBLEM

Reduce the fraction 48/64 to lowest terms.

SOLUTION

1. Divide 4, a common divisor, into the numerator and
    denominator:

$$\frac{48 \div 4}{64 \div 4} = \frac{12}{16}$$

2. The fraction 12/16 can be reduced further. Divide 4
    again into the numerator and the denominator:

$$\frac{12 \div 4}{16 \div 4} = \frac{3}{4}$$

We cannot reduce 3/4 any further.

Note: We could have divided 16, the highest common
divisor, into the numerator and the denominator to solve
the problem in one step.

ANSWER

$$\frac{48}{64} = \frac{3}{4}$$

Exercises: Changing Fractions by Dividing

Reduce the following to lowest terms:

2-11. 20/5	2-12. 49/7	2-13. 120/8
2-14. 80/5	2-15. 979/11	2-16. 3/6
2-17. 9/27	2-18. 32/64	2-19. 180/240
2-20. 720/1440		

Procedure for Changing Fractions
to Mixed Numbers

• •

1. Divide the numerator by the denominator to obtain
    the whole number.
2. Write the remainder over the divisor to obtain the
    fraction.

• •

Example 2–6

PROBLEM

Reduce 47/8 to a mixed number.

SOLUTION

Divide the numerator, 47, by the denominator, 8, and write the remainder, 7, over the divisor:

$$\begin{array}{r} 5\frac{7}{8} \\ 8\overline{)47} \\ \underline{40} \\ 7 \end{array}$$

ANSWER

$$\frac{47}{8} = 5\frac{7}{8}$$

Exercises: Changing Fractions to Mixed Numbers

Reduce the following to mixed numbers:

2-21. 25/4	2-22. 39/6	2-23. 74/8
2-24. 145/16	2-25. 389/32	2-26. 1495/35

Procedure for Changing Mixed Numbers to Improper Fractions

1. Multiply the denominator of the fraction by the whole number and add the numerator to obtain the numerator of the improper fraction.
2. Write the sum over the denominator of the fraction in the mixed number.

Example 2–7

PROBLEM

Convert 3-7/8 to an improper fraction.

SOLUTION

1. Multiply the denominator of the fraction by the whole number:

    8 x 3 = 24

2. Add the numerator of the fraction to the product:

    24 + 7 = 31

3. Write the sum over the denominator of the fraction
    in the mixed number:

    $$\frac{31}{8}$$

ANSWER

$$3\frac{7}{8} = \frac{31}{8}$$

Exercises: Changing Mixed Numbers
to Improper Fractions

Change the following to improper fractions:

2-27. 2-1/2   2-28. 3-1/3   2-29. 4-7/8

2-30. 12-5/6   2-31. 14-11/32   2-32. 125-11/36

2–3   Multiplication of Fractions

Multiplication is an easy fundamental operation to
perform with fractions. We start with multiplication
because it will also help us in the division, addition,
and subtraction of fractions. The procedure is illustrated in the examples that follow.

The multiplication of fractions can be simplified
by cancellation. _Cancellation_ is the process of dividing
by the same number both the numerator and the denominator
of the fractions being multiplied. The number chosen
must divide evenly into both the numerator and the
denominator. A simple rule to remember is that you can
divide "top into bottom" and "bottom into top" (see
Examples 2-8 and 2-9).

Before multiplying fractions, attempt to reduce them
through cancellation. Reducing fractions makes multiplication easier because the numbers being multiplied are
smaller. Multiply the numerators ("top times top"), then
the denominators ("bottom times bottom"), and reduce the
product to lowest terms or to a mixed number. The same
procedures apply when more than two fractions are being
multiplied.

To multiply mixed numbers, change them to improper
fractions and proceed as in the multiplication of proper
fractions. Whole numbers should be written over a
denominator of 1.

Procedure for Multiplying Fractions and Whole Numbers

1. Cancel or reduce all numerators and denominators. Remember that in cancellation, you can divide "top into bottom" and "bottom into top" only.
2. Multiply all numerators.
3. Multiply all denominators.
4. Write the product as a proper fraction or a mixed number and reduce to lowest terms.

Example 2–8

PROBLEM

Multiply 3/4 by 7/15.

SOLUTION

1. Reduce the fractions by dividing "top into bottom":

$$\frac{\cancel{3}^{1}}{4} \times \frac{7}{\cancel{15}_{5}}$$

2. Multiply first the numerators and then the denominators:

$$\frac{1 \times 7}{4 \times 5} = \frac{7}{20}$$

ANSWER

$$\frac{3}{4} \times \frac{7}{15} = \frac{7}{20}$$

Example 2–9

PROBLEM

Multiply 7/8 by 48.

SOLUTION

1. When you write the problem, show 48 with a denominator of 1:

$$\frac{7}{8} \times \frac{48}{1}$$

2. Reduce the numerator and the denominator by dividing "bottom into top":

$$\frac{7}{\cancel{8}_{1}} \times \frac{\cancel{48}^{6}}{1}$$

3. Multiply first the numerators and then the denominators:

$$\frac{7 \times 6}{1 \times 1} = \frac{42}{1} = 42$$

ANSWER

$$\frac{7}{8} \times 48 = 42$$

Exercises: Multiplying Fractions and Whole Numbers

Multiply the following:

2-33.   1/2 x 5/8   2-34.   7/16 x 1/3

2-35.   2/3 x 12   2-36.   3/4 x 24

2-37.   16 x 7/8   2-38.   88 x 3/4

2-39.   72/125 x 25/81   2-40.   144/148 x 37/156

Procedure for Multiplying Fractions and Mixed Numbers

●●

1. Convert all mixed numbers to improper fractions.
2. Cancel or reduce all numerators and denominators.
    Remember that in cancellation, you can divide "top
    into bottom" and "bottom into top" only.
3. Multiply all numerators.
4. Multiply all denominators.
5. Write the resulting product as a proper fraction or
    a mixed number and reduce to lowest terms.

●●

Example 2–10

PROBLEM

Multiply 1/2 by 3-3/4.

SOLUTION

1. Convert 3-3/4 to an improper fraction:

$$4 \times 3 + 3 = 15 \quad \text{so} \quad 3\frac{3}{4} = \frac{15}{4}$$

2. Write as a multiplication problem:

$$\frac{1}{2} \times \frac{15}{4}$$

3. Attempt to cancel. However, no quantity will divide "top to bottom" or "bottom to top." Therefore,

4. Multiply the numerators and the denominators:

$$\frac{1}{2} \times \frac{15}{4} = \frac{1 \times 15}{2 \times 4} = \frac{15}{8}$$

5. Convert 15/8 to a mixed number:

$$\frac{15}{8} = 1\frac{7}{8}$$

ANSWER

$$\frac{1}{2} \times 3\frac{3}{4} = 1\frac{7}{8}$$

Example 2–11

PROBLEM

Multiply 3-1/4 by 2-5/8.

SOLUTION

1. Convert 3-1/4 and 2-5/8 to improper fractions:

$$3\frac{1}{4} = \frac{4 \times 3 + 1}{4} = \frac{13}{4}$$

$$2\frac{5}{8} = \frac{8 \times 2 + 5}{8} = \frac{21}{8}$$

2. Write in multiplication form:

$$\frac{13}{4} \times \frac{21}{8}$$

3. Attempt to cancel. However, the fractions cannot be reduced because no quantity will evenly divide into the numerators and the denominators.

4. Multiply the numerators and the denominators:

$$\frac{13}{4} \times \frac{21}{8} = \frac{13 \times 21}{4 \times 8} = \frac{273}{32}$$

5. Convert to a mixed number:

$$\frac{273}{32} = 32\overline{)273} = 8\frac{17}{32}$$
$$\underline{256}$$
$$17$$

ANSWER

$$3\frac{1}{4} \times 2\frac{5}{8} = 8\frac{17}{32}$$

Exercises: Multiplying Fractions and Mixed Numbers

Multiply the following:

2-41.   1/4 x 3-1/2   2-42.   1/2 x 7-1/2

2-43.   2/3 x 3-4/5   2-44.   3/4 x 16-1/4

2-45.   5/9 x 20-2/3   2-46.   1-3/4 x 1-5/8

2-47.   2-1/2 x 2-1/2   2-48.   3-7/8 x 10-2/5

2-49.   9-6/15 x 8-1/8   2-50.   15-5/16 x 7-7/8

Procedure for Multiplying Fractions, Mixed Numbers, and Whole Numbers

• •

1. Convert all mixed numbers to improper fractions.
2. Place all whole numbers over a denominator of 1.
3. Cancel or reduce all numerators and denominators.
    Remember that in cancellation, you can divide "top
    into bottom" and "bottom into top" only.
4. Multiply all numerators.
5. Multiply all denominators.
6. Write the resulting product as a proper fraction or
    a mixed number and reduce to lowest terms.

• •

Example 2–12

PROBLEM

Multiply 3/4 by 6-7/8 by 60.

SOLUTION

Note: The procedures used in the previous examples also
apply to the multiplication of several fractions.

1. Write 6-7/8 as an improper fraction:

$$6\frac{7}{8} = \frac{8 \times 6 + 7}{8} = \frac{55}{8}$$

2. When you write the problem, show 60 with a denominator
    of 1:

$$\frac{3}{4} \times \frac{55}{8} \times \frac{60}{1}$$

3. First look for cancellations. Note that 4 divides
    evenly into 4 and 60:

$$\frac{3}{\cancel{4}_1} \times \frac{55}{8} \times \frac{\cancel{60}^{15}}{1}$$

4. Multiply the numerators and the denominators:

$$\frac{3 \times 55 \times 15}{1 \times 8 \times 1} = \frac{2475}{8}$$

5. Convert 2475/8 to a mixed number:

```
        309 3/8
    8) 2475
       24
       ‾‾‾
        075
         72
        ‾‾‾
          3
```

ANSWER

$$\frac{3}{4} \times 6\frac{7}{8} \times 60 = 309\frac{3}{8}$$

Exercises: Multiplying Fractions, Mixed Numbers, and Whole Numbers

Multiply the following:

2-51.   1/2 x 3-1/2 x 12   2-52.   1/4 x 2-1/8 x 24

2-53.   7/8 x 4-1/4 x 32   2-54.   11/12 x 12-5/8 x 42

2-55.   15/16 x 14-7/16 x 96

2–4   Division of Fractions

The division of fractions is similar to the multiplication
of fractions and is almost as easy. To divide by a
fraction, invert the divisor and multiply. To _invert_
means to reverse the positions of the numerator and the
denominator or, as some say, "tip the fraction upside
down." Remember that to divide fractions, you invert the
fraction on the right (the divisor) and proceed as in
multiplication. After inverting, always cancel whenever
possible. The examples that follow illustrate the proper
procedures.

Procedure for Dividing Fractions

1. Place all whole numbers over a denominator of 1.
2. Invert the divisor (the fraction to the right of
    the division sign).
3. Cancel or reduce all numerators and denominators.
    Remember that in cancellation, you can divide "top
    into bottom" and "bottom into top" only.
4. Multiply all numerators and denominators.
5. Reduce the answer to lowest terms. Convert to a
    mixed number if necessary.

Example 2–13

PROBLEM

Divide 1/2 by 3/4.

SOLUTION

1. Invert the second fraction (the fraction to the right
    of the division sign) and write the problem as a
    multiplication problem:

$$\frac{1}{2} \div \frac{3}{4} = \frac{1}{2} \times \frac{4}{3}$$

2. Cancel by dividing 2 into 2 and 4:

$$\frac{1}{\cancel{2}} \times \frac{\cancel{4}}{3}$$

3. Multiply the numerators and the denominators:

$$\frac{1 \times 2}{1 \times 3} = \frac{2}{3}$$

ANSWER

$$\frac{1}{2} \div \frac{3}{4} = \frac{2}{3}$$

Example 2–14

PROBLEM

Divide 28 by 4/5.

SOLUTION

1. When you write the problem, show 28 with a denominator
    of 1. Then invert 4/5 and write the problem as a
    multiplication problem:

$$\frac{28}{1} \div \frac{4}{5} = \frac{28}{1} \times \frac{5}{4}$$

2.   Cancel by dividing 4 into 28:

$$\frac{\overset{7}{\cancel{28}}}{1} \times \frac{5}{\cancel{4}}$$

3.   Multiply the numerators and the denominators:

$$\frac{7 \times 5}{1 \times 1} = \frac{35}{1} = 35$$

ANSWER

$$28 \div \frac{4}{5} = 35$$

Example 2–15

PROBLEM

Divide 5/16 by 144.

SOLUTION

1.   When you write the problem, show 144 with a denominator
     of 1. Then invert 144/1 to change the problem to a
     multiplication problem:

$$\frac{5}{16} \div \frac{144}{1} = \frac{5}{16} \times \frac{1}{144}$$

2.   Since the fractions cannot be reduced, multiply the
     numerators and the denominators:

$$\frac{5}{16} \times \frac{1}{144} = \frac{5 \times 1}{16 \times 144} = \frac{5}{2304}$$

ANSWER

$$\frac{5}{16} \div 144 = \frac{5}{2304}$$

Exercises: Dividing Fractions

Divide the following:

2-56. 3/4 ÷ 7/8   2-57. 1/2 ÷ 3/4   2-58. 8/9 ÷ 2/3

2-59. 2/5 ÷ 3/8   2-60. 8 ÷ 2/3   2-61. 7/40 ÷ 21/25

2-62. 12 ÷ 3/4   2-63. 48 ÷ 4/5   2-64. 4 ÷ 3/5

2-65. 288 ÷ 9/16   2-66. 3/4 ÷ 48   2-67. 7/16 ÷ 144

Procedure for Dividing Fractions, Mixed Numbers, and Whole Numbers

••

1.   Convert all mixed numbers to improper fractions.

2. Place all whole numbers over a denominator of 1.
3. Invert the divisor (the fraction to the right of the
    division sign).
4. Cancel or reduce all numerators and denominators.
    Remember that in cancellation, you can divide "top
    into bottom" and "bottom into top" only.
5. Multiply all numerators and denominators.
6. Reduce the answer to lowest terms. Convert to a
    mixed number if necessary.

• •

Example 2–16

PROBLEM

Divide 5-1/2 by 3/4.

SOLUTION

1. Change 5-1/2 to an improper fraction:

$$5\frac{1}{2} = \frac{11}{2}$$

2. Invert 3/4, try to reduce, and multiply:

$$\frac{11}{2} \times \frac{4}{3}$$

3. Cancel by dividing 2 into 2 and 4:

$$\frac{11}{\cancel{2}_1} \times \frac{\cancel{4}^2}{3}$$

4. Multiply the numerators and the denominators and reduce
    the answer to a mixed number:

$$\frac{11 \times 2}{1 \times 3} = \frac{22}{3} = 7\frac{1}{3}$$

ANSWER

$$5\frac{1}{2} \div \frac{3}{4} = 7\frac{1}{3}$$

Example 2–17

PROBLEM

Divide 12-2/3 by 24-4/5.

SOLUTION

1. Convert each fraction to an improper fraction:

$$12\frac{2}{3} = \frac{3 \times 12 + 2}{3} = \frac{38}{3}$$

$$24\frac{4}{5} = \frac{5 \times 24 + 4}{5} = \frac{124}{5}$$

2. Invert the divisor (the fraction to the right of the division sign):

$$\frac{38}{3} \div \frac{124}{5} = \frac{38}{3} \times \frac{5}{124}$$

3. Reduce by dividing 2 into 38 and 124:

$$\frac{\overset{19}{\cancel{38}}}{3} \times \frac{5}{\underset{62}{\cancel{124}}}$$

4. Multiply the numerators and the denominators and reduce if possible:

$$\frac{19 \times 5}{3 \times 62} = \frac{95}{186}$$

Note: 95/186 cannot be reduced, so the product remains 95/186.

ANSWER

$$12\frac{2}{3} \div 24\frac{4}{5} = \frac{95}{186}$$

Exercises: Dividing Fractions, Mixed Numbers, and Whole Numbers

Divide the following:

2-68. 6-1/2 ÷ 7/16	2-69. 7-3/4 ÷ 5/8
2-70. 12-1/2 ÷ 15/16	2-71. 1-3/5 ÷ 29/32
2-72. 2-1/2 ÷ 1-2/3	2-73. 6-3/13 ÷ 16-2/5
2-74. 120-1/2 ÷ 12-3/4	2-75. 25-19/30 ÷ 15-3/5

2-5   Addition of Fractions

The addition of fractions is more involved than the multiplication or division of fractions because the fractions must have the same denominator in order to be added. If they do not have the same denominator, we must find the lowest common denominator and convert the fractions to equivalent fractions (fractions with the same denominator) before adding. The lowest common denominator (LCD) is the smallest number that the denominators will divide into and yield a whole number.

When adding fractions, be sure the denominators are equal. Then add the numerators and reduce. If the denominators are not equal, find the LCD, convert the fractions to equivalent fractions with equal denominators, then add the numerators, and reduce.

Finding the LCD is easy if you use the following
procedure.

Procedure for Finding The Lowest Common Denominator

1. Arrange the denominators of the fractions in an
    inverted division sign.
2. Divide each denominator by the smallest number that
    evenly divides into at least one of the denominators.
    If a denominator does not evenly divide by that
    number, bring that denominator down. Write the
    quotient beneath the inverted division sign.
3. Continue the process by dividing each number in the
    quotient by the next larger divisor that will evenly
    divide into at least one of the numbers.
4. When the numbers can no longer be divided, multiply
    all the divisors and the final quotient to obtain the
    lowest common denominator (LCD).

Example 2–18

PROBLEM

Find the LCD for the following fractions: 1/3, 3/5,
and 5/6.

SOLUTION

Note: Find the smallest number that 3, 5, and 6 will
divide into evenly. If you do not immediately see that
the LCD is 30, use the following method:

1. Arrange the denominators in an inverted division sign:

)3 5 6

2. Divide each number by 2. Write the quotient below
    the number being divided, as shown. Bring down
    numbers that do not divide evenly:

    2) 3 5 6
        3 5 3

3. Again attempt to divide by 2. If you cannot do so,
    use the next larger number, 3, as a divisor. Make
    another inverted division sign and divide by 3:

    2) 3 5 6
      3)3 5 3
          1 5 1

4. The numbers in the quotient can no longer be divided.

Multiply all the divisors and the quotient to obtain the LCD:

2 x 3 x 1 x 5 x 1 =

6   x   5   = 30

ANSWER

30 is the LCD for 1/3, 3/5, and 5/6.

Procedure for Adding Fractions

1. When necessary, change all fractions to equivalent fractions having the same denominator by finding the lowest common denominator (LCD).
2. Add the numerators.
3. Write the sum of the numerators over the common denominator and reduce to lowest terms.
4. If the sum is an improper fraction, convert to a mixed number.

Example 2–19

PROBLEM

Add 1/4, 3/4, and 2/4.

SOLUTION

1. Because the denominators of the fractions are the same, simply add the numerators:

$$\frac{1 + 3 + 2}{4} = \frac{6}{4}$$

2. Reduce by dividing 2 into the numerator and the denominator:

$$\frac{\overset{3}{\cancel{6}}}{\underset{2}{\cancel{4}}} = \frac{3}{2}$$

3. Convert to a mixed number:

$$\frac{3}{2} = 1\frac{1}{2}$$

ANSWER

$$\frac{1}{4} + \frac{3}{4} + \frac{2}{4} = 1\frac{1}{2}$$

Example 2–20

PROBLEM

Add 1/8, 3/4, and 7/12.

SOLUTION

<u>Note</u>: In order to add the fractions, we must write
them with the same denominator. Observe that the common
denominator is 24. If you do not see that 24 is the
LCD, use the suggested procedure for finding the LCD:

$$
\begin{array}{r}
2 \,)\, \underline{8 \quad 4 \quad 12} \\
2 \,)\, \underline{4 \quad 2 \quad 6} \\
2 \quad 1 \quad 3
\end{array}
$$

$$2 \times 2 \times 2 \times 1 \times 3 = 24$$

1. Convert each fraction to an equivalent fraction with
    24 as the denominator:

$$\frac{1 \times 3}{8 \times 3} = \frac{3}{24}$$

$$\frac{3 \times 6}{4 \times 6} = \frac{18}{24}$$

$$\frac{7 \times 2}{12 \times 2} = \frac{14}{24}$$

2. Arrange the equivalent fractions and add the
    numerators:

$$\frac{3}{24} + \frac{18}{24} + \frac{14}{24} = \frac{3 + 18 + 14}{24} = \frac{35}{24}$$

3. Convert to a mixed number:

$$\frac{35}{24} = 1\frac{11}{24}$$

ANSWER

$$\frac{1}{8} + \frac{3}{4} + \frac{7}{12} = 1\frac{11}{24}$$

Example 2–21

PROBLEM

Add 3/8, 5/6, and 7/32.

SOLUTION

<u>Note</u>: In order to add the fractions, we must write them
with the same denominator. The common denominator of

these fractions is difficult to observe. Therefore, use
the suggested procedure for finding the LCD:

```
2 ) 8  6  32
 2 ) 4  3  16
  2) 2  3   8
     1  3   4
```

2 x 2 x 2 x 1 x 3 x 4 = 96

1. Convert the fractions to equivalent fractions with 96
    as the denominator:

$$\frac{3 \times 12}{8 \times 12} = \frac{36}{96}$$

$$\frac{5 \times 16}{6 \times 16} = \frac{80}{96}$$

$$\frac{7 \times 3}{32 \times 3} = \frac{21}{96}$$

2. Arrange the equivalent fractions and add the
    numerators:

$$\frac{36}{96} + \frac{80}{96} + \frac{21}{96} = \frac{36 + 80 + 21}{96} = \frac{137}{96}$$

3. Convert to a mixed number:

$$\frac{137}{96} = 1\frac{41}{96}$$

ANSWER

$$\frac{3}{8} + \frac{5}{6} + \frac{7}{32} = 1\frac{41}{96}$$

Exercises: Adding Fractions

Add the following and reduce when possible:

2-76.   3/8 + 2/8

2-77.   7/15 + 4/15 + 4/15

2-78.   5/16 + 7/16 + 15/16

2-79.   3/32 + 7/32 + 5/32

2-80.   3/4 + 5/8 + 7/16

2-81.   1/2 + 1/4 + 1/16

2-82.   61/54 + 13/16 + 7/8

2-83.   2/9 + 7/16 + 3/4

2-84.   5/6 + 6/15 + 7/8 + 3/16

2-85.   1/6 + 3/10 + 3/4 + 11/32

Procedure for Adding Fractions, Mixed Numbers, and Whole Numbers

1. When necessary, change all fractions to equivalent fractions having the same denominator.
2. Add the numerators.
3. Write the sum of the numerators over the denominator and reduce to lowest terms.
4. If the result is an improper fraction, convert to a mixed number.
5. Add the whole numbers.
6. Add the mixed number (or fraction) to the sum of the whole numbers.

Example 2–22

PROBLEM

Add 12-1/2, 7-7/8, and 6-5/16.

SOLUTION

1. Arrange the mixed numbers vertically:

$$12\frac{1}{2}$$
$$7\frac{7}{8}$$
$$+\ 6\frac{5}{16}$$

2. Find the common denominator of the fractions by observation. The common denominator is 16.

3. Convert the fractions to equivalent fractions with 16 as the denominator:

$$12\frac{1}{2} = 12\frac{1 \times 8}{2 \times 8} = 12\frac{8}{16}$$

$$7\frac{7}{8} = 7\frac{7 \times 2}{8 \times 2} = 7\frac{14}{16}$$

$$6\frac{5}{16} = 6\frac{5 \times 1}{16 \times 1} = 6\frac{5}{16}$$

4. Add the numerators of the equivalent fractions and then add the whole numbers:

$$\frac{8 + 14 + 5}{16} = \frac{27}{16}$$

$$12 + 7 + 6 = 25$$

5. Convert the improper fraction to a mixed number:

$$\frac{27}{16} = 1\frac{11}{16}$$

6. Add 1-11/16, the sum of the fractions, to 25, the sum
    of the whole numbers:

$$1\frac{11}{16} + 25 = 26$$

ANSWER

$$12\frac{1}{2} + 7\frac{7}{8} + 6\frac{5}{16} = 26\frac{11}{16}$$

Example 2–23

PROBLEM

Add 18-5/8, 7/12, 5-7/10, and 136-2/5.

SOLUTION

1. Arrange the quantities vertically:

$$18\frac{5}{8}$$
$$\frac{7}{12}$$
$$5\frac{7}{10}$$
$$+\ 136\frac{2}{5}$$

2. Find the lowest common denominator:

```
2 )   8  12  10  5
 2 )  4   6   5  5
  5 ) 2   3   5  5
      2   3   1  1
```

$$2 \times 2 \times 5 \times 2 \times 3 \times 1 \times 1 = 120$$

3. Convert the fractions to equivalent fractions with
    120 as the denominator:

$$18\frac{5 \times 15}{8 \times 15} = 18\frac{75}{120}$$

$$\frac{7 \times 10}{12 \times 10} = \frac{70}{120}$$

$$5\frac{7 \times 12}{10 \times 12} = 5\frac{84}{120}$$

$$136\frac{2 \times 24}{5 \times 24} = 136\frac{48}{120}$$

4. Add the numerators of the fractions and then add the
    whole numbers:

$$\frac{75 + 70 + 84 + 48}{120} = \frac{277}{120}$$

$$18 + 5 + 136 = 159$$

5. Convert 277/120 to a mixed number:

$$\frac{277}{120} = 2\frac{37}{120}$$

6. Add the sum of the fraction, 2-37/120, to the sum of the whole number, 159:

$$2\frac{37}{120} + 159 = 161\frac{37}{120}$$

ANSWER

$$18\frac{5}{8} + \frac{7}{12} + 5\frac{7}{10} + 136\frac{2}{5} = 161\frac{37}{120}$$

Exercises: Adding Fractions, Mixed Numbers, and Whole Numbers

Add the following and reduce when possible:

2-86.   1-3/8 + 2-5/16 + 3-5/8

2-87.   6-6/8 + 1-4/5 + 4-1/20

2-88.   1-3/4 + 5-3/10 + 4-2/5

2-89.   5-5/12 + 1-3/4 + 14

2-90.
$$\begin{aligned}
11&\frac{7}{10}\\
28&\frac{11}{30}\\
+\ 2&\frac{2}{15}\\
\hline
\end{aligned}$$

2-91.
$$\begin{aligned}
29&\\
3&\frac{9}{10}\\
+\ 6&\frac{23}{30}\\
\hline
\end{aligned}$$

2-92.
$$\begin{aligned}
11&\frac{5}{6}\\
7&\frac{5}{8}\\
+\ 16&\frac{11}{32}\\
\hline
\end{aligned}$$

2-93.
$$\begin{aligned}
45&\frac{7}{8}\\
18&\frac{5}{9}\\
+\ 7&\frac{5}{6}\\
\hline
\end{aligned}$$

2-94.
$$\begin{aligned}
21&\frac{3}{4}\\
39&\frac{3}{7}\\
+\ 10&\frac{11}{28}\\
\hline
\end{aligned}$$

2-95.
$$\begin{aligned}
18&\frac{3}{5}\\
1&\frac{7}{8}\\
47&\frac{3}{4}\\
+\ 6&\frac{3}{16}\\
\hline
\end{aligned}$$

2–6   Subtraction of Fractions

The procedure for subtracting fractions is very similar to the procedure for adding fractions. Fractions that are to be subtracted must have the same denominator. Therefore, we may need to find the lowest common denominator (LCD), just as we do when adding fractions. To subtract fractions, subtract the numerators and reduce the answer.

When subtracting mixed numbers, we must sometimes "borrow" from the whole number to make the numerator of

the top fraction greater than the numerator of the bottom
fraction. To complete the subtraction process, reduce
the answer and convert any improper fraction to a mixed
number. This procedure is shown in the examples.

Procedure for Subtracting Fractions

1. When necessary, change all fractions to equivalent
    fractions having the same denominator (the lowest
    common denominator).
2. Subtract the numerator of the bottom fraction from
    the numerator of the top fraction and write the
    result over the denominator.
3. Reduce the answer to lowest terms and convert to a
    mixed number when necessary.

Example 2–24

PROBLEM

Subtract 5/16 from 9/16.

SOLUTION

Because these fractions have the same denominator, simply
subtract the numerators and reduce the answer:

$$\frac{9}{16} - \frac{5}{16} = \frac{9-5}{16} = \frac{\cancel{4}}{\cancel{16}} = \frac{1}{4}$$

ANSWER

$$\frac{9}{16} - \frac{5}{16} = \frac{1}{4}$$

Example 2–25

PROBLEM

Subtract 5/8 from 29/32.

SOLUTION

1. Arrange the fractions vertically:

$$\begin{array}{r} \frac{29}{32} \\ -\frac{5}{8} \\ \hline \end{array}$$

2. Find the LCD, which is 32.

3. Convert the fractions to equivalent fractions with the same denominator, subtract the numerators, and write the result over the denominator:

$$\frac{29}{32} = \qquad \frac{29}{32} = \frac{29}{32}$$

$$-\frac{5}{8} = \frac{5 \times 4}{8 \times 4} = \frac{20}{32}$$

Therefore,

$$\begin{array}{r} \frac{29}{32} \\ -\frac{20}{32} \\ \hline \frac{9}{32} \end{array}$$

4. Reduce the answer if possible (9/32 cannot be reduced).

ANSWER

$$\frac{29}{32} - \frac{5}{8} = \frac{9}{32}$$

Exercises: Subtracting Fractions

Subtract the following and reduce when possible:

2-96. 3/4 - 1/4	2-97. 7/8 - 5/8
2-98. 15/16 - 13/16	2-99. 39/64 - 23/64
2-100. 1/2 - 1/4	2-101. 7/8 - 7/16
2-102. 3/8 - 3/10	2-103. 9/10 - 8/9
2-104. 5/12 - 3/10	2-105. 21/50 - 1/8

Procedure for Subtracting Mixed Numbers

1. When necessary, change all fractions to equivalent fractions having the same denominator (the lowest common denominator).
2. Subtract the numerator of the bottom fraction from the numerator of the top fraction and write the result over the denominator.
3. If the numerator of the top fraction is smaller than the numerator of the bottom fraction, borrow one unit from the whole number.
4. After subtracting the fractions, subtract the whole numbers.

Example 2–26

PROBLEM

Subtract 2-7/12 from 4-5/8.

SOLUTION

1. Arrange the mixed numbers as shown, find the LCD (24),
   and convert the fractions to equivalent fractions
   with a denominator of 24:

$$4\frac{5}{8} \qquad \frac{5 \times 3}{8 \times 3} = \frac{15}{24} \qquad 4\frac{15}{24}$$

$$2\frac{7}{12} \qquad \frac{7 \times 2}{12 \times 2} = \frac{14}{24} \qquad 2\frac{14}{24}$$

2. Subtract the fractions and the whole numbers:

$$
\begin{array}{r}
4\frac{15}{24} \\
-\ 2\frac{14}{24} \\
\hline
2\frac{1}{24}
\end{array}
$$

ANSWER

$$4\frac{5}{8} - 2\frac{7}{12} = 2\frac{1}{24}$$

Example 2–27

PROBLEM

Subtract 7-3/4 from 9-1/9.

SOLUTION

1. Arrange the mixed numbers as shown, find the LCD (36),
   and convert the fractions to equivalent fractions with
   a denominator of 36:

$$9\frac{1}{9} = 9\frac{1 \times 4}{9 \times 4} = 9\frac{4}{36} = 9\frac{4}{36}$$

$$7\frac{3}{4} = 7\frac{3 \times 9}{4 \times 9} = 7\frac{27}{36} = 7\frac{27}{36}$$

2. Because 27 cannot be subtracted from 4, borrow 1, or
   36/36, from 9 and add it to 4/36:

$$9\frac{4}{36} = 8\frac{4}{36} + \frac{36}{36} = 8\frac{40}{36}$$

$$
\begin{array}{r}
-\ 7\frac{27}{36} = \qquad\qquad 7\frac{27}{36} \\
\hline
\end{array}
$$

3. Subtract the fraction from the fraction and subtract
    the whole number from the whole number:

$$
\begin{array}{r}
8\frac{40}{36} \\
-\ 7\frac{27}{36} \\
\hline
1\frac{13}{36}
\end{array}
$$

ANSWER

$$9\frac{1}{9} - 7\frac{3}{4} = 1\frac{13}{36}$$

Exercises: Subtracting Mixed Numbers

Subtract the following and reduce:

2-106.   4-1/4 - 2-1/2   2-107.   12-5/9 - 3-1/3

2-108.   32-1/32 - 23-23/64   2-109.   45-7/9 - 3-7/27

2-110.   12 - 7-5/8   2-111.   23-7/10 - 15-7/8

2-112.   43 - 26-3/16   2-113.   156-3/4 - 96-9/32

2-114.   15-7/9 - 10-7/11   2-115.   301 - 289-3/35

2–7   Simplifying Complex Fractions

A <u>complex</u> <u>fraction</u> is a fraction that contains one or
more fractions in the numerator, in the denominator, or
in both. Many applied problems require the simplification
of complex fractions.
     To simplify complex fractions, first perform all
the operations in the numerator and then perform all
the operations in the denominator. Finally, divide the
denominator into the numerator. The examples show how
to simplify fractions.

Procedure for Simplifying Complex Fractions

• •

1. Perform all the operations in the numerator.
2. Perform all the operations in the denominator.
3. Divide the numerator by the denominator.

• •

Example 2–28

PROBLEM

Simplify the following complex fraction:

$$\frac{\frac{2}{3} \times \frac{5}{6}}{4\frac{3}{4} - 2\frac{1}{2}} \left.\begin{array}{l}\\\end{array}\right\} \text{numerator} \atop \left.\begin{array}{l}\\\end{array}\right\} \text{denominator}$$

SOLUTION

Note: To simplify the fraction, you must perform the multiplication in the numerator and then perform the subtraction in the denominator.

1. First convert the fractions in the denominator into equivalent fractions with 4 as the LCD:

$$\frac{\frac{2}{3} \times \frac{5}{6}}{4\frac{3}{4} - 2\frac{2}{4}} = \frac{\frac{2}{3} \times \frac{5}{6}}{2\frac{1}{4}} = \frac{\frac{5}{9}}{2\frac{1}{4}}$$

2. Then divide the denominator 2-1/4 = 9/4 into the numerator 5/9. Remember to invert and multiply:

$$\frac{\frac{5}{9}}{\frac{9}{4}} = \frac{5}{9} \div \frac{9}{4}$$

$$= \frac{5}{9} \times \frac{4}{9}$$

$$= \frac{20}{81}$$

ANSWER

$$\frac{\frac{2}{3} \times \frac{5}{6}}{4\frac{3}{4} - 2\frac{1}{2}} = \frac{20}{81}$$

Example 2–29

PROBLEM

Simplify the following complex fraction:

$$\frac{1}{\frac{1}{2} + \frac{2}{3} + \frac{3}{4}}$$

SOLUTION

1. To add the fractions in the denominator, convert them to equivalent fractions with a denominator of 12:

$$\frac{1}{\frac{1}{2} + \frac{2}{3} + \frac{3}{4}} = \frac{1}{\frac{1 \times 6}{2 \times 6} + \frac{2 \times 4}{3 \times 4} + \frac{3 \times 3}{4 \times 3}}$$

2. Add the fractions:

$$= \frac{1}{\frac{6 + 8 + 9}{12}}$$

$$= \frac{1}{\frac{23}{12}}$$

3. Divide the denominator into the numerator:

$$= 1 \div \frac{23}{12}$$

$$= 1 \times \frac{12}{23}$$

$$= \frac{12}{23}$$

ANSWER

$$\frac{1}{\frac{1}{2} + \frac{2}{3} + \frac{3}{4}} = \frac{12}{23}$$

Exercises: Simplifying Complex Fractions

Simplify the following:

2-116. $\dfrac{\frac{2}{3} \div \frac{2}{3}}{1\frac{1}{2} + \frac{1}{2}}$

2-117. $\dfrac{13\frac{1}{3} - 5\frac{5}{6}}{7\frac{1}{2}}$

2-118. $\dfrac{1}{\frac{2}{3} + \frac{3}{4} + \frac{5}{6}}$

2-119. $\dfrac{3\frac{1}{2} \times 4 \times 5\frac{1}{2}}{7\frac{1}{2}}$

2-120. $\dfrac{\left(2\frac{3}{4} \times 12\right) - 4\frac{1}{2}}{8\frac{3}{8} \div 4\frac{3}{16}}$

2–8   Practical Applications of Fractions

The coming of the metric system was supposed to greatly reduce the use of fractions. However, the examples and the problems illustrate how important fractions still are in many everyday applications.

Example 2–30

PROBLEM

How many square feet are in a 1/4-acre lot?   (1 acre = 43,560 sq ft.)

SOLUTION

Multiply 43,560 by 1/4:

$$43560 \times \frac{1}{4} = 10890$$

ANSWER

There are 10,890 sq ft in a 1/4-acre lot.

Example 2–31

PROBLEM

Find the width of floor space covered by 24 2x4s if each 2x4 is 3-1/2" wide and 3/8" is allowed between each piece.

SOLUTION

1. Draw a sketch like the one in Figure 2-2.

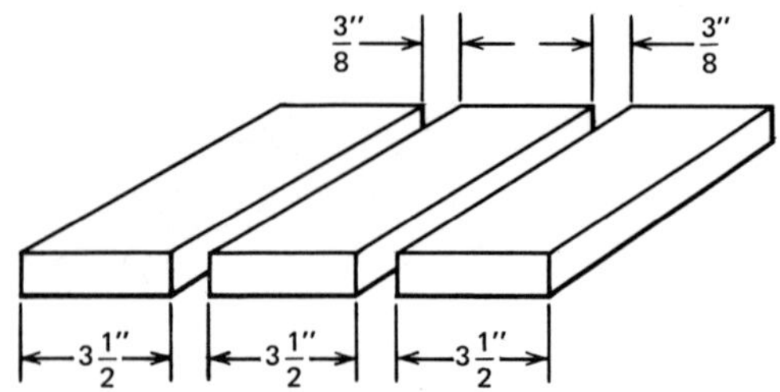

Figure 2–2

2. Observe that each 2x4 is 3-1/2" wide and there is a space of 3/8" between each pair.

3. Multiply 24 by the sum of 3-1/2 and 3/8. To do so, enclose the 3-1/2 + 3/8 in parentheses to keep them as one quantity:

$$24\left(3\frac{1}{2} + \frac{3}{8}\right) = 24\left(3\frac{4}{8} + \frac{3}{8}\right) = 24\left(3\frac{7}{8}\right)$$

$$= \frac{24}{1} \times \frac{31}{8}$$

$$= 93$$

ANSWER

The width of floor space covered by 24 2x4s is 93".

Exercises: Practical Applications of Fractions

Solve the following:

2-121. How far will you walk in 15 steps if you walk 30-1/2" in each step?

2-122. What is the cost of 2-1/2 dozen donuts at $2 a dozen?

2-123. Find the cost of 5-3/4 lb of walleye fillets if they cost $6 per pound.

2-124. Find the change in Minnetonka Laboratories, Inc. stock if it drops from 16-3/8 to 11-1/4 during a week.

2-125. How many pizzas are needed for a softball team if each of the 12 members eats 3/4 of a pizza?

2-126. 12 pizzas were ordered for a class party of 30 students. If the teacher had 3/4 of a pizza, how much could each student have?

2-127. Which paneling is thicker: 1/4" or 7/64"?

2-128. How much time is spent for travel 5 days per week if it takes 3/4 hour to drive one way to work?

2-129. If it takes 20 hours to drive 1010 mi, determine the rate of speed per hour.

2-130. If it takes 50 minutes to drive one way to work, how many hours are spent driving in 49-1/4 weeks, 5 days per week?

2-131. Find the cost of carpeting a living room 8-1/2 yd by 5-1/3 yd if the carpet costs $27 per sq yd. (Area = length x width.)

2-132. A cubic foot contains about 7-1/2 gal. How many cubic feet are there in a 55-gal barrel?

2-133. One spouse offers the other $20 for every pound they lose. How much will the reward be if the weight drops from 141-1/2 lb to 126-1/4 lb?

2-134. How much would you pay for a $300 lawn mower and a $40 trimmer if there were discounts of 1/3 on the lawn mower and 1/5 on the trimmer?

2-135. Find the total air fare to Florida for a family if the father pays full fare; the mother, 2/3 fare; the children under 18, 1/2 fare; and the children under 12, 1/3 fare. The full fare is $288. The children's ages are 17, 15, 12, 10, and 6.

Chapter 2

Fractions

Solving Applied Welding Problems

Now that you have mastered the fundamental operations on fractions, you are ready to apply these math skills to typical problems that welders encounter on the job. After studying the examples that follow, complete the applied welding problems.

Example 2–32

PROBLEM

What is the total height of the bracket in Figure 2-3?

SOLUTION

Arrange the quantities in proper columns, convert the fractions to equivalent fractions with a denominator of 4, and add:

$$1\frac{1}{4} = \frac{1}{4}$$
$$4\frac{1}{2} = \frac{2}{4}$$
$$+\ 2\frac{1}{2} = \frac{2}{4}$$
$$\overline{\qquad\qquad}$$
$$7 \qquad \frac{5}{4} = 1\frac{1}{4}$$
$$+\ 1\frac{1}{4}$$
$$\overline{\qquad\qquad}$$
$$8\frac{1}{4}$$

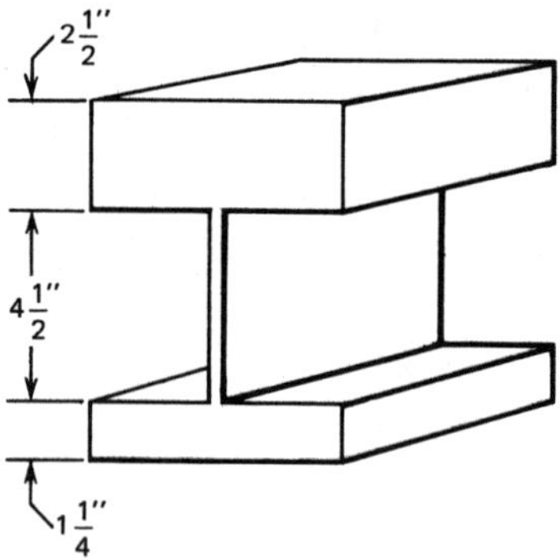

Figure 2–3

ANSWER

$$1\frac{1}{4}" + 4\frac{1}{2}" + 2\frac{1}{2}" = 8\frac{1}{4}"$$

Example 2–33

PROBLEM

A 12-1/2' length of tubing has a 4-3/8' piece cut from it.
How much remains?

SOLUTION

Subtract the 4-3/8' piece from the 12-1/2' length:

$$
\begin{array}{r}
12\frac{1}{2} = \frac{4}{8} \\
- \ 4\frac{3}{8} = \frac{3}{8} \\
\hline
8 \qquad \frac{1}{8}
\end{array}
$$

ANSWER

$$12\frac{1}{2}' - 4\frac{3}{8}' = 8\frac{1}{8}'$$

Example 2–34

PROBLEM

A bracket requires 6-3/4" of weld. How many inches of
weld are required for 20 brackets?

SOLUTION

Multiply the number of brackets by the number of inches
of weld:

$$6\frac{3}{4} \times 20 = \frac{27}{4} \times \frac{20}{1}$$

$$= 135$$

ANSWER

$$6\frac{3}{4}" \times 20 = 135"$$

Example 2–35

PROBLEM

The 8-3/4' long section of angle iron in Figure 2-4 is
cut into 5 equal parts. How long should each part be?

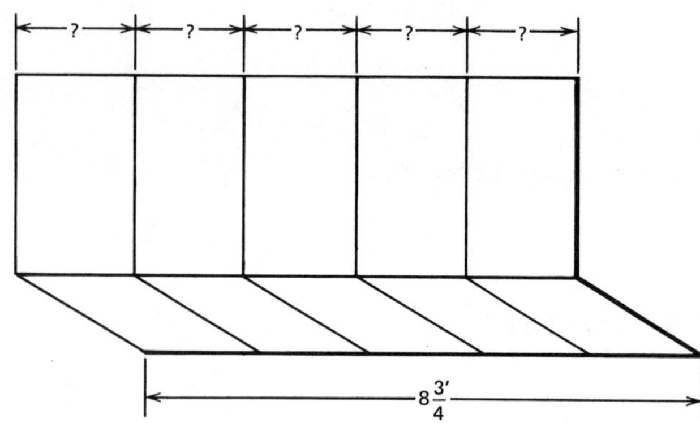

Figure 2—4

SOLUTION

Divide the number of parts into the total length:

$$8\frac{3}{4} \div 5 = \frac{\overset{7}{\cancel{35}}}{4} \times \frac{1}{\cancel{5}}$$

$$= \frac{7}{4}$$

$$= 1\frac{3}{4}$$

ANSWER

$$8\frac{3}{4}' \div 5 = 1\frac{3}{4}'$$

Applied Welding Problems

2-136.   What is the outside diameter of the pipe in Figure 2-5?

2-137.   What is the length of the end cap in Figure 2-6?

2-138.   What is the length of the pipe spacer plate in Figure 2-7?

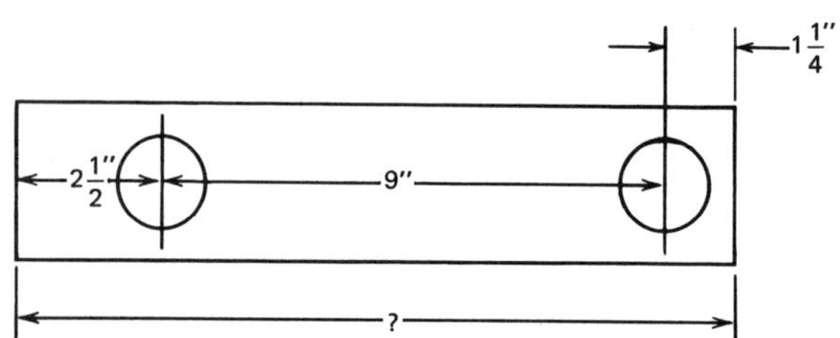

Figure 2— 7

2-139.   What is the total length of the spacers in Figure 2-8?

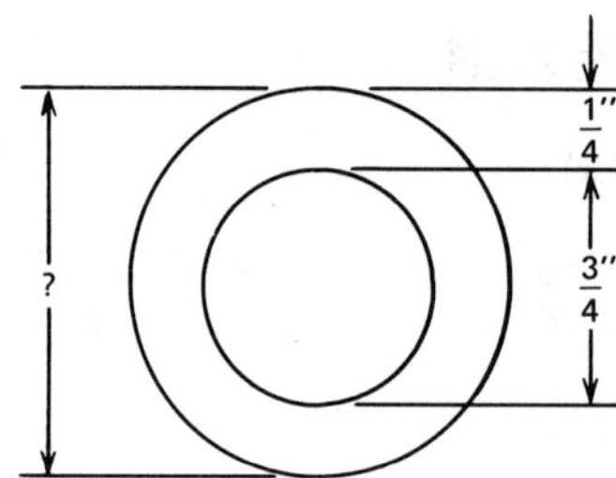

Figure 2—5

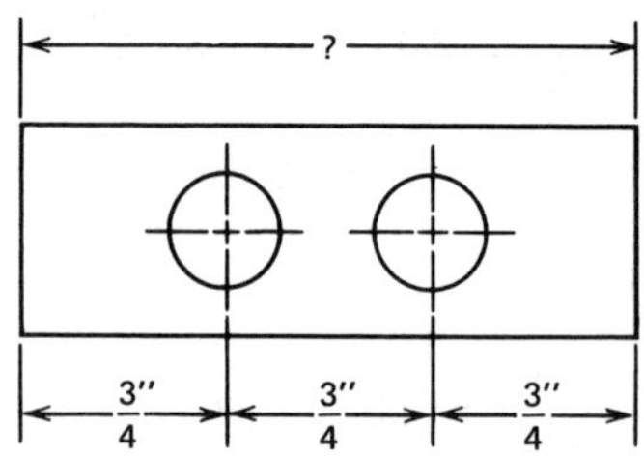

Figure 2—6

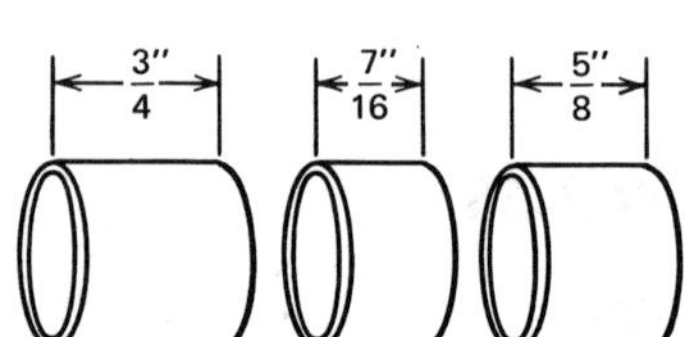

Figure 2—8

2-140. Figure 2-9 shows metal spacers.
    a. What is the total height of spacers A, B, and C?
    b. What is the combined height of B and C?
    c. What is the combined height of A and C?

2-141. Use the measurements of the turnbuckle given in Figure 2-10.
    a. Find dimension A.
    b. Find dimension B.
    c. Find dimension C.
    d. Find dimension D.

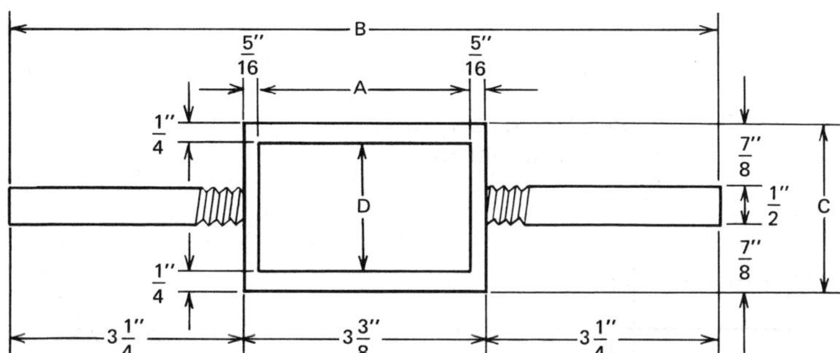

Figure 2—9

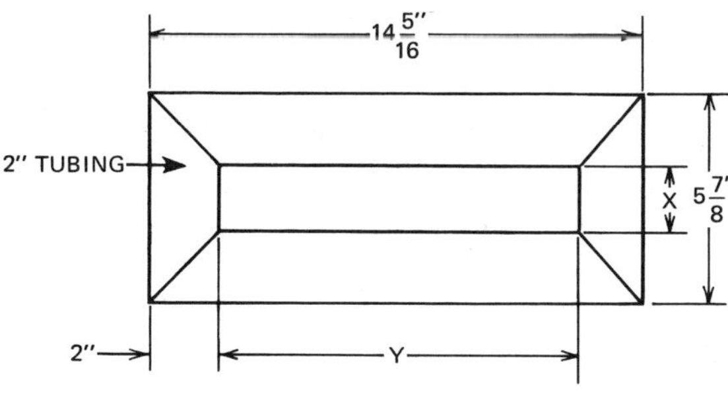

Figure 2—10

2-142. Figure 2-11 shows a grate.
    a. What is the total amount of 2" tubing needed to build the grate?
    b. How wide is dimension X?
    c. How long is dimension Y?
    d. How many inches of 2" tubing are needed to build 5 grates?

Figure 2—11

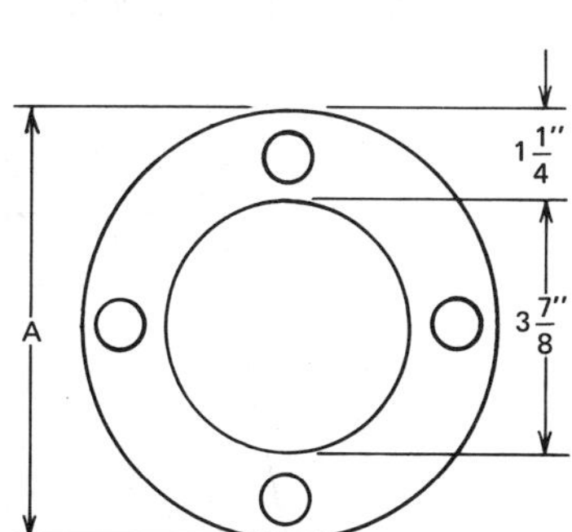

Figure 2—12

2-143. How long is dimension A of the flange in Figure 2-12?

2-144. Figure 2-13 shows two pipes and flanges.
    a. Find dimension X.
    b. Find dimension Y.
    c. Find dimension Z.

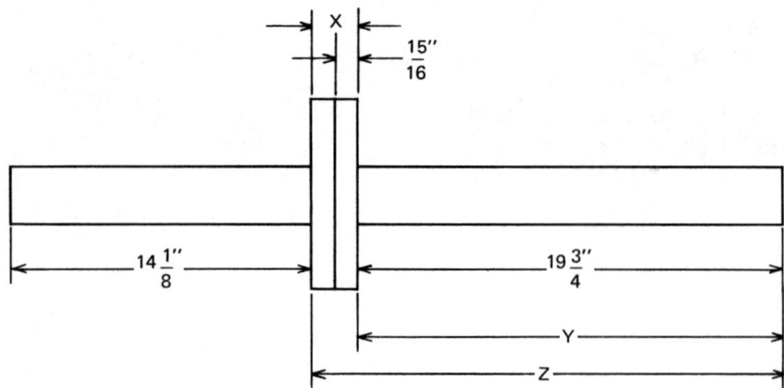

Figure 2—13

2-145. A 7-7/8" section of an 11-1/16" section is cut
from the 20" pipe in Figure 2-14. How much
material remains?

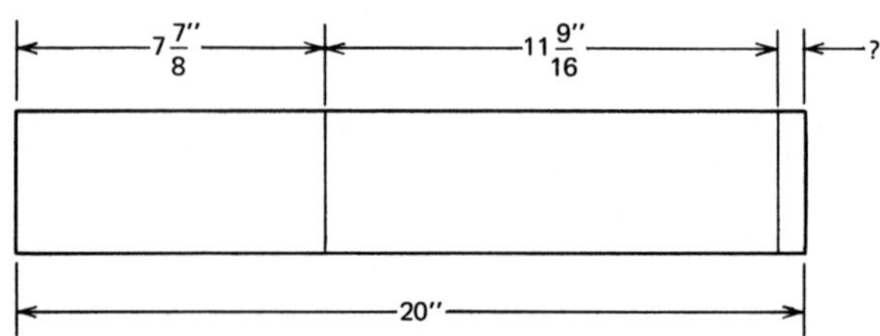

Figure 2—14

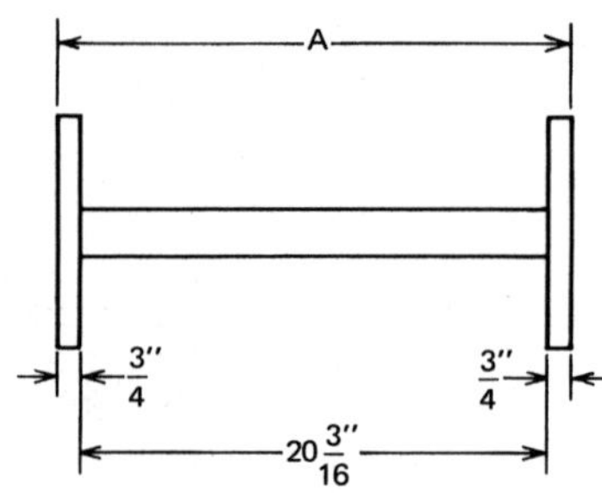

Figure 2—15

2-146. How long is dimension A of the foot scraper in
Figure 2-15?

2-147. A reinforcing web is shown in Figure 2-16.
a. Find dimension W.
b. Find dimension X.
c. Find dimension Y.
d. Find dimension Z.

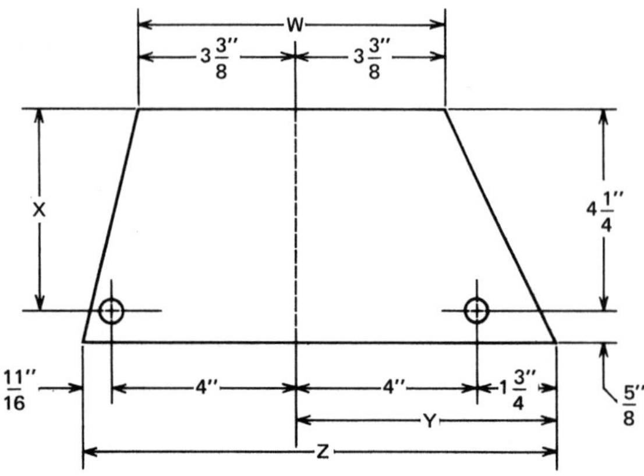

Figure 2—16

2-148.   Use the measurements of the slide rail given in
         Figure 2-17.
         a.   Find dimension A.
         b.   Find dimension B.
         c.   Find dimension C.

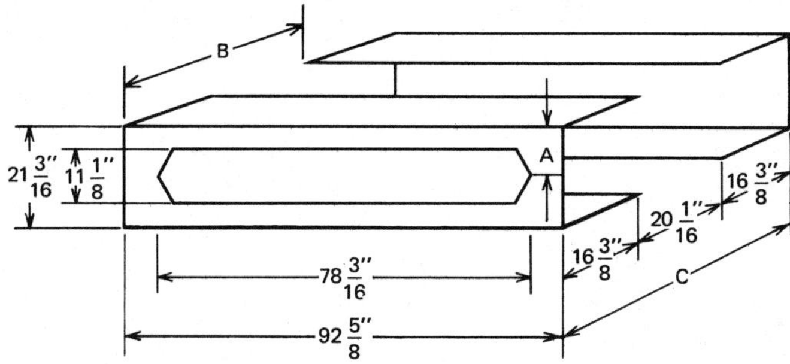

Figure 2—17

2-149.   The reinforcing channel X in Figure 2-18 is
         3/16" too long on one end and 1-1/4" too long
         on the other end.
         a.   What is the length of the channel when it
              is cut to size?
         b.   What is dimension Y?
         c.   What is dimension Z?

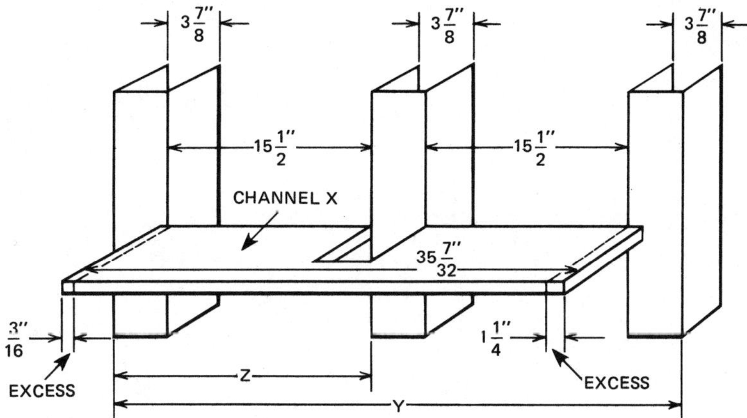

Figure 2—18

2-150.   Use the measurements of the vapor recovery
         adaptor plate given in Figure 2-19.

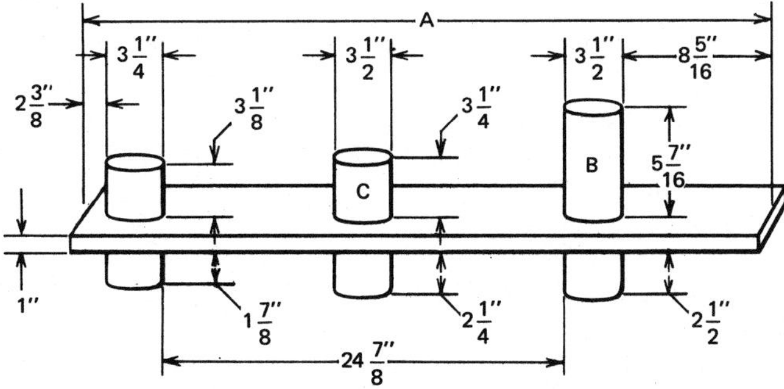

Figure 2—19

a. Find dimension A.
b. How many inches of pipe are needed to build
    the vapor recovery unit?
c. How much longer is pipe B than pipe C?

2-151. Figure 2-20 shows a pipe mounting basket.
        a. What is the diameter of pipe A?
        b. What is the overall length of pipe B?
        c. What is the outside diameter of pipe C?
        d. What is the distance between the centers of
            pipes D and A?
        e. What is dimension E?
        f. What is dimension F?

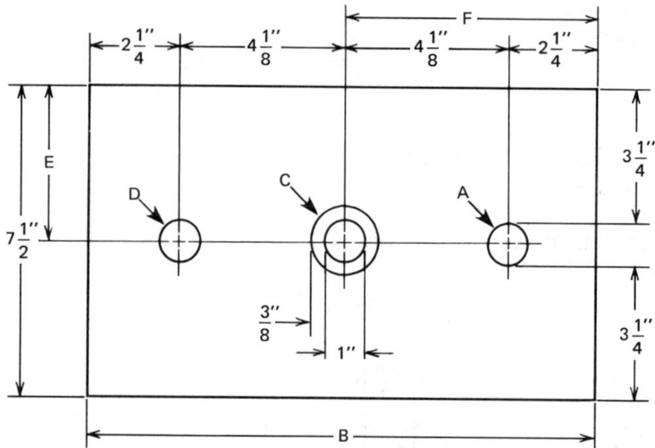

Figure 2—20

2-152. A welding shop manufactures the hinge plate
        shown in Figure 2-21.
        a. How many inches of 3/8" material are required
            for 9 hinge plates?
        b. If each hinge plate weighs 1-3/4 lb, how
            many pounds will 20 plates weigh?
        c. Parts A and B are welded all around. How
            many inches of weld are on each hinge plate?
        d. How many inches of weld are required for 9
            hinge plates?

2-153. Figure 2-22 shows a length of pipe.
        a. How many 6-1/4" lengths of pipe can be cut
            from a 65" length of pipe?
        b. The pipe base is 8-7/8" square. What length
            of 8-7/8" material is needed for 10 bases?
        c. Each unit requires 6-1/2" of weld. How
            many inches of weld are needed for 10 units?

2-154. How much does a 60' long piece of steel weigh
        if each foot weighs 1-3/4 lb?

2-155. A length of metal weighs 150 lb. How long is it
        if it weighs 2-1/2 lb per foot?

2-156. What are the center dimensions of the stiffener
        plate in Figure 2-23?

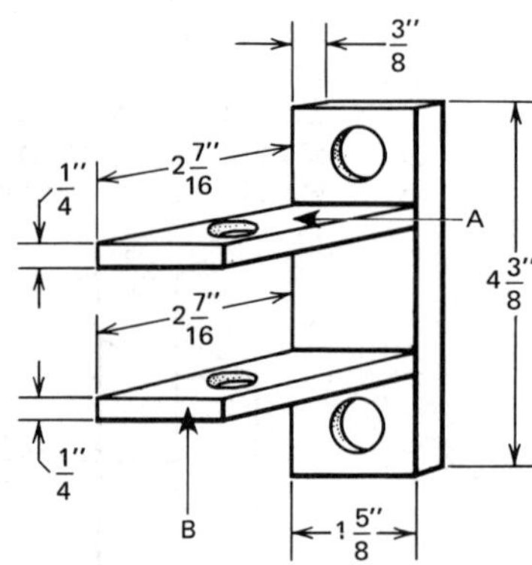

Figure 2—21

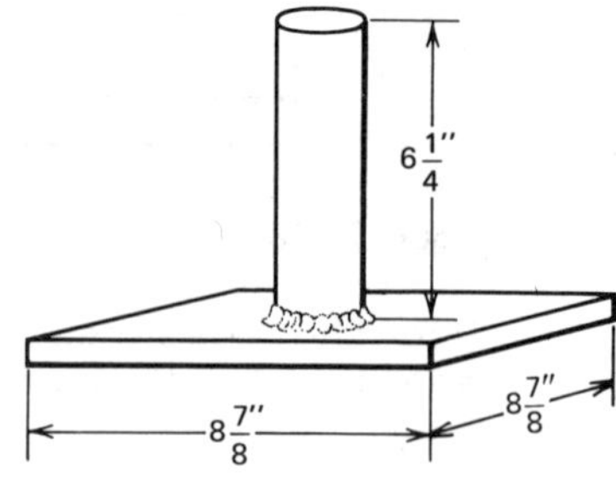

Figure 2—22

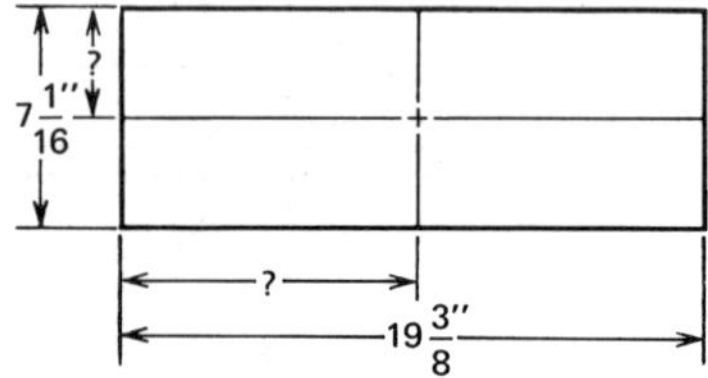

Figure 2—23

2-157.   What are the inside dimensions of the tubing air
          duct frame in Figure 2-24?

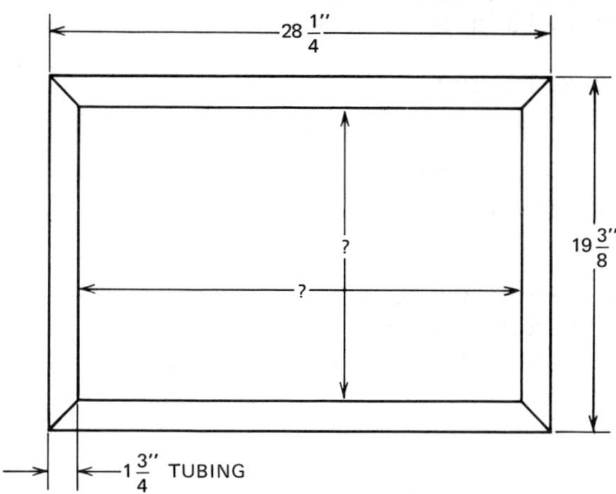

Figure 2—24

2-158.   The tank in Figure 2-25 measures 63-5/16" in
          diameter one way and 61-7/32" in diameter the
          other way.
          a.   How far out of round is the tank?
          b.   The tank has 12 support channels evenly
                spaced at 22-3/16". What is the length of the
                tank?

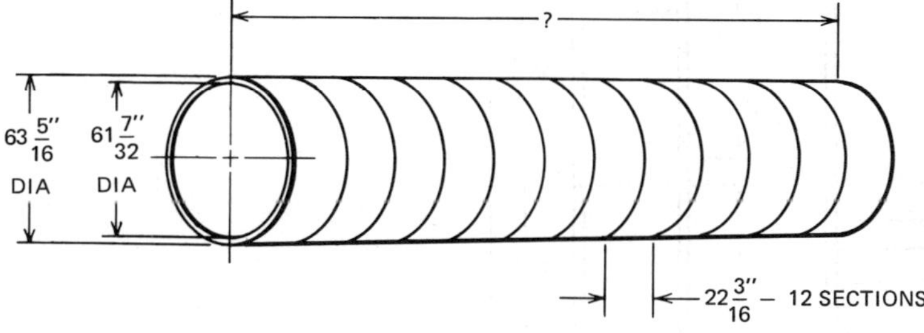

Figure 2—25

2-159.   9 sections of a 4-3/16" square tubing are cut
          from a 60" length.
          a.   How much tubing remains? (Disregard waste.)
          b.   There is 1/8" waste per cut. What is the
                total amount of waste?

2-160.   A slot is cut in the slide plate in Figure 2-26.
          What is the length of the slot?

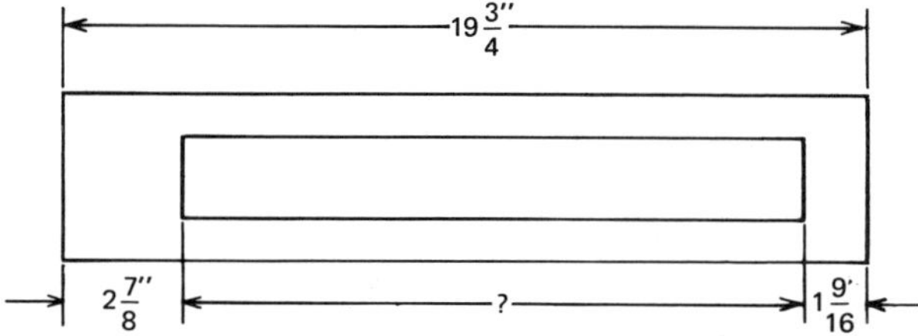

Figure 2—26

2-161. A reinforcing pad requires 16-1/4" of 3/8"
        fillet weld and 7-7/16" of 1/4" fillet weld.
        a. How many inches of 3/8" fillet weld are
            required for 42 pads?
        b. How many inches of 1/4" fillet weld are
            required for 42 pads?

2-162. If 63 sections of pipe 11-7/8" long are welded
        to a water supply tank, what is the total amount
        of pipe needed?

2-163. One-half of an order of 110 hanger brackets has
        been rejected. The brackets need an additional
        1-5/8" of weld to pass inspection. What is the
        total amount of weld needed?

2-164. Figure 2-27 shows a storage cart.
        a. Find dimension A.
        b. Find dimension B.
        c. Find dimension C.
        d. Find dimension D.
        e. Find dimension E.
        f. Find dimension F.
        g. Find dimension G.

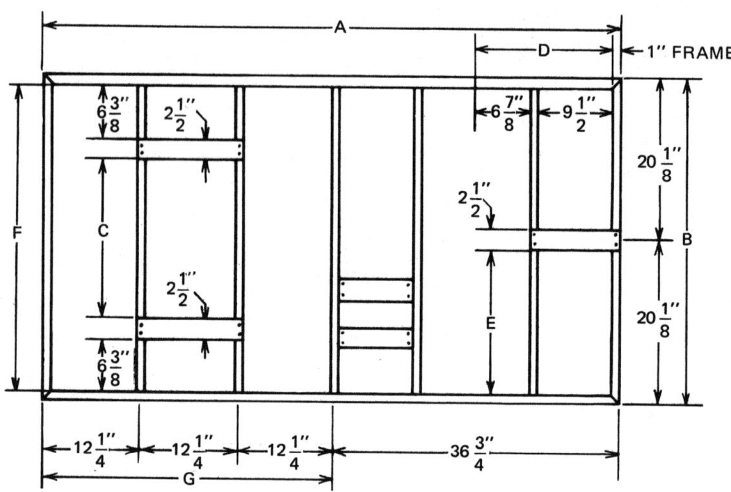

Figure 2—27

Chapter 2: Fractions

Applied Welding Problems
Survey Test

Name ____

Course/Sec. ________________   Date _____________________

2-1.   Three sheets of metal are stacked together. The   2-1. _______________
        sheets have the following thicknesses: 5/32",
        9/16", and 3/4". What is the total thickness of
        the stack?

2-2.   Find the total length of the angle iron sections   2-2. _______________
        in Figure 2-28.

$29\frac{1}{4}''$ $20\frac{1}{8}''$ $36\frac{3}{16}''$

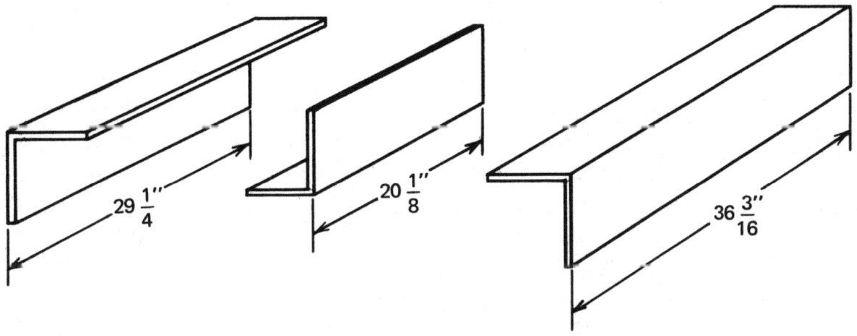

Figure 2-28

2-3.   To reinforce a farm loader, a welder used the   2-3. _______________
        following lengths of pipe: 4'5/8", 28-3/16",
        32-1/4", 37-3/4", and 44-1/8". What is the total
        length of pipe used?

2-4.   A casting was 3-7/8" long before grinding. If   2-4. _______________
        3/32" was ground off each end, determine the length
        of the casting after grinding.

2-5.   The outside diameter of a pipe is 2-1/8", and the   2-5. _______________
        wall is 3/16" thick. Find the inside diameter of
        the pipe.

2-6.   Find A in Figure 2-29.

2-6. _______________

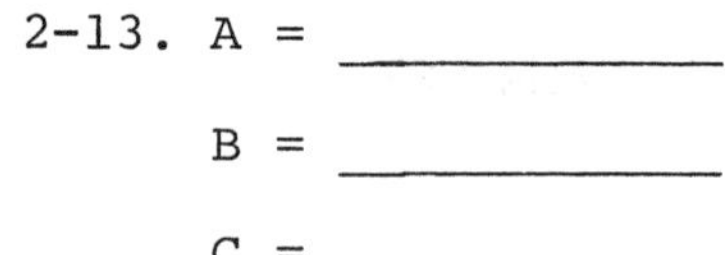

Figure 2—29

2-7.   8 pieces 2-5/8" long are to be cut from a length of round stock. How many inches of round stock are needed?

2-7. _______________

2-8.   If 9" of silver solder weighs 27/32 oz, how much does 1" weigh?

2-8. _______________

2-9.   How many pounds of welding wire are needed for a job if each unit requires 14-1/2 lb of welding wire and there are 28 units?

2-9. _______________

2-10.   How many pieces of sheet metal strips, each 3/4" wide, can be cut from a 96" piece? (Assume nothing is lost in cutting.)

2-10. _______________

2-11.   A drill bit will drill through 48" of iron per sharpening. How many 3/8" deep holes can be drilled in one sharpening?

2-11. _______________

2-12.   A welder is to place 8 reinforcement rods on a 48-1/2" angle iron. If the rods are equally spaced, how far apart are they from center to center?

2-12. _______________

2-13.   The three pipes in Figure 2-30 have a radius (distance from center to curve) of 9-1/4", 22-1/8", and 31-15/16". Find A, B, and C.

2-13. A = _______________

B = _______________

C = _______________

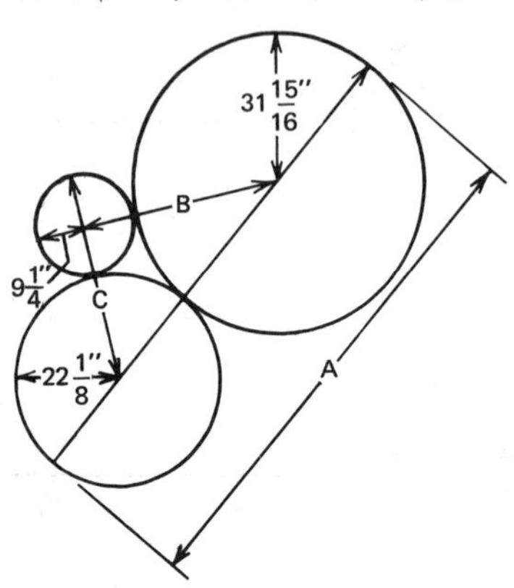

Figure 2—30

2-14. The table frame in Figure 2-31 is made from 2"
       square tubing.
       a. Find the total amount of material needed for   2-14a.________________
           the legs.
       b. Find the total amount of material needed for   2-14b.________________
           the top frame.
       c. Find the total amount of bracing material   2-14c.________________
           needed for the legs.

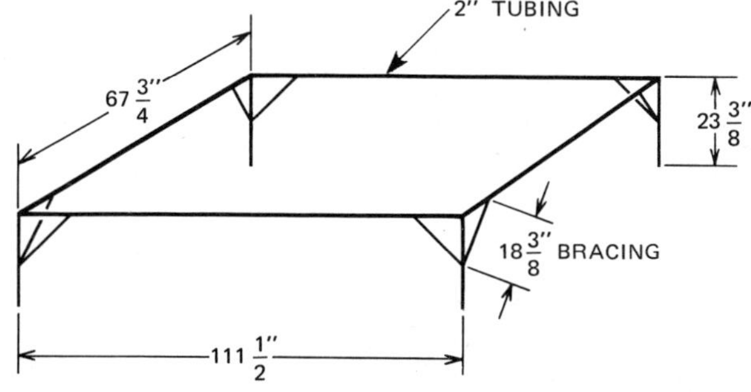

Figure 2—31

2-15. The metal steps in Figure 2-32 are constructed
       from 1/4" diamond plate and are welded into place
       on channel iron supports.
       a. How much material would be needed for 12   2-15a.________________
           treads?
       b. What would the total rise be for 12 steps?   2-15b.________________
       c. How much depth would be necessary for 12   2-15c.________________
           steps?

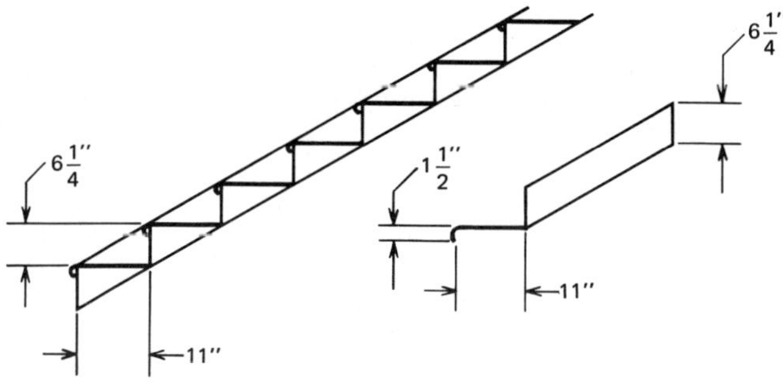

Figure 2—32

2-16. A welder can produce an arc-welded bracket every   2-16.________________
       5/8 hour. How many brackets can the welder
       complete in 10 hours?

2-17. A welder cuts the following strips from a piece of   2-17.________________
       sheet metal: 7-5/8", 9-5/32", 16-1/16", 2-3/16",
       and 11-3/4". How much is left if the original
       sheet was 48" wide and each cut wastes 1/8"?

2-18. How many pieces of square tubing 6-3/4" long can   2-18.________________
       be cut from a 260" length of tubing?

2-19. If 1' of angle iron weighs 2-3/8 lb, how much will   2-19.________________
       96' of angle iron weigh?

Chapter 3: Decimals and Decimal Fractions
Preview and Self-Test

· ·

Name ____ ____________________

Course/Sec. _______________   Date ____________________

Chapter 3 covers decimals and decimal fractions and the
fundamental operations on decimals. Before you turn to
the chapter, complete the Self-Test to determine which
sections in the chapter you need to study carefully.

Self-Test

PROPERTIES OF DECIMALS

3-1. Write fifty-two thousandths as a decimal.   3-1. ________________

3-2. Write four and four-hundredths as a fraction.   3-2. ________________

                     Properties of decimals score   ________________

CHANGING FRACTIONS TO DECIMALS

3-3. Change 3/8 to a decimal.   3-3. ________________

3-4. Change 5/16 to a decimal.   3-4. ________________

3-5. Change 14-7/32 to a decimal.   3-5. ________________

               Changing fractions to decimals score   ________________

CHANGING DECIMALS TO FRACTIONS

3-6. Change 0.125 to a fraction.   3-6. ________________

3-7. Change 32.8125 to an improper fraction.   3-7. ________________

3-8. Write 8.09375 as a mixed number.   3-8. ________________

               Changing decimals to fractions score   ________________

ROUNDING DECIMALS

3-9. Round 0.8245 to the nearest thousandth.   3-9. ________________

3-10. Round 68476.437 to the nearest hundredth.   3-10. ________________

3-11. Write 7/16 as a decimal and round to the nearest 3-11. ________________
       hundredth.

                       Rounding decimals score   ________________

ADDITION AND SUBTRACTION OF DECIMALS

3-12. Find the total rainfall for a three-day rain   3-12. ________________
       during which the following measurements were
       obtained: first day, 1.87"; second day, 3.27";
       and third day, 0.23".

3-13.   A piece of laminated wood is made up of the
        following layers of wood: 0.125", 0.0625",
        0.1875", 0.0625", and 0.125". What is its thick-
        ness if there is 0.03125" of glue between each
        layer?   3-13. _______________

3-14.   A piston measures 4.165". It is oversized by   3-14. _______________
        0.035". What is the correct size?

                Addition and subtraction of decimals score   _______________

MULTIPLICATION OF DECIMALS

3-15.   An apartment building uses 21.37 cu ft of gas per   3-15. _______________
        day. How many cubic feet would be used in 30.4
        days?

3-16.   The area of a rectangle is found by multiplying   3-16. _______________
        the length by the width. What is the area of a
        room 12.5' x 14.75'?

3-17.   A paper clip weighs 0.03524 oz. How many ounces   3-17. _______________
        will 150 paper clips weigh (to the nearest tenth
        of an ounce)?

                Multiplication of decimals score   _______________

DIVISION OF DECIMALS

3-18.   You paid $23.60 for 16.4 gal of gasoline. How   3-18. _______________
        much did you pay per gallon (to the nearest tenth
        of a cent)?

3-19.   No. 8 gage sheet metal is 0.1285" thick. How many   3-19. _______________
        sheets are in a stack 5.5255" high?

                Division of decimals score   _______________

PRACTICAL APPLICATIONS OF DECIMALS

3-20.   Find the cost per mile (to the nearest thousandth)   3-20. _______________
        for gasoline if gas costs $1.389 per gallon and
        you get 28.78 miles per gallon.

3-21.   A wood screw is 1.875" long. Write the length as   3-21. _______________
        a fraction.

3-22.   No. 14 gage wire is 0.064084" in diameter.   3-22. _______________
        Convert this figure to the nearest fraction of
        16ths.

3-23.   The weight of air is 0.0795 lb per cubic foot.   3-23. _______________
        How many cubic feet of air would weigh 5 lb (to
        the nearest hundredth)?

3-24.   If a drill feeds into the work at the rate of   3-24. _______________
        0.63 centimeters (cm) per revolution, how many
        revolutions are needed to drill a hole 7.43cm
        deep?

3-25.   High-quality North Dakota spring wheat contains   3-25. _______________
        0.158 lb of protein per pound of wheat. How many
        pounds of wheat would you need in order to obtain
        0.553 lb of protein?

                Practical applications of decimals score   _______________

After you have checked your answers on the Self-Test,
transfer your scores to the Chapter 3 Objectives.

Chapter 3

••

Decimals and Decimal Fractions

Upon successful completion of this chapter, you will
know the properties of decimals. You will be able to
perform fundamental operations on decimals, including
changing fractions to decimals, changing decimals to
fractions, and rounding decimals; and you will be able to
make practical applications of decimals and decimal
fractions.

Objectives

	Self-Test Scores	If you did poorly on the Self-Test, turn to:
• Properties of decimals	__________	Section 3-2
• Changing fractions to decimals	__________	Section 3-3
• Changing decimals to fractions	__________	Section 3-4
• Rounding decimals	__________	Section 3-5
• Addition and subtraction of decimals	__________	Section 3-6
• Multiplication of decimals	__________	Section 3-7
• Division of decimals	__________	Section 3-8
• Practical applications of decimals	__________	Section 3-9

3-1   Introduction to Decimals and Decimal Fractions

Quantities that contain a decimal point are generally called decimals. Decimals or decimal fractions make calculation much easier than do common fractions. As the hand-held calculator, which uses decimals, grows in popularity, the use of decimals will become more essential in our daily lives. Increasing use of the metric system will also result in greater use of decimals. By learning the basic procedures for working with decimals, you should be able to calculate much more accurately and easily than before.

3-2   Properties of Decimals

Fractions were developed because whole numbers were not adequate for all calculations. A decimal fraction is a special kind of fraction. Its denominator is 10 or some multiple of 10--that is, 10, 100, 1000, and so on. Some examples of decimal fractions are:

$$\frac{7}{10}, \quad \frac{31}{100}, \quad \frac{126}{1000}$$

We write decimal fractions without a denominator by placing a period called a decimal point in the numerator. There are as many figures to the right of the decimal point as there are zeros in the denominator. Thus, 7/10 is written 0.7, and 31/100 is written 0.31. You should place a zero to the left of the decimal point when there is no whole number. Thus, .7 = 0.7, and .31 = 0.31.

The fact that the number of figures to the right of the decimal equals the number of zeros in the denominator can be shown by dividing the denominator into the numerator:

$$\frac{7}{10} = 10\overline{)7.0} = 0.7$$

$$\frac{31}{100} = 100\overline{)31.00} = 0.31$$

Thus, 7/10 and 31/100 are the fractional forms, and 0.7 and 0.31 are the decimal forms, of the same quantities.

Whole numbers and fractions may also be written in decimal form:

$$\frac{4326}{1000} \quad \text{is written as} \quad 4.326$$

The decimal form can be found by dividing the
denominator into the numerator:

```
              4.326
    1000)4326.000
         4000
         ----
          3260
          3000
          ----
           2600
           2000
           ----
            6000
            6000
            ----
```

When reading a number like 4.326, we read "four
point three two six" or "four and three hundred twenty-
six thousandths." Note the word _and_ is used in place of
the decimal point. The ability to read decimal fractions
in written form and to change written decimal fractions
to numerical form is important.
The chart shows the place values for decimals:

Millions	Hundred thousands	Ten thousands	Thousands	Hundreds	Tens	Units	Decimal point	Tenths	Hundredths	Thousandths	Ten-thousandths	Hundred-thousandths	Millionths
0	0	0	0	0	0	0	.	0	0	0	0	0	0

Procedure for Writing Decimal Fractions in Decimals and in Words

1. To write a decimal fraction in decimals, place a
   decimal point in the numerator so that as many figures
   appear to the right of the decimal as there are zeros
   in the denominator.
2. If the decimal fraction is part of a mixed number,
   write the whole number to the left of the decimal
   point.
3. If there is no whole number with the decimal fraction,
   place a zero to the left of the decimal point.
4. To write a decimal fraction in words, refer to the
   chart showing the place values for decimals. Use the
   word _and_ in place of the decimal point.

Example 3–1

PROBLEM

Write the following fractions as decimals and in words:

$$\frac{5}{10}, \quad \frac{32}{100}, \quad \text{and} \quad \frac{57}{1000}$$

SOLUTION AND ANSWER

$$\frac{5}{10} = 0.5 = \text{five-tenths}$$

$$\frac{32}{100} = 0.32 = \text{thirty-two hundredths}$$

$$\frac{57}{1000} = 0.057 = \text{fifty-seven thousandths}$$

Note: There should always be as many decimal places in the decimal form as there are zeros in the denominator of the fraction. In writing 57/1000 in decimal form, we had to add a zero in front of the 57 as a place holder.

Example 3–2

PROBLEM

Write the following mixed numbers as decimals and in words:

$$3\frac{4}{100}, \quad 16\frac{41}{1000}, \quad \text{and} \quad 301\frac{301}{10,000}$$

SOLUTION AND ANSWER

$$3\frac{4}{100} = 3.04 = \text{three and four-hundredths}$$

$$16\frac{41}{1000} = 16.041 = \text{sixteen and forty-one thousandths}$$

$$301\frac{301}{10,000} = 301.0301 = \text{three hundred one and three hundred one ten-thousandths}$$

Example 3–3

PROBLEM

The pistons of a rebuilt motor were forty-thousandths inch oversize. Express this figure as a decimal and as a fraction.

SOLUTION

Forty-thousandths is written 0.040 as a decimal and 40/1000 as a fraction.

Exercises: Writing Decimal Fractions in Decimals and in Words

Write the following as decimals and in words:

3-1.   9/10   3-2.   75/100   3-3.   33/1000

3-4.   7/10,000   3-5.   7/1000   3-6.   527-17/1000

3-7.   42-17/10,000   3-8.   29-927/10,000

3-9.   376-86/1000   3-10. 144-1246/100,000

Write the following as decimals and as fractions:

3-11.   nine-tenths

3-12.   sixty-five thousandths

3-13.   ten and ten-hundredths

3-14.   ninety-nine and sixty-four ten-thousandths

3-15.   four hundred forty-seven thousand and forty-nine
        millionths

Make the following changes:

3-16.   Write thirty-six hundredths inch as a decimal and
        as a fraction.

3-17.   No. 28 gage copper wire has a diameter of 0.320".
        Write this figure as a fraction and in words.

3-18.   The width of a compact car is listed as ninety-
        three and twenty-five hundredths inches. Write
        this figure as a mixed number and as a decimal.

3-19.   Eight-thousandths inch is planed off the cylinder
        head of a motorcycle. Write this figure as a
        decimal and as a fraction.

3-20.   The thickness of a plastic bag is given as 1.5
        mils. (1000 mils = 1".) Write this figure as a
        decimal and as a fraction.

3-3   Changing Fractions to Decimals

Not all fractions have denominators that are multiples of
10. However, all fractions can be changed to decimals by
dividing the denominator into the numerator. Sometimes a
repeating decimal may result (the same digit repeats).
In this case, we place a small bar or dot above the
repeating digit or add three dots after the digit.

Procedure for Changing Fractions to Decimals

1. Place a decimal point to the right of the numerator
    and add one or more zeros.

2. Divide the numerator by the denominator.
3. If the result is not even, carry out the division to the desired number of places.
4. If the last digit in the quotient repeats, place a small bar or dot above the digit or add three dots after the digit.

•••

Example 3–4

PROBLEM

Change 1/4 to a decimal.

SOLUTION

1. Divide the denominator (4) into the numerator (1).

2. Place the decimal point to the right of the 1 and add two zeros.

3. Divide the denominator into the numerator. Remember to bring the decimal point up into the quotient:

```
     0.25
  4)1.00
     8
    ‾20
     20
```

Note: The numerator always goes inside the division sign.

ANSWER

$$\frac{1}{4} = 0.25$$

Example 3–5

PROBLEM

Change 2/3 to a decimal.

SOLUTION

1. Divide the denominator (3) into the numerator (2).

2. Place the decimal point after the 2 and directly above in the quotient.

3. Add zeros and then divide the denominator into the numerator:

```
     0.666
  3)2.000
     1 8
     ‾20
      18
      ‾20
       18
```

4. The result is called a repeating decimal. To indicate that it is repeating, place a bar or a dot above the last digit or three dots after it.

ANSWER

$$\frac{2}{3} = 0.66\overline{6} \quad \text{or} \quad 0.66\dot{6} \quad \text{or } 0.666 \ldots$$

Note: A repeating decimal can also be written as a combination of a decimal and a fraction:

$$\frac{2}{3} = 0.66\frac{2}{3}$$

Example 3–6

PROBLEM

Change 5/16 to a decimal.

SOLUTION

1. Divide the denominator (16) into the numerator (5).

2. Place the decimal point after the 5 and directly above in the quotient.

3. Add zeros and then divide the numerator into the denominator:

```
      0.3125
16 ) 5.0000
     4.8
      20
      16
       40
       32
        80
        80
```

ANSWER

$$\frac{5}{16} = 0.3125$$

Exercises: Changing Fractions to Decimals

Change the following to decimals:

3-21. 1/2   3-22. 3/4   3-23. 1/8

3-24. 3/8   3-25. 7/16   3-26. 1/32

3-27. 15/32   3-28. 5/12   3-29. 43/64

3-30. 45/128

Procedure for Changing Mixed Numbers to Decimals

●●

1. Write the whole number to the left of the decimal
    point.
2. Change the fraction to a decimal by dividing the
    denominator into the numerator.

●●

Example 3–7

PROBLEM

Change 12-5/8 to a decimal.

SOLUTION

1. Write 12 to the left of the decimal point.

2. Divide the denominator into the numerator. Remember
    to place the decimal point after the 5 and to bring
    the decimal point up into the quotient, add zeros, and
    then divide:

```
        0.6 25
    8 ) 5.000
        4 8
          20
          16
           40
           40
```

ANSWER

    12-5/8 = 12.625

Exercises: Changing Mixed Numbers to Decimals

Change the following to decimals:

3-31.   7-1/4   3-32.   10-1/2   3-33.   9-3/4

3-34.   25-7/8   3-35.   144-5/32

3–4   Changing Decimals to Fractions and Mixed Numbers

Converting a decimal into a fraction or a mixed number
may sometimes be necessary.

Procedure for Changing Decimals to Fractions

• •

1. Write the quantity as a fraction with 10 or a multiple of 10 as the denominator.
2. Reduce to lowest terms.
3. If there is a whole number, write the whole number and then convert the decimal part to a fraction.

• •

Example 3–8

PROBLEM

Write 0.75 as a fraction.

SOLUTION (Method 1)

1. Read 0.75 as "seventy-five hundredths."

2. Write it as a fraction: 75/100.

3. Reduce by dividing the numerator and the denominator by 25:

$$\frac{75 \div 25}{100 \div 25} = \frac{3}{4}$$

ANSWER

$$0.75 = \frac{3}{4}$$

Example 3–9

PROBLEM

Change 0.3125 to a fraction.

SOLUTION (Method 2)

1. Write 0.3125 as a fraction by using 3125 as the numerator and 10,000 as the denominator. Note that the numerator is written as a whole number and that the denominator has as many zeros as there are digits to the right of the decimal point. You can also think of this step as multiplying the numerator and the denominator by 10,000 or as moving the decimal point to the right four places in the numerator and in the denominator:

$$0.3125 = \frac{3125}{10,000}$$

2. Reduce by dividing by 25 two times:

$$\frac{3125 \div 25}{10,000 \div 25} = \frac{125 \div 25}{400 \div 25} = \frac{5}{16}$$

ANSWER

$$0.3125 = \frac{5}{16}$$

Example 3–10

PROBLEM

Change 25.875 to a mixed number.

SOLUTION

1. Write 25 as the whole number.

2. Change 0.875 to a fraction by writing 875 as the numerator and 1000 as the denominator.

3. Reduce by dividing the numerator and the denominator by 25 and by 5:

$$25\frac{875 \div 25}{1000 \div 25} = 25\frac{35 \div 5}{40 \div 5} = 25\frac{7}{8}$$

ANSWER

$$25.875 = 25\frac{7}{8}$$

Exercises: Changing Decimals to Fractions and Mixed Numbers

Change the following to mixed numbers:

3-36. 7.5	3-37. 9.75	3-38. 10.25
3-39. 15.375	3-40. 49.625	3-41. 42.1875
3-42. 60.4375	3-43. 79.6875	3-44. 8.15625
3-45. 144.8125		

3–5   Rounding Decimals

When telling someone the number of miles on your car, you may say the mileage gauge shows 27,000 when the number is actually 27,437.4. This process is called rounding, or rounding off. Sometimes it is desirable to "round off" a number when the given number of places of accuracy (digits) is not necessary.

To round off a number, first decide how accurate the answer should be, or how many places are desired. Then look at the digit to the right of the number in the place of accuracy desired. If the digit is 5 or more, round the number to the left by increasing it by 1. If it is less than 5, do not change the number. The examples illustrate this procedure.

Procedure for Rounding Decimals

• •

1. Decide how many places of accuracy are desired.
2. Look at the digit to the right of the number in the
    place of accuracy desired.
3. If the digit to the right of the number is 5 or more,
    increase the number by 1.
4. If the digit to the right of the number is less than
    5, do not change the number, but replace all the
    digits to the right of the number by zeros.

• •

Example 3–11

PROBLEM

Round 4358.5 (a) to the nearest hundred, then (b) to the
nearest ten, and finally (c) to the nearest unit.

SOLUTION (a)

1. To round to the nearest hundred, look at the digit to
    the right of the 3, the number in the hundreds place.

2. Since that digit is 5, increase the 3 in the hundreds
    place by 1.

ANSWER (a)

4358.5 rounded to the nearest hundred is 4400.

SOLUTION (b)

1. To round to the nearest ten, look at the digit to
    the right of the 5, the number in the tens place.

2. Since that digit is 8 and is greater than 5, increase
    the 5 in the tens place by 1.

ANSWER (b)

4358.5 rounded to the nearest ten is 4360.

SOLUTION (c)

1. To round to the nearest unit, look at the digit to
    the right of the 8, the number in the units place.

2. Since that digit is 5, increase the 8 in the units
    place by 1.

ANSWER (c)

4358.5 rounded to the nearest unit is 4359.

Example 3–12

PROBLEM

Round 716.3295 (a) to the nearest hundred, (b) to the nearest tenth, (c) to the nearest hundredth, and (d) to the nearest thousandth.

SOLUTION (a)

1 is to the right of the 7 in the hundreds place and is less than 5, so replace the digits to the right of the 7 by zeros.

ANSWER (a)

716.3295 rounded to the nearest hundred is 700.

SOLUTION (b)

2 is to the right of the 3 in the tenths place and is less than 5, so replace the digits to the right of the 3 by zeros.

Note: If the zeros are decimal numbers to the right of the desired place of accuracy, they may be eliminated from the answer.

ANSWER (b)

716.3295 rounded to the nearest tenth is 716.3.

SOLUTION (c)

9 is to the right of the 2 in the hundredths place and is greater than 5, so increase the 2 by 1.

ANSWER (c)

716.3295 rounded to the nearest hundredth is 716.33.

SOLUTION (d)

5 is to the right of the 9 in the thousandths place, so increase the 9 by 1. Because the sum is 10, carry the 1 to the 2 in the hundredths place.

ANSWER (d)

716.3295 rounded to the nearest thousandth is 716.330.

Exercises: Rounding Decimals

Round the following:

3-46.   Round 1.387 to the nearest hundredth.

3-47.   Round 3929 to the nearest hundred.

3-48.   Round 5.273 to the nearest tenth.

3-49.   Round 29,846.5 to the nearest thousand.

3-50.   Round 4,569,346.59 to the nearest ten thousand.

3-51.   Round 617.3395 to the nearest ten.

3-52.   Round 555.55 to the nearest tenth.

3-53.   Round 4545.45 to the nearest unit.

3-54.   Round .0498345 to the nearest ten-thousandth.

3-55.   Round 5.54449 to the nearest thousandth.

3–6   Addition and Subtraction of Decimals

Adding and subtracting decimals are operations similar to
adding and subtracting whole numbers. But placing the
decimal point in the correct position is important. To
add or subtract decimals, arrange them so that the decimal
points are aligned in a straight, vertical column.
    Add and carry from one column to the other as in
adding whole numbers and place the decimal point in the
sum, directly beneath the decimal points above.
    To subtract decimals, write the decimals in vertical
form and subtract as whole numbers. Borrow the same way
you do with whole numbers. The decimal point in the
difference is placed directly below the decimal points in
the subtrahend and minuend.

Procedure for Addition and Subtraction
of Decimals

••

1.   Arrange the numbers so that the decimal points are
     aligned in a vertical column.
2.   To add, proceed as in adding whole numbers and bring
     the decimal point down to the sum.
3.   To subtract, proceed as in subtracting whole numbers
     and bring the decimal point down to the difference.

••

Example 3–13

PROBLEM

Add 10.46, 5.006, 144.82, 5280.6, and 0.00492.

SOLUTION

1.   Arrange the addends in columns.

2. Make certain the decimals are aligned vertically and
    add:

```
        10.46
         5.006
       144.82
      5280.6
    +     0.00492
      5440.89092
```

ANSWER

    10.46 + 5.006 + 144.82 + 5280.6 + 0.00492 =
    5440.89092

CHECK

Add upward to obtain the same sum.

Example 3–14

PROBLEM

Subtract 12.398 from 37.1.

SOLUTION

1. Arrange the larger quantity (minuend) above the
    smaller quantity (subtrahend), with the decimals in
    a vertical column. You may wish to add zeros to make
    subtraction easier.

2. Borrow as necessary and subtract:

```
      37.100
    - 12.398
      24.702
```

ANSWER

    37.1 - 12.398 = 24.702

CHECK

Add the difference and the subtrahend to obtain the
minuend:

```
      12.398
    + 24.702
      37.100
```

Example 3–15

PROBLEM

A plumber cut the following length from a 10' (120")
length of 3/4" copper pipe: 12.375", 25.625", and
73.125". How much pipe is left?

SOLUTION

1. Add the amounts used:

        12.375
        25.625
      + 73.125
       111.125

2. Subtract this sum from the original amount (if
    there is no decimal in a number, the decimal point
    is placed after the last digit):

        120.000
      - 111.125
          8.875

ANSWER

    120" - 111.125" = 8.875"

Exercises: Addition and Subtraction of Decimals

Perform the indicated operations:

3-56.   23.97 + 14.34 + 5.98

3-57.   25.37 + 3.16 + 125.08

3-58.   6.284 + 14.3 + 291.0006

3-59.   0.007 + 0.070 + 0.777 + 7.794

3-60.   2.9834 + 0.5627 + 739.246 + 39.0007 + 576.321

3-61.   0.47 - 0.34   3-62.   4.392 - 3.967

3-63.   7.205 - 6.3976   3-64.   100 - 49.875

3-65.   0.00574 - 0.00093

3-66.   Find the total cash income of a small business
        for a week if the daily income is as follows:
        $198.57, $276.43, $204.29, $187.56, and $294.63.

3-67.   Find the total number of hours of a part-time
        worker who worked 3.25 hours on Monday, 2.23 hours
        on Tuesday, 5.46 hours on Wednesday, 2.15 hours on
        Thursday, and 6.17 hours on Friday.

3-68.   How much fence will be needed to enclose an odd-
        shaped lot with the following dimensions: 110.5',
        132.75', 120.25', and 122.75'?

3-69.   A family spends $39.79 and $83.46 for groceries.
        How much money is left if they have budgeted $150
        for groceries?

3-70.   Determine the outside diameter of the pipe shown
        in Figure 3-1.

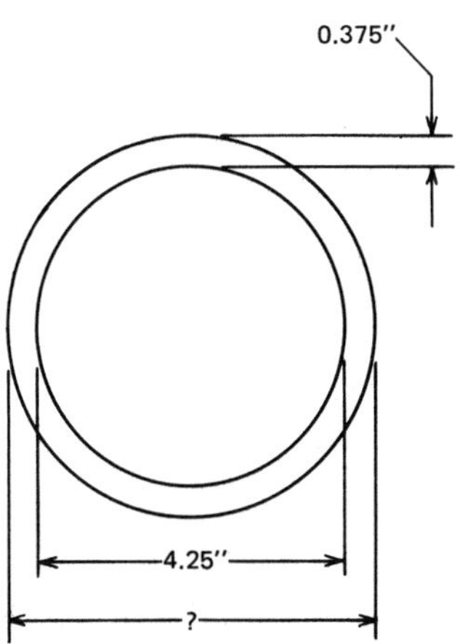

Figure 3—1

3–7   Multiplication of Decimals

Multiplying decimals is similar to multiplying whole
numbers. Write the decimal numbers under each other just
as you do when working with whole numbers. Multiply the
two decimal numbers. Then count the total number of
decimal places (places to the right of the decimal point)
in the two quantities that you multiplied and count off
the same number of places in the product. Check by
reversing the two decimals and multiplying.

Procedure for Multiplication of Decimals

•••

1.   Arrange vertically the decimal numbers to be
     multiplied.
2.   Multiply the two numbers as if they were whole
     numbers.
3.   Count the total number of decimal places in the two
     numbers.
4.   Count off the same number of places in the product,
     from right to left, and insert a decimal point in the
     product to mark that number of places.
5.   Check by reversing the two decimal numbers and
     multiplying.

•••

Example 3–16

PROBLEM

Multiply 33.5 by 6.37.

SOLUTION

1.   Arrange vertically the decimal numbers to be multi-
     plied and multiply as if they were whole numbers.

2.   Count the total number of decimal places in the two
     decimal numbers.

3.   Count off the same number of places in the product
     and insert a decimal point:

```
     33.5    ( 1 place)
   x 6.37    (+ 2 places)
     2345
     1005
     2010
   213.395   ( 3 places)
```

ANSWER

     33.5 x 6.37 = 213.395

Example 3–17

PROBLEM

Multiply 42.9 by 0.000086.

SOLUTION

1. Arrange vertically the numbers to be multiplied and
    multiply as if they were whole numbers.

2. Count the total number of decimal places in the two
    decimal numbers.

3. Count off this total in the product and insert a
    decimal point:

```
        42.9      ( 1 place)
   x  .000086     (+ 6 places)
       2574
       3432
   0.0036894      ( 7 places)
```

Note: Two zeros must be added in the product to obtain
the same number of places that are in the two decimals
that were multiplied.

ANSWER

    42.9 x 0.000086 = 0.0036894

Exercises: Multiplication of Decimals

Perform the indicated operation:

3-71.   0.73 x 6   3-72.   8.4 x 0.9

3-73.   7.89 x 1.2   3-74.   0.684 x 98

3-75.   8.324 x 0.67   3-76.   14.84 x 0.84

3-77.   1000 x 0.874   3-78.   1001 x 39.6

3-79.   17,000 x 0.135   3-80.   9876.46 x 0.105

3–8   Division of Decimals

Dividing decimals is similar to dividing whole numbers,
except that placing the decimal point must be considered.

Procedure for Division of Decimals

•••

1. Place the number to be divided (dividend) inside the
    division sign and place the divisor in front of it.

2. If the divisor contains a decimal point, move the
    decimal point in the divisor to the right of the last
    digit.
3. Move the decimal point in the dividend the same
    number of places to the right. If necessary, add
    zeros.
4. Bring the decimal point in the dividend directly up
    into the quotient; divide as a whole number and add
    zeros as needed.
5. Check by multiplying the divisor by the quotient to
    obtain the dividend.

••

Example 3–18

PROBLEM

Divide 1.440 by 24.

SOLUTION

1. Place 1.440 (dividend) inside the division sign and
    24 (divisor) in front of it.

2. Place the decimal point in the quotient and divide.

3. Since 24 will not divide into 14, place a zero in the
    quotient and determine how many times 24 will divide
    into 144 (6 times):

```
        0.06
    24 )1.44
        144
```

ANSWER

    1.440 ÷ 24 = 0.06

CHECK

    24 x 0.06 = 1.44

Example 3–19

PROBLEM

Divide 1776 by 71.04.

SOLUTION

1. Place 1776 (dividend) inside the division sign and
    71.04 (divisor) in front of it; place a decimal point
    after 1776:

```
    71.04 )1776.
```

2. Move the decimal point to the right of the last digit
    in the divisor.

3. Move the decimal point in the dividend the same

number of places (2) to the right. To do so, add two
zeros to the dividend:

$$71.04, \overline{)177600,}$$

4. Place the decimal point in the quotient and divide as
   a whole number:

```
              25.
71.04, )177600,
        14208
        35520
        35520
```

ANSWER

   $1776 \div 71.04 = 25$

CHECK

   $71.04 \times 25 = 1776$

Example 3–20

PROBLEM

Divide 0.07832 by 0.0016.

SOLUTION

1. Place 0.07832 (dividend) inside the division sign and
   0.0016 (divisor) in front of it:

   $$0.0016 \overline{)0.07832}$$

2. Move the decimal point to the right 4 places in the
   divisor and in the dividend:

   $$00016, \overline{)00783,20}$$

3. Place the decimal point in the quotient and divide as
   a whole number:

```
         48.95
16. )783.20
     64
     143
     128
     152
     144
      80
      80
```

ANSWER

   $0.07832 \div 0.0016 = 48.95$

CHECK

   $0.0016 \times 48.95 = 0.07832$

Exercises: Division of Decimals

Perform the indicated operation:

3-81.   0.16 ÷ 2

3-82.   0.18 ÷ 0.6

3-83.   38.6 ÷ 0.2

3-84.   76.4 ÷ 0.02

3-85.   123.57 ÷ 137.4

3-86.   29.145 ÷ 0.000067

3-87.   2576 ÷ 2.24

3-88.   5280 ÷ 2993.76

3-89.   Divide 0.25956 by 0.412.

3-90.   Divide 4683 by 3947.769.

3-9   Practical Application of Decimals

We make numerous applications of decimals in our daily
lives. The use of computers, hand-held calculators, and
the metric system will result in even greater use of
decimals in the future. (Of course, after obtaining a
decimal solution, we must sometimes round the result to
the nearest cent, dollar, mile, and so on.) Some applications of decimals are illustrated by the following
examples.

Example 3-21

PROBLEM

A checkout clerk earns $261.20 (before taxes) in a
40-hour week. How much does the clerk earn per hour?

SOLUTION

1. Divide the hours worked (40) into the total earned.

2. Place the decimal in the quotient and divide:

```
        6.53
40 )261.20
    240
    ‾‾‾
     212
     200
     ‾‾‾
      120
      120
      ‾‾‾
```

ANSWER

The clerk earns $6.53 per hour:

    $261.20 ÷ 40 = $6.53

Example 3–22

PROBLEM

Find the cost of 145.5' of speaker wire at 3.7¢ per foot.

SOLUTION

1. Each foot costs 3.7¢ or $0.037, so multiply $0.037
    by 145.5:

```
      145.5
    x 0.037
      10185
      4365
      5.3835
```

2. Round to the nearest hundredth:

    5.3835 = 5.38

ANSWER

The cost of the speaker wire is $5.38:

    $0.037 x 145.5 = $5.38

Example 3–23

PROBLEM

Find the miles per gallon if you use 25.6 gal to drive
371.2 mi.

SOLUTION

1. Divide the gallons (25.6) into the miles driven
    (371.2) to find the miles per gallon:

    25.6) 371.2

2. Move the decimal point one place to the right in the
    divisor and in the dividend, place the decimal point
    in the quotient, and divide. Add zeros as needed:

```
            14.5
    256. ) 3712.0
           256
           1152
           1024
           1280
           1280
```

ANSWER

    371.2 mi ÷ 25.6 gal = 14.5 mpg

Exercises: Practical Applications of Decimals

Solve the following:

3-91.   A weekly time card reads 8.10, 7.94, 8.32, 8.14,
and 8.68 hours. Find the total number of hours
worked.

3-92.   Three cartons on a delivery van weigh 106.4
kilograms (kg), 283.5kg, and 421.7kg. What is
their total weight?

3-93.   A stainless steel rod is 48" long. Three pieces,
each 4.875" long, are cut off. Each cut wastes
0.125". How much of the rod is left?

3-94.   A discount of $14.83 is allowed on a purchase of
$494.33 if paid in 10 days. What is the bill
after the discount?

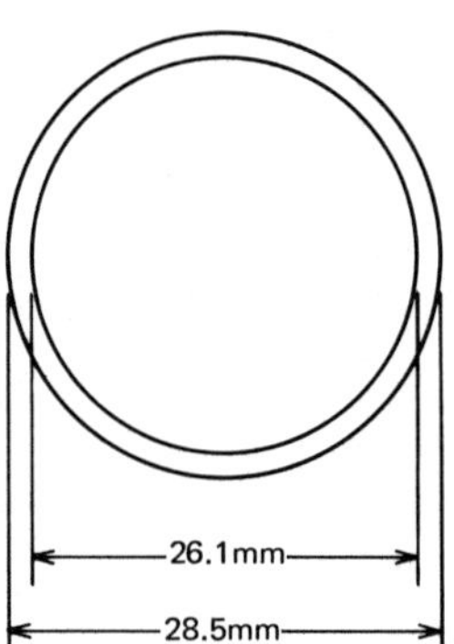

3-95.   Find the thickness of the pipe shown in Figure 3-2. **Figure 3—2**

3-96.   A machinist earns $8.37 per hour. She works 43
hours one week. Find her total check if she is
paid time and a half for any time over 40 hours.

3-97.   A lot sells for $25,412.63 and contains 14,520
sq ft. How much is paid per square foot?

3-98.   An oil well flows at 1.85 barrels per hour. The
price of crude is $15.37 per barrel. How many
dollars will the well produce per 24-hour day?

3-99.   How many rivets are in a 25 lb box if each rivet
weighs 0.0625 lb?

3-100. Each revolution of a wheel takes 7.3304'. How
many revolutions (to the nearest whole revolution)
are made when driving 6.4 miles (5280' = 1 mi)?

Chapter 3

Decimals and Decimal Fractions

Solving Applied Welding Problems

Now that you have mastered the fundamental operations on decimals and decimal fractions, you are ready to apply these math skills to typical problems that welders encounter on the job. After studying the examples that follow, complete the applied welding problems.

Example 3–24

PROBLEM

What is the total amount of steel tubing needed to fabricate the work table top shown in Figure 3-3?

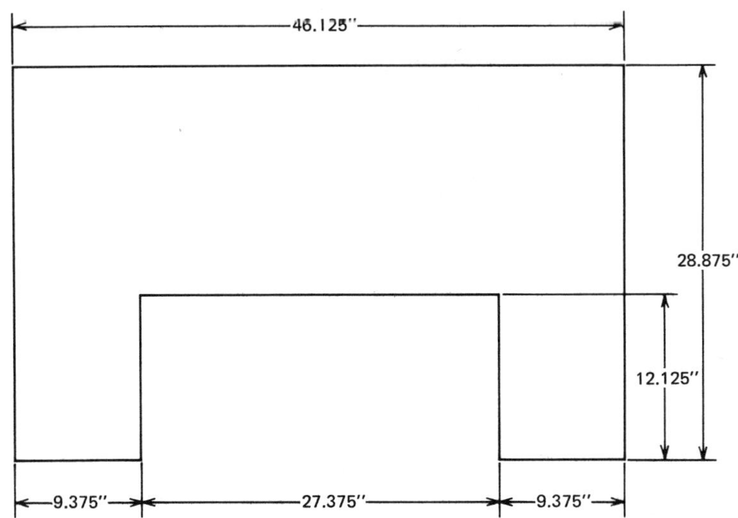

Figure 3—3

SOLUTION

Arrange in proper columns the lengths of material needed for each side and add:

```
      28.875
      28.875
      46.125
       9.375
       9.375
      27.375
      12.125
   +  12.125
     174.250
```

ANSWER

174.25" of tubing are needed.

Example 3–25

PROBLEM

What is the total length of the 6 slots cut in the adapter plate shown in Figure 3-4?

SOLUTION

Multiply the length of each slot by the number of slots:

```
   2.3125
   x    6
  13.8750
```

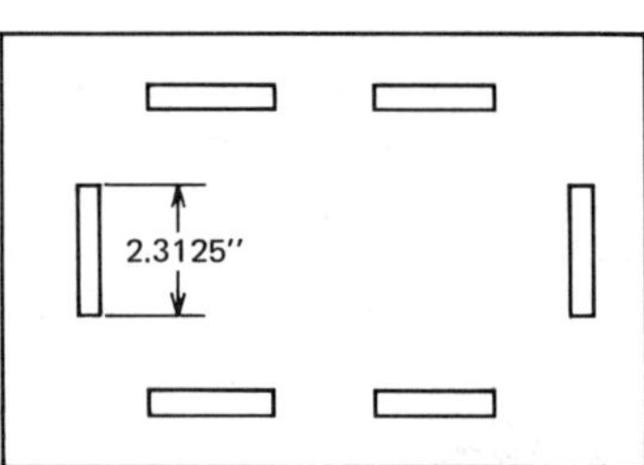

Figure 3—4

ANSWER

The total length of the 6 slots is 13.875".

Example 3–26

PROBLEM

The total weight of 152 trailer frames is 38,000 lb. What is the weight of each frame?

SOLUTION

Divide the total weight of the frames by the number of units:

```
          250
  152 )38000
        304
        760
        760
        000
        000
```

ANSWER

The weight of each frame is 250 lb.

Example 3–27

PROBLEM

The cost per pound of a trailer frame is $1.25. How much does a completed 250 lb frame cost?

SOLUTION

Multiply the weight of the trailer frame by the cost per pound:

```
    250
  x 1.25
   1250
    500
   250
  31250
```

ANSWER

The cost of a completed frame is $312.50.

Applied Welding Problems

3-101. Figure 3-5 shows a table frame.
  a. What is the total amount of square tubing needed to build the table frame?
  b. What is the amount of tubing used for the legs?

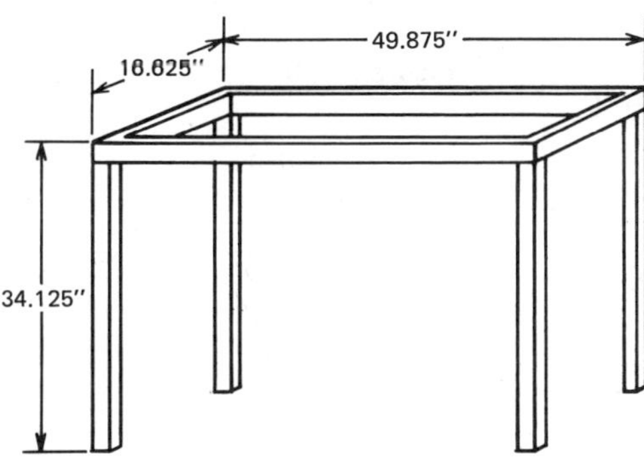

Figure 3—5

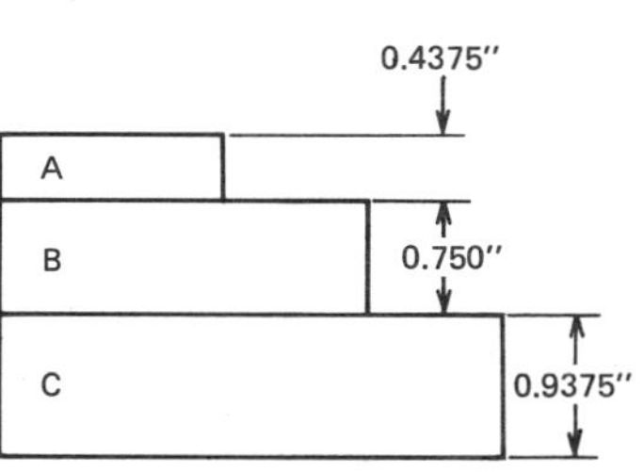

Figure 3—6

3-102. Refer to Figure 3-6.
  a. What is the total height of shims A, B, and C?
  b. What is the combined height of shims B and C?
  c. What is the combined height of shims A and B?

3-103. A motor mount bracket is shown in Figure 3-7.
  a. What is its overall length?
  b. What is its overall width?
  c. What is the total length of the 4 slot openings in the bracket?
  d. What is dimension A?
  e. What is dimension B?

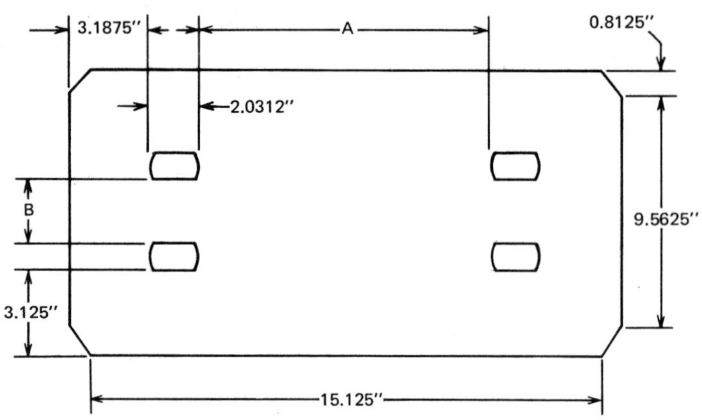

Figure 3—7

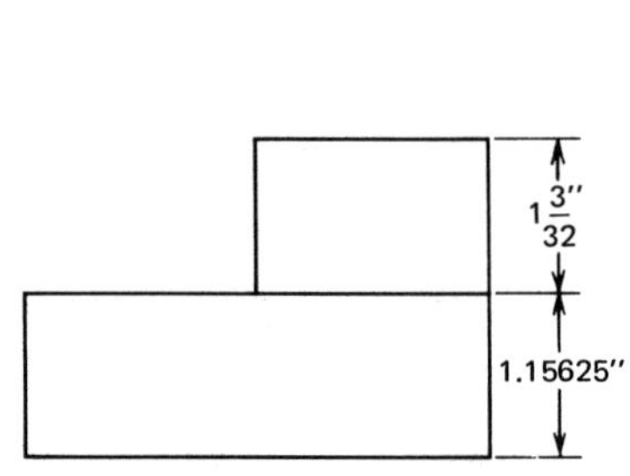

Figure 3—8

3-104. After consulting a blueprint, a welder knows that the 34-5/8" channel he has is 4.6878" too long. How long should the channel be?

3-105. What is the total height of the two shims in Figure 3-8?

3-106. Refer to Figure 3-9.
    a. Find dimension A.
    b. Find dimension B.

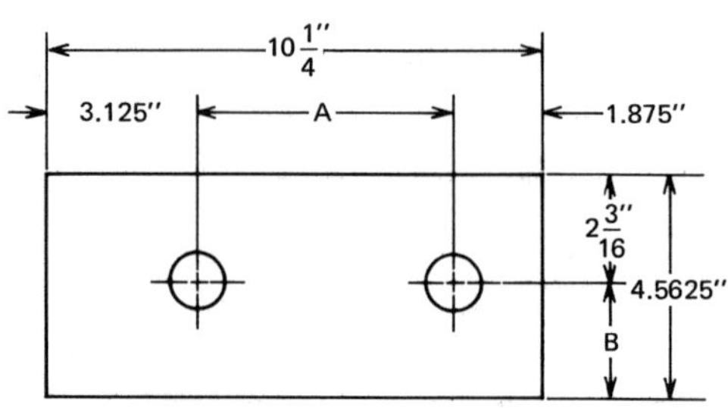

Figure 3—10

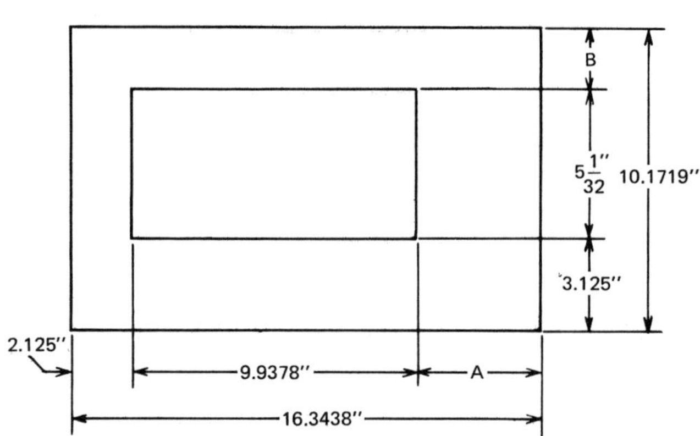

Figure 3—9

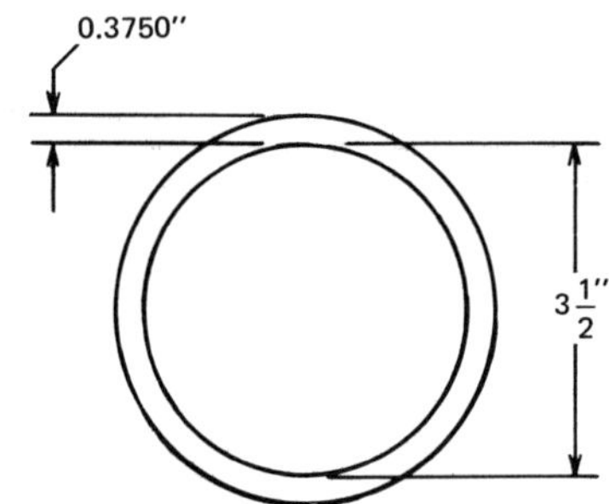

Figure 3—11

3-107. Refer to Figure 3-10.
    a. What is distance A between the centers of the holes?
    b. What is dimension B?

3-108. What is the outside diameter of the pipe in Figure 3-11?

3-109. How much larger is the gear tooth at the bottom than at the top in Figure 3-12?

3-110. Figure 3-13 shows a spacer plate.
    a. Find dimension A.
    b. Find dimension B.

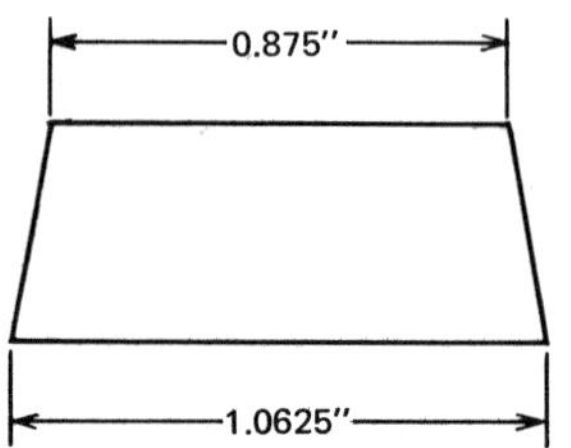

Figure 3—12

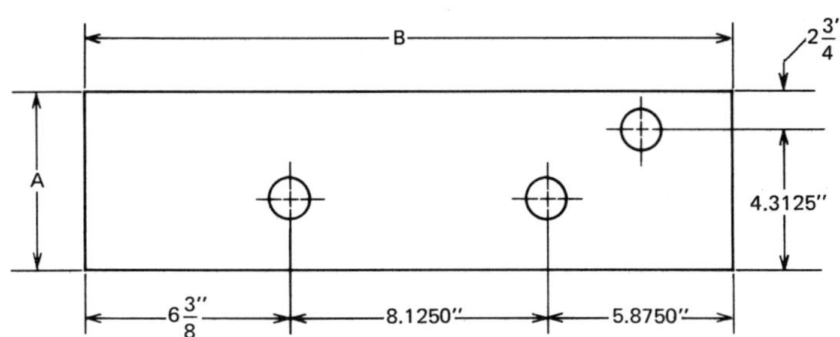

Figure 3—13

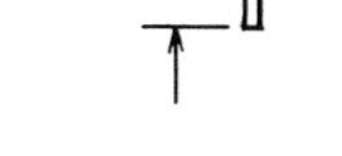

3-111. How many inches of weld are needed to weld a
        3-1/2" square frame to a plate?

3-112. What is the overall length of the flange and pipe
        in Figure 3-14?

Figure 3—14

3-113. Refer to Figure 3-15.
        a. What is the overall length of the flanges,
            gasket, and pipe?
        b. What is the overall length of the pipe, not
            including the gasket and flanges?
        c. What is the overall width of the flanges and
            gasket?

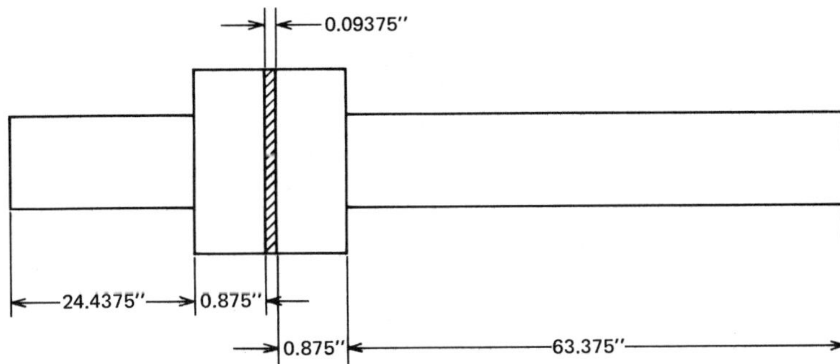

Figure 3—15

3-114. 74 sheets of stainless steel are stacked together.
        Each sheet is 0.375" thick. What is the total
        height of the stack?

3-115. A base plate has 8 anchor bolt slots that are
        each 8.7500" long. What is the total length of
        the slots?

3-116. A single base plate weighs 248.625 lb. How much
        do 248 base plates weigh?

3-117. 168 frame gussets require 16.6875" of weld each.
        What is the total amount of weld needed?

3-118. What size sheet of metal is needed to cut 16
        plates that are each 12.9375" x 4.875"?

3-119. Figure 3-16 shows a frame.
   a. What is the total amount of 1" metal tubing needed to build the frame?
   b. How much material is needed to build 19 frames?
   c. If 19 frames weigh a total of 826.5 lb, how much does each frame weigh?
   d. What is the total amount of space A between parts of the frame?

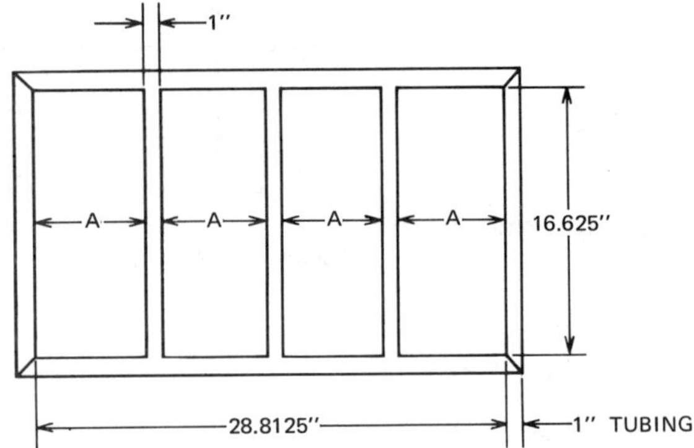

Figure 3—16

3-120. The tolerance for the plate in Figure 3-17 is ±1/64". (This tolerance means that the plate can be 1/64" larger or 1/64" smaller than the given dimension.)
   a. What is the largest limit of dimension A?
   b. What is the smallest limit of dimension B?

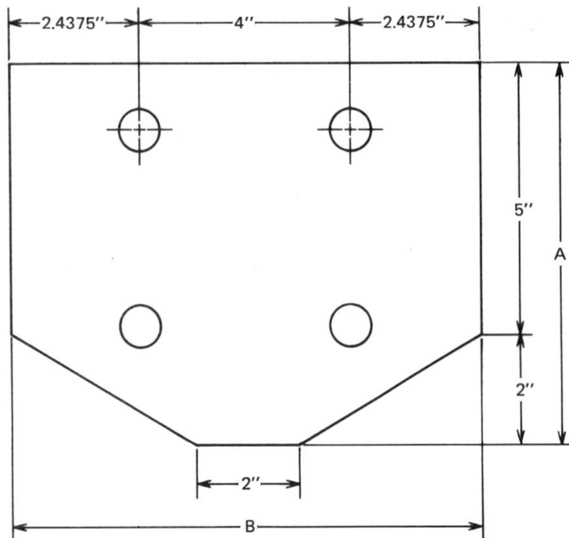

Figure 3—17

3-121. How many pipe spacers 12.25" long can be cut from a 245" length, allowing 0.125" of waste per cut?

3-122. Refer to Figure 3-18.
    a. Find dimension A (1/2 the length).
    b. Find dimension B (1/2 the width).

3-123. The anchor bolt in Figure 3-19 weighs 3-1/2 lb.
    a. If 655 lb of anchor bolts are in the warehouse, how many anchor bolts are there?
    b. How many anchor bolts can be made from 526.25" of round stock, allowing 0.125" per cut for waste?

3-124. A quantity of 1/2" cold roll flat bar stock weighs 5377.32 lb at 4.68 lb per foot. How many feet of bar stock are there?

3-125. The bar in Figure 3-20 is to be cut into 6 equal parts. How long will each part be?

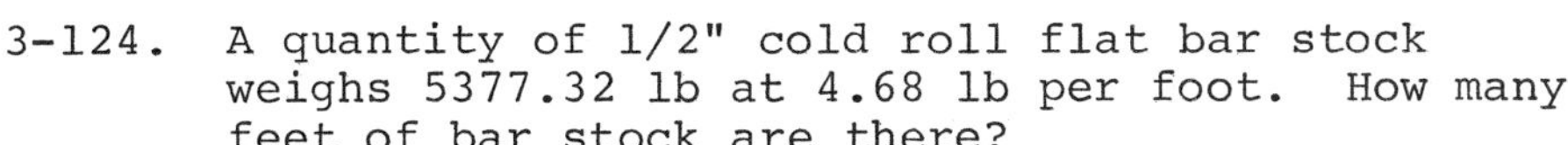
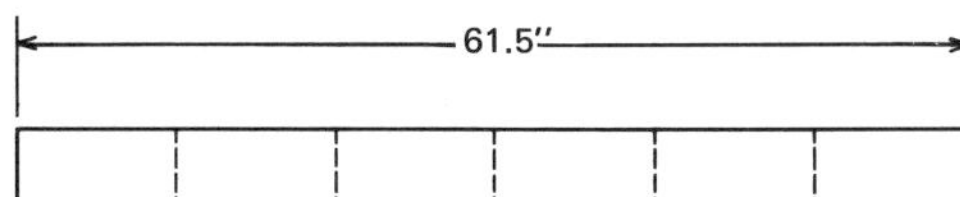

Figure 3—20

3-126. A piece of tubing 3/8" square weighs 2.478 lb per foot. How much does a 20' length weigh?

3-127. A hanger bracket is shown in Figure 3-21.
    a. What length of material is needed for one hanger bracket?
    b. How many hanger brackets can be cut from 200" of flat stock?
    c. How much material will be left?

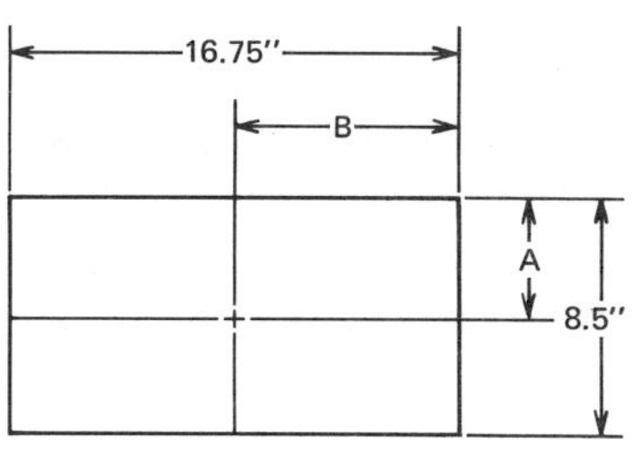

Figure 3—18

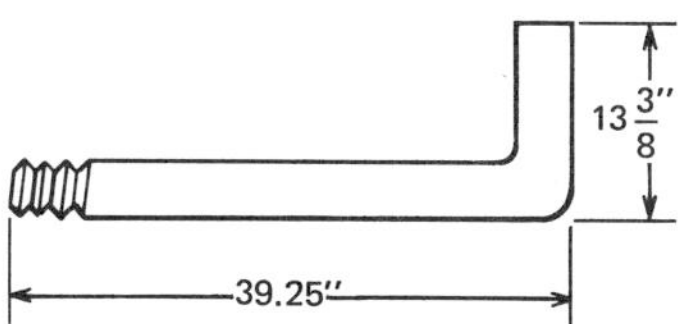

Figure 3—19

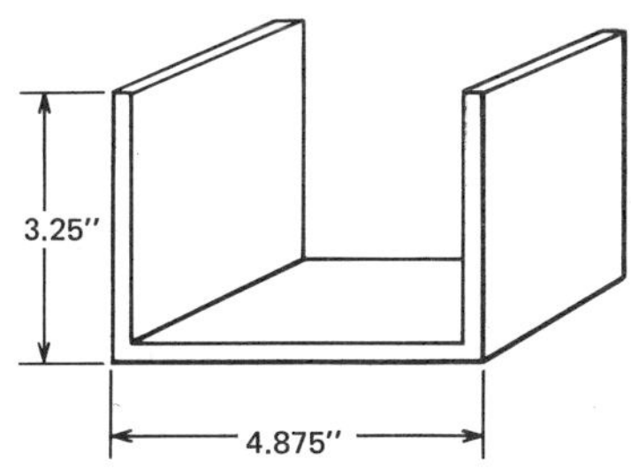

Figure 3—21

Chapter 3: Decimals and Decimal Fractions

Applied Welding Problems Survey Test

Name ______________________________

Course/Sec. ______________ Date ______________

3-1. Figure 3-22 shows plates A, B, and C. 3-1a. ____________

a. What is the combined length of the plates? 3-1b. ____________

b. Which plate is the longest? 3-1c. ____________

c. Which plate is the shortest?

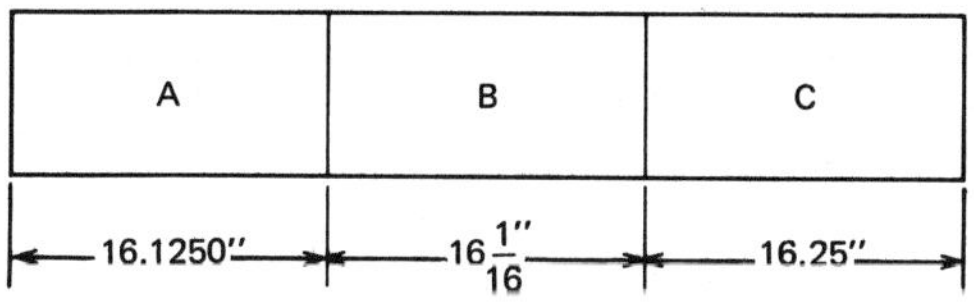

A
B
C
16.1250"
$16\frac{1}{16}$"
16.25"

Figure 3–22

3-2. What length remains if 4.8750" are cut from a 9-3/8" piece of steel? 3-2. ____________

3-3. Determine distance A in Figure 3-23. 3-3. ____________

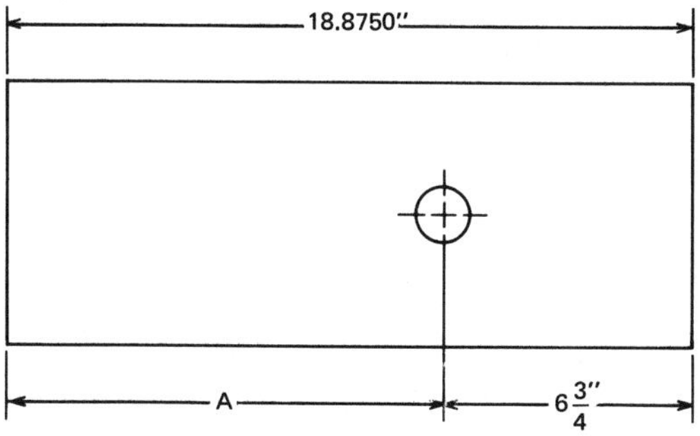

18.8750"
A
$6\frac{3}{4}$"

Figure 3–23

3-4.   How much smaller is pipe B than pipe A in Figure   3-4. ______________
        3-24?

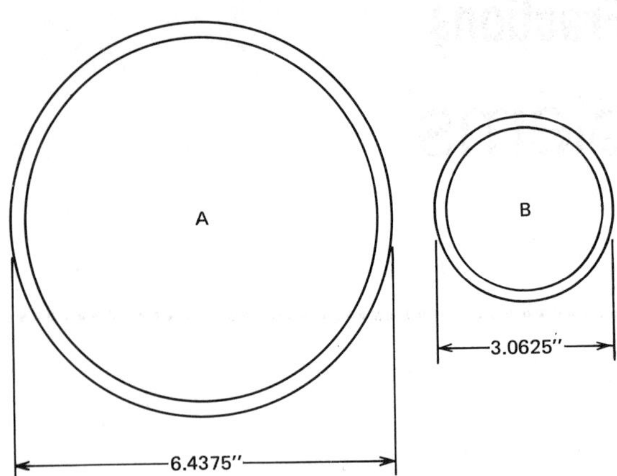

Figure 3—24

3-5.   Fireplace grates require 80 lengths of rod. Each
        rod is 12.625" in length and weighs 1.25 lb.

        a. What is the total length of rod needed?   3-5a. ______________

        b. How many pounds of rod are needed?   3-5b. ______________

3-6.   A welder earns $12.42 per hour. Last week he
        worked 40 hours of regular time and 6 hours of
        overtime. The overtime rate is 1-1/2 times the
        hourly rate.

        a. What was his pay for 40 hours of work?   3-6a. ______________

        b. What was his overtime rate?   3-6b. ______________

        c. How much did he earn on overtime?   3-6c. ______________

        d. What was his total pay for the 46 hours worked? 3-6d. ______________

3-7.   Figure 3-25 shows an end cap.

        a. What is dimension A of the joining collar on   3-7a. ______________
            the end cap?

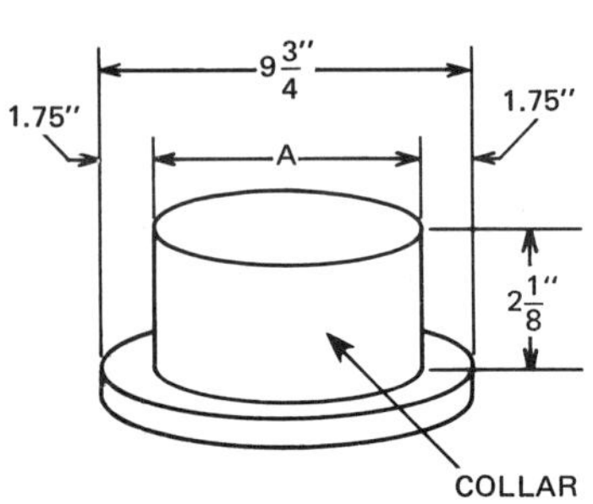

Figure 3—25

b. What length of material is needed to build 26 collars?

3-7b. ______________________

3-8. How many 9.1250" lengths can be cut from a bar 182-1/2" long?

3-8. ______________________

3-9. What is the equal distance between the 1" pipes coming out of the header pipe shown in Figure 3-26?

3-9. ______________________

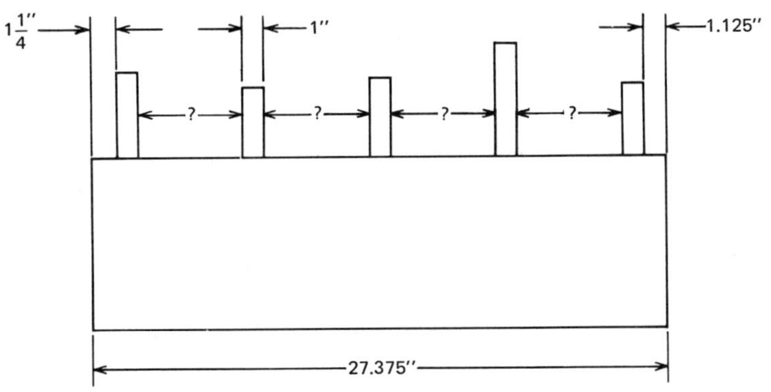

Figure 3—26

3-10. A bearing support weighs 10.5 lb. How many supports can be made from 420 lb of steel?

3-10. ______________________

Chapter 4: Percent

Preview and Self-Test

●●●

Name ____

Course/Sec. ________________   Date ___________________

Chapter 4 covers the properties of percent, the solution
of percent problems, and practical applications of
percent. Before you turn to the chapter, complete the
Self-Test to determine which sections in the chapter you
need to study carefully.

Self-Test

PROPERTIES OF PERCENT

4-1. Change 45% to a decimal.   4-1. ________________

4-2. Change 8-1/2% to a decimal.   4-2. ________________

4-3. Write 7.5 as a percent.   4-3. ________________

4-4. Write 0.0875 as a percent.   4-4. ________________

                    Properties of percent score   ________________________

CHANGING FRACTIONS TO PERCENTS

4-5. Change 1/8 to a percent.   4-5. ________________

4-6. Change 7/15 to a percent.   4-6. ________________

4-7. Change 5-5/32 to a percent.   4-7. ________________

                Changing fractions to percents score   ________________________

CHANGING PERCENTS TO FRACTIONS

4-8. Change 18-1/2% to a fraction.   4-8. ________________

4-9. Change 17.5% to a fraction.   4-9. ________________

4-10. Change 138-1/2% to a mixed number.   4-10. ________________

                Changing percents to fractions score   ________________________

SOLVING BASIC PERCENT PROBLEMS

4-11.   Find 9% of 6879.   4-11. ______________

4-12.   Find 9-3/4% of 76.89.   4-12. ______________

4-13.   40 is 39% of what number?   4-13. ______________

4-14.   Find 0.125% of 450.56.   4-14. ______________

4-15. $15,000 is 66-2/3% of how many dollars?   4-15. ______________

                    Solving basic percent problems score   ______________________

PRACTICAL APPLICATIONS OF PERCENT

4-16.   How much would you pay for a $298.95 lawn mower   4-16. ______________
        if you received a 15% discount?

4-17.   Find the interest on $8575 at 9.5% for a year.   4-17. ______________

4-18.   A load of wheat contains 14.8% water. How many   4-18. ______________
        pounds of water are in a 380-bushel load? (Wheat
        weighs 60 lb per bushel.)

4-19.   A landscaping company sold its tillers after the   4-19. ______________
        season for $189.75. This price was 60% of the
        original cost. Find the original cost.

4-20.   There were 9.2 sq yd of waste on a carpet-laying   4-20. ______________
        job. If this amount was 2-1/2% of the total, how
        much carpet was ordered?

4-21.   If 83% of a job can be completed in 87.15 hours,   4-21. ______________
        how long will it take to do the whole job?

4-22.   The tolerance for a 1-3/8" drive shaft is ±0.05.%   4-22. ______________
        Find the largest and the smallest acceptable sizes
        to the nearest ten-thousandth.

4-23.   How much would you pay for a $279.95 water heater   4-23. ______________
        including a 7% sales tax if you were given a 20%
        discount?

4-24.   An interior decorating studio had a gross income   4-24. ______________
        of $84,397. The following year there was a 24%
        increase in business. Find the gross income.

4-25.   Inflation rates for three consecutive years were   4-25. ______________
        9.7%, 12.9%, and 17.9%. How much would you need
        to be earning at the end of the third year to keep
        up with inflation if you earned $25,000 the first
        year?

                    Practical applications of percent score   ______________________

After you have checked your answers on the Self-Test,
transfer your scores to the Chapter 4 Objectives.

Chapter 4

••

Percent

Upon successful completion of this chapter, you will know
the properties of percent. You will be able to change
fractions to percents and percents to fractions, solve
basic percent problems, and make practical applications
of percent.

Objectives

	Self-Test Scores	If you did poorly on the Self-Test, turn to:
• Properties of percent	_________	Section 4-2
• Changing fractions to percents	_________	Section 4-3
• Changing percents to fractions	_________	Section 4-4
• Solving basic percent problems	_________	Sections 4-5 and 4-6
• Practical applications of percent	_________	Section 4-7

4–1   Introduction to Percent

The word <u>percent</u> comes from the Latin words <u>per</u> <u>centum</u>,
which mean "by the hundred" or "for every hundred." When
you are working with percent, remember that you are
dealing with so many out of 100 or with a part of the
whole that can be expressed as a percent.

Percents are used frequently in our everyday lives.
Understanding their use and calculation is very important.

4–2   Properties of Percent

A percent refers to a given number of parts of the whole,
which is equal to one hundred percent. "Twenty percent"
can be written 20% and means "20 out of 100." The % is
the symbol that means percent.

Figure 4-1 shows a "whole" divided into 100 equal
squares. The 100 squares are equal to 100%. The 20
shaded squares in the figure equal 20 out of 100, or
20/100, or 20%.

Similarly, Figure 4-2 shows a "whole" divided into
3 equal parts. 3 out of 3, or 3/3, equal 100%. The part
that is shaded equals 1 out of 3, or 1/3, or 33-1/3%.
The 33-1/3% is obtained by changing the fraction 1/3 to a
decimal. (Follow the procedure for changing fractions to
decimals given in Chapter 3.)

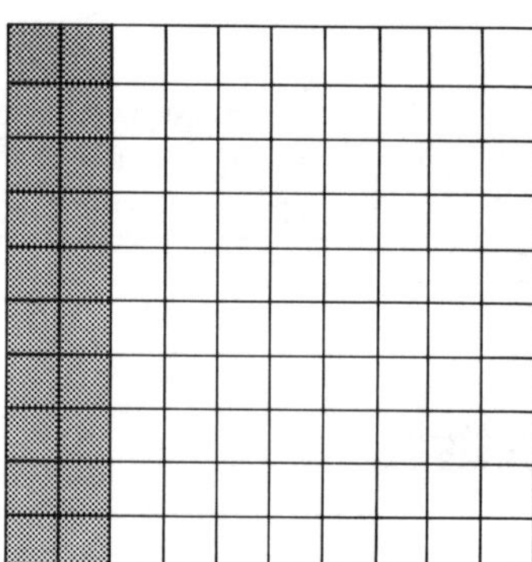

Figure 4–1

Procedure for Changing Decimals to Percents

• •

Move the decimal point two places to the right and add a
percent symbol (%).

• •

Example 4–1

PROBLEM

Change the following decimals to percents: 0.45, 0.08,
0.705, 0.004, and 4.76.

SOLUTION AND ANSWER

To change a decimal to a percent, move the decimal point
two places to the right and add the percent symbol:

$$0.45 = 45\%$$
$$0.08 = 8\%$$
$$0.705 = 70.5\%$$
$$0.004 = 0.4\%$$
$$4.76 = 476\%$$

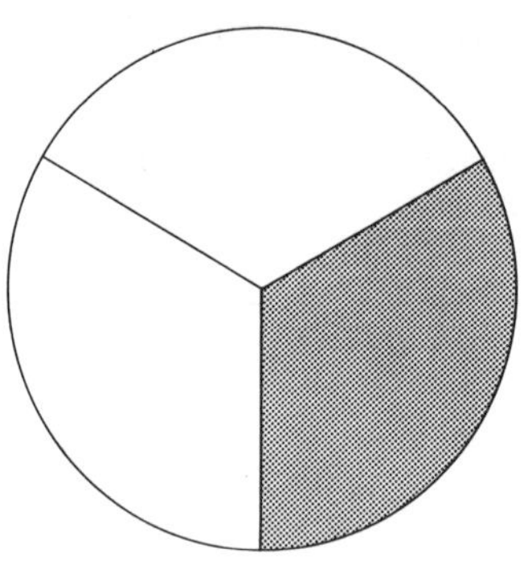

Figure 4–2

Exercises: Changing Decimals to Percents

Change the following to percents:

4-1.   0.75   4-2.   0.68   4-3.   0.03

4-4.   0.09   4-5.   0.006   4-6.   0.0095

4-7.   0.4005   4-8.   3.46   4-9.   2.2

Procedure for Changing Percents to Decimals

●●

1.   Remove the percent symbol (%) and move the decimal
     point two places to the left.
2.   Add zeros if necessary.

●●

Example 4-2

PROBLEM

Change the following percents to decimals: 8%, 15%,
0.5%, 7.5%, and 872%.

SOLUTION AND ANSWER

To change a percent to a decimal, remove the percent
symbol and move the decimal point two places to the left:

$$8\% = 08.\% = 0.08$$
$$15\% = 15.\% = 0.15$$
$$0.5\% = 00.5\% = 0.005$$
$$7.5\% = 07.5\% = 0.075$$
$$872\% = 872.\% = 8.72$$

Exercises: Changing Percents to Decimals

Change the following to decimals:

4-10.   7%   4-11.   1%   4-12.   25%

4-13.   33%   4-14.   0.2%   4-15.   0.8%

4-16.   610%   4-17.   4263%   4-18.   0.25%

4-3   Changing Fractions to Percents

To solve certain applied problems, changing a fraction
to a percent or a percent to a fraction is sometimes
necessary. To change a fraction to a percent, first

change the fraction to a decimal by dividing the denominator into the numerator. Then move the decimal point two places to the right and add the percent symbol.

The following figures illustrate the equivalence among fractions, decimals, and percent. Figure 4-3 shows 42 shaded squares out of 100 squares, or 42/100. To change this fraction to a decimal, divide the denominator, 100, into the numerator, 42:

$$
\begin{array}{r}
0.42 \\
100\ \overline{)42.00} \\
\underline{400} \\
200 \\
\underline{200}
\end{array}
$$

Then move the decimal point in the quotient two places to the right and add the percent symbol: 42%.

Figure 4-4 shows 15 shaded squares out of 40 squares, or 15/40. To change this fraction to a decimal, divide the denominator into the numerator:

$$
\begin{array}{r}
0.375 \\
40\ \overline{)15.000} \\
\underline{120} \\
300 \\
\underline{280} \\
200 \\
\underline{200}
\end{array}
$$

Then move the decimal point in the quotient two places to the right and add the percent symbol: 37.5%.

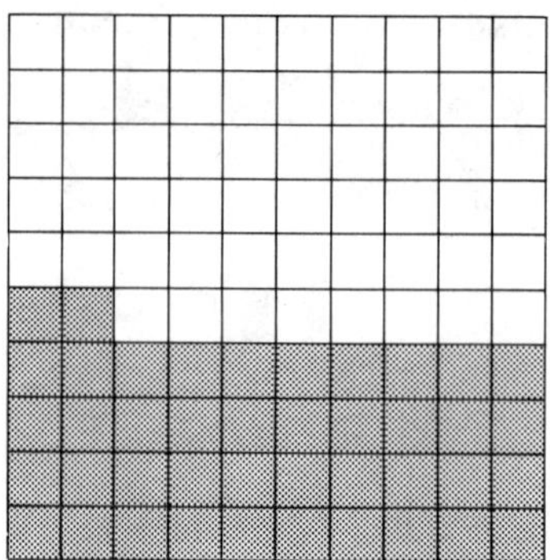

42 SHADED SQUARES OUT OF 100 SQUARES = $\frac{42}{100}$ = 0.42 = 42%

Figure 4—3

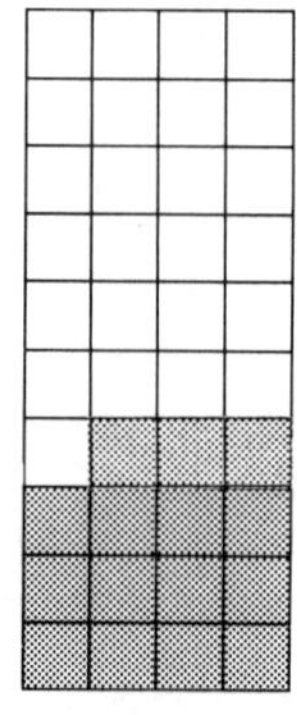

15 SHADED SQUARES OUT OF 40 SQUARES = $\frac{15}{40}$ = 0.375 = 37.5%

Figure 4—4

Procedure for Changing Fractions to Percents

• •

1. Change the fraction to a decimal by dividing the denominator into the numerator. Remember to insert the decimal point in the proper places.
2. Then move the decimal point in the quotient two places to the right and add the percent symbol (%).

• •

Example 4–3

PROBLEM

Change 5/16 to a percent.

SOLUTION

1. Convert 5/16 to a decimal:

$$
\frac{5}{16} =
\begin{array}{r}
0.3125 \\
16\ \overline{)5.0000} \\
\underline{4\,8} \\
20 \\
\underline{16} \\
40 \\
\underline{32} \\
80 \\
\underline{80}
\end{array}
$$

2. Move the decimal point two places to the right and
    add the percent symbol:

    0.3125 = 31.25%

ANSWER

$$\frac{5}{16} = 31.25\%$$

Exercises: Changing Fractions to Percents

Convert the following to percents:

4-19. 1/2   4-20. 1/4   4-21. 1/3

4-22. 5/8   4-23. 7/20   4-24. 31/32

4—4   Changing Percents to Fractions

To convert a percent to a fraction, change the percent
to a decimal by removing the percent symbol and moving
the decimal point two places to the left. Change the
decimal to a decimal fraction (remember that the denominator of a decimal fraction is 10 or a multiple of 10)
and reduce.

Figure 4-5, like the previous two figures, illustrates the equivalence of percents, decimals, and
fractions. In Figure 4-5, 35% of the squares are shaded.
In other words, 35 out of 100 squares, or 35/100, are
shaded. (This decimal fraction could also be written as
a decimal: 0.35.) To complete the operation of changing
35% to a fraction, reduce 35/100 by 5:

$$\frac{35 \div 5}{100 \div 5} = \frac{7}{20}$$

Table 4-1 gives the decimal equivalents of many
common fractions.

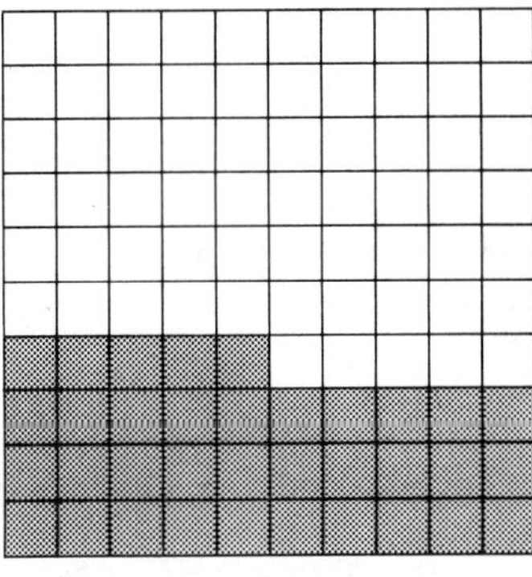

35 SHADED SQUARES OUT OF
100 SQUARES = $\frac{35}{100}$ = 0.35 = 35%

Figure 4—5

Procedure for Changing Percents to Fractions

1. Remove the percent symbol and move the decimal point
    two places to the left.
2. Change the number to a decimal fraction and reduce.

Example 4—4

PROBLEM

Change 87-1/2% to a fraction.

SOLUTION

1. Change the percent to a decimal by removing the
    percent symbol and moving the decimal point two
    places to the left (remember that 1/2 = 0.5):

    $87\frac{1}{2}\% = 87.5\% = 0.875$

2. Read 0.875 as "eight hundred seventy-five thousandths"
    and write it as a fraction:

    $$\frac{875}{1000}$$

    (Remember that there should be as many zeros in the
    denominator as there are digits in the decimal.)

3. Reduce the fraction:

    $$\frac{875 \div 25}{1000 \div 25} = \frac{35 \div 5}{40 \div 5} = \frac{7}{8}$$

ANSWER

$87\frac{1}{2}\% = \frac{7}{8}$

Table 4-1   Decimal Equivalents

$\frac{1}{64} = 0.0156$	$\frac{17}{64} = 0.2656$	$\frac{33}{64} = 0.5156$	$\frac{49}{64} = 0.7656$
$\frac{1}{32} = 0.0312$	$\frac{9}{32} = 0.2812$	$\frac{17}{32} = 0.5312$	$\frac{25}{32} = 0.7812$
$\frac{3}{64} = 0.0468$	$\frac{19}{64} = 0.2968$	$\frac{35}{64} = 0.5468$	$\frac{51}{64} = 0.7968$
$\frac{1}{16} = 0.0625$	$\frac{5}{16} = 0.3125$	$\frac{9}{16} = 0.5625$	$\frac{13}{16} = 0.8125$
$\frac{5}{64} = 0.0781$	$\frac{21}{64} = 0.3281$	$\frac{37}{64} = 0.5781$	$\frac{53}{64} = 0.8281$
$\frac{3}{32} = 0.0937$	$\frac{11}{32} = 0.3437$	$\frac{19}{32} = 0.5937$	$\frac{27}{32} = 0.8437$
$\frac{7}{64} = 0.1093$	$\frac{23}{64} = 0.3593$	$\frac{39}{64} = 0.6093$	$\frac{55}{64} = 0.8593$
$\frac{1}{8} = 0.1250$	$\frac{3}{8} = 0.3750$	$\frac{5}{8} = 0.6250$	$\frac{7}{8} = 0.8750$
$\frac{9}{64} = 0.1406$	$\frac{25}{64} = 0.3906$	$\frac{41}{64} = 0.6406$	$\frac{57}{64} = 0.8906$
$\frac{5}{32} = 0.1562$	$\frac{13}{32} = 0.4062$	$\frac{21}{32} = 0.6562$	$\frac{29}{32} = 0.9062$
$\frac{11}{64} = 0.1718$	$\frac{27}{64} = 0.4218$	$\frac{43}{64} = 0.6718$	$\frac{59}{64} = 0.9218$
$\frac{3}{16} = 0.1875$	$\frac{7}{16} = 0.4375$	$\frac{11}{16} = 0.6875$	$\frac{15}{16} = 0.9375$
$\frac{13}{64} = 0.2031$	$\frac{29}{64} = 0.4531$	$\frac{45}{64} = 0.7031$	$\frac{61}{64} = 0.9531$
$\frac{7}{32} = 0.2187$	$\frac{15}{32} = 0.4687$	$\frac{23}{32} = 0.7187$	$\frac{31}{32} = 0.9687$
$\frac{15}{64} = 0.2343$	$\frac{31}{64} = 0.4843$	$\frac{47}{64} = 0.7343$	$\frac{63}{64} = 0.9843$
$\frac{1}{4} = 0.25$	$\frac{1}{2} = 0.5$	$\frac{3}{4} = 0.75$	$1 = 1.0$

Example 4-5

PROBLEM

Change 66-2/3% to a fraction.

SOLUTION

1. Recall that percent means part of 100. Write the
    percent as a decimal fraction with a denominator of
    100:

$$66\tfrac{2}{3}\% = \frac{66\tfrac{2}{3}}{100}$$

2. Change 66-2/3 to an improper fraction and divide by
    100/1:

$$\frac{66\tfrac{2}{3}}{100} = \frac{\tfrac{200}{3}}{\tfrac{100}{1}} = \frac{200}{3} \div \frac{100}{1}$$

$$= \frac{200}{3} \times \frac{1}{100}$$

$$= \frac{2}{3}$$

Note: When dividing fractions, remember to invert the
fraction to the right of the division sign ($\div$) and
multiply.

ANSWER

$$66\tfrac{2}{3}\% = \frac{2}{3}$$

Exercises: Changing Percents to Fractions

Change the following to fractions:

4-25.   75%   4-26.   45%   4-27.   12.5%

4-28.   43.75%   4-29.   7-1/2%   4-30.   33-1/3%

4-5   Solving Basic Percent Problems

There are three basic types of percent problems:
(1) finding the _percentage_, or part of the whole amount
given; (2) finding the _base_, or the whole amount; and
(3) finding the _rate_, or the percent. All percent
problems can be "arranged" into one of these three types
of problems and solved.
     The procedure for calculating percent problems is to
use the following formula:

    base x rate = percentage

Say, for example, that you want to purchase a $20.00
fishing rod. If there is a 4% sales tax, you should pay
$0.80 for tax, or $20.80 total:

$20.00 x 4% = $20.00 x 0.04 = $0.80

In the above, $20.00 is the base; 4%, the rate; and
$0.80, the percentage.
     All percent problems are a variation of the fact
that base x rate = percentage. If you are given two of
these quantities, then you can solve for the third.
     The following suggestions will help you solve
percentage problems:

- Use base x rate = percentage (B x R = P) to find
  the percentage when you are given the base and
  the rate.
- Use base = percentage/rate (B = P/R) to find the
  base when you are given the percentage and the
  rate.
- Use rate = percentage/base (R = P/B) to find the
  rate when you are given the percentage and the
  base.

     Identifying the rate, the base, and the percentage
is easy if you remember the following facts:

- The <u>rate</u> includes a percent symbol or the word
  "percent."
- The <u>base</u> is the quantity that follows the word "of"
  in a stated problem and refers to the whole quantity.
- The <u>percentage</u> is the result of finding the rate of
  a given base.

Procedure for Finding the Percentage

••

When you are given the base and the rate, use the
following formula to find the percentage:

base x rate = percentage   or   B x R = P

••

Example 4–6

PROBLEM

Find 34% of $560.

SOLUTION

1. 34% is the rate, and $560 is the base, so you are
    looking for percentage, or P.

2. Use the following formula:

    B x R = P

3. Change 34% to 0.34 and multiply the base by the rate:

```
      560
    x 0.34
     2240
    1680
    190.40
```

ANSWER

P = 34% of $560 = $190.40

Example 4–7

PROBLEM

Find 7-1/2% of $2564.

SOLUTION

1. 7-1/2% is the rate, and $2564 is the base, so you are
    looking for percentage, or P.

2. Use the following formula:

    B x R = P

3. Change 7-1/2% to a decimal (1/2 = 0.50) and multiply
    the base by the rate:

$$7\tfrac{1}{2}\% = 7.5\% = 0.075$$

```
      2564
    x 0.075
     12820
    17948
    192.300
```

ANSWER

$$P = 7\tfrac{1}{2}\% \text{ of } \$2564 = \$192.30$$

Exercises: Finding the Percentage

Solve the following problems:

4-31. 7% of 200   4-32. 8% of 326

4-33. 12% of 829   4-34. 17% of 663

4-35. 28% of 146.7   4-36. 5.3% of 98.74

4-37. 225% of $446.70   4-38. 337% of $215.97

4-39. Find 39.75% of 37,864   4-40. Find 0.55% of 15,367

Procedure for Finding the Base

When you are given the percentage and the rate, use the
following formula to find the base:

$$\text{base} = \frac{\text{percentage}}{\text{rate}} \quad \text{or} \quad B = \frac{P}{R}$$

Example 4–8

PROBLEM

The quantity 160 is 62-1/2% of what number?

SOLUTION

1. 160 is the percentage, and 62-1/2% is the rate, so
    you want to find the base.

2. Use the following formula:

    $$B = \frac{P}{R}$$

3. Change 62-1/2% to a decimal and divide the rate into
    the percentage:

    $$62\tfrac{1}{2}\% = 62.5\% = 0.625$$

    ```
              256.
    .625) 160000
          1250
          3500
          3125
          3750
          3750
    ```

ANSWER

$$B = \frac{160}{0.625} = 256$$

Exercises: Finding the Base

Solve the following problems:

4-41.   180 is 30% of what number?

4-42.   425 is 25% of what number?

4-43.   12.5 is 12.5% of what number?

4-44. $47.30 is 4% of what number?

4-45.   5280 is 33-1/3% of what number?

Procedure for Finding the Rate

When you are given the percentage and the base, use the
following formula to find the rate:

$$\text{rate} = \frac{\text{percentage}}{\text{base}} \quad \text{or} \quad R = \frac{P}{B}$$

Example 4–9

PROBLEM

What percent is 1216 of 25,789, to the nearest tenth of
a percent?

SOLUTION

1. 1216 is the percentage, and 25,789 is the base, so
    you want to find the rate.

2. Use the following formula:

$$R = \frac{P}{B}$$

3. Divide the base into the percentage:

$$R = \frac{1216}{25,789} = 0.0472 = 4.7\%$$

4. Carry the quotient out four places so that you can
    round it to the nearest tenth of a percent:

$$0.0472 = 4.72\% = 4.7\%$$

ANSWER

$$R = \frac{1216}{25,789} = 0.0472 = 4.7\%$$

Example 4–10

PROBLEM

Find what percent 199.15 is of 569.

SOLUTION

1. 199.15 is the percentage, and 569 the base, so you
    want to find the rate.

2. Use the following formula:

$$R = \frac{P}{B}$$

3. Divide the base into the percentage:

$$
\begin{array}{r}
0.35 \\
569\,\overline{)199.15} \\
1707 \\
\overline{2845} \\
2845
\end{array}
$$

4. Change 0.35 to a percent:

    0.35 = 35%

ANSWER

$$
R = \frac{199.15}{569} = 0.35 = 35\%
$$

Exercises: Finding the Rate

Solve the following problems:

4-46.   10 is 15% of what number?

4-47.   18 is 75% of what number?

4-48. $4.89 is 4% of what number?

4-49.   16.46 is 1.5% of what number?

4-50.   2020 is 15.6% of what number?

4-6   Jacobs's Law

Percent problems are difficult for many people because
they do not understand the language of percent--that is,
they have trouble identifying the rate, the percentage,
and the base. A procedure called Jacobs's Law* makes
this translation easier. The procedure can best be
explained by examples.

Example 4-11

PROBLEM

What is 43% of 250?

TRANSLATION TO JACOBS'S LAW

    ? = 0.43 x 250

Note: "What" = ?, "is" = equals sign, "43%" = 0.43, and
"of" = x.

*Jacobs's Law is named after Scotty Jacobs, a former fellow
instructor who used this procedure to teach his students. It is
used here with his permission and his blessing.

SOLUTION

Multiply 0.43 by 250:

$$0.43 \times 250 = 107.5$$

ANSWER

43% of 250 is 107.5

Example 4–12

PROBLEM

45 is what percent of 360?

TRANSLATION TO JACOBS'S LAW

$$45 = ? \times 360$$

SOLUTION

1. Divide each side by 360 (note that the numerator and
    denominator on the right-hand side cancel each other):

$$\frac{45}{360} = \frac{? \times 360}{360}$$

$$= 0.125$$

2. Convert 0.125 to a percent:

$$0.125 = 12.5\%$$

ANSWER

45 is 12.5% of 360.

Example 4–13

PROBLEM

$14.85 is 5% of what?

TRANSLATION TO JACOBS'S LAW

$$\$14.85 = 0.05 \times ?$$

SOLUTION

Divide both sides by 0.05:

$$\frac{14.85}{0.05} = \frac{0.05 \times ?}{0.05}$$

$$= \$297$$

ANSWER

$14.85 is 5% of $297.

4–7   Practical Applications of Percent

Numerous applications of percent are made in industry.
One type of basic problem in industry is to find the
percentage of a number--for example, sales tax, interest,
discounts, or commissions. Another is to find the rate
of change--for example, profit, loss, increase, or
decrease. Jacobs's Law can help solve such problems.
The examples that follow illustrate some common applications of percent.

Example 4–14

PROBLEM

Find the interest for 3 years on a new subcompact car
that sells for $5450. The simple interest rate is 10.5%
per year.

SOLUTION

Determine the interest by multiplying the base ($5450),
called the principal, by the rate (10.5%), by the time in
years, or $I = prt$:

    I = prt
    I = 5450 x 0.105 x 3
      = 572.25 x 3
      = 1716.75

ANSWER

The interest for 3 years is $1716.75.

Example 4–15

PROBLEM

Find the cost of a $450 dishwasher after a 30% discount.

SOLUTION

1. Find the discount by multiplying the rate (30%) by
    the base ($450) and then subtracting the amount of
    the discount from the regular price to find the sale
    price:

        discount = regular price x rate of discount
                 = 450 x 30% = 450 x 0.30
                 = 135

2. Subtract the discount from the regular price to find
    the sale price:

        450   (regular price)
      - 135   (discount)
        315   (sale price)

ANSWER

The cost of a $450 dishwasher after a 30% discount is
$315.

Example 4–16

PROBLEM

The number of fans in a National Hockey League city
increased from 9846 to 11,392. What was the rate of
increase, to the nearest tenth of a percent?

SOLUTION

1. Find the amount of change (the increase):

    11392 - 9846 = 1546

2. Divide the amount of change by the original number
    of fans:

    $$\frac{\text{amount of change}}{\text{original number}} = \text{\% of change}$$

    $$\frac{1546}{9846} = 0.1570$$

3. Round to the nearest tenth and change to a percent:

    0.1570 = 15.7 = 15.7%

ANSWER

The number of fans increased by 15.7%.

Note: A similar procedure is used for various applications of percent in finding profit, loss, increase,
decrease, and so on. Use common sense in applying the
formula:

$$\text{rate of change} = \frac{\text{amount of change}}{\text{original number}}$$

Example 4–17

PROBLEM

Some quality controls indicate the percent of tolerance
of certain parts. Find the high and low acceptable
levels for a 1-1/4" car axle if the level of tolerance is
±0.5%.

SOLUTION

1. Because the tolerance level is ±0.5%, you must find
    0.5% of 1-1/4 (recall that 1-1/4 = 1.25, and 0.5% =
    0.005):

    1.25 x 0.005 = 0.00625

2. To find the upper limit of acceptance, add the per-
    centage of tolerance to 1.25:

        1.25000
      + 0.00625
        ─────────
        1.25625

3. To find the lower limit of acceptance, subtract the
    percentage of tolerance from 1.25:

        1.25000
      - 0.00625
        ─────────
        1.24375

ANSWER

The upper and lower limits of acceptance are 1.25625" and
1.24375".

Exercises: Practical Applications of Percent

Solve the following problems:

4-51. Find the 7.5% interest on $2500 for 1 year.

4-52. Find the 4% sales tax on a bill for $2376.49.

4-53. Good spring wheat contains 15.8% protein. How
       much protein is in 15,000 lb of good spring wheat?

4-54. Find the interest on $250,000 for 1 year at 8.33%.

4-55. Find the 7% sales tax on a new $297.89 lawn mower.

4-56. What is the commission on a $69,750 house if the
       rate of commission is 3.5%?

4-57. What do you pay (before sales tax) for a $297.89
       lawn mower if the store offers a discount of 15%?

4-58. A well-advertised soap claims to be 99.55% pure.
       What part of a 4 oz bar is impure?

4-59. Insurance charged by a truck line is 1/2% of the
       cargo value. What is the insurance charge for 6
       new compact cars valued at $7984 each?

4-60. Some companies offer discounts if accounts are
       paid within 10 days of delivery. Find the amount
       paid on a $7386.48 bill paid early, if the dis-
       count is 1-1/4%.

4-61. Highway fatalities were reduced from 580 to 435 in
       one year with a safety campaign. Find the percent
       of decrease.

4-62. The 1983 Pontiac claims 17 mpg in the city and
       25 mpg on the highway. Find the percent of
       increase from city to highway driving.

4-63. From 1970 to 1980, the population of Endsville
       changed from 4397 to 7646 people. Find the per-
       cent of increase (to the nearest tenth) during the
       10 years.

4-64. The price of gasoline in 1978 was 56.9¢; in 1979,
       it was 98.9¢. What was the percent of increase?

4-65. A farmer found that by spraying for weeds and
       insects, his wheat yield per acre changed from
       28.6 to 32.4 bushels. Find the percent of
       increase.

4-66. Radial tires are supposed to increase mileage by
       8-1/2%. If you can drive 343.2 mi on a tank of
       gasoline with regular tires, how much farther can
       you drive on radials?

4-67. Unpaid accounts are charged a 1-1/2% service
       charge per month. Find the total service charge
       for a year for an interior decorator whose cus-
       tomers have an average of $889.76 in unpaid bills
       per month.

4-68. The rent for a one-bedroom apartment was increased
       from $169 to $180. What was the percent of
       increase (to the nearest tenth)?

4-69. The protein content of North Dakota hard spring
       wheat averages 14.28%. How many pounds of protein
       are in a 353-bushel load of this wheat? (1 bushel
       = 60 lb.)

4-70. The approximate depreciation rates for a full-size
       car are 18.5% of the original price the first
       year, 12.5% of the original price the second year,
       and 7.5% of the original price the third year.
       What is the value of a $9546 car (full size) at
       the end of 3 years?

Chapter 4

Percent

••

Solving Applied Welding Problems

Now that you have mastered the fundamental operations
of percent, you are ready to apply these math skills to
typical problems that welders encounter on the job.
After studying the examples that follow, complete the
applied welding problems.

Example 4–18

PROBLEM

The total weight of a gate valve is 19 lb. 33% of the
total weight is stainless steel; 65% of the total weight
is mild steel; and 2% of the total weight is welded
metal. Find the number of pounds of stainless steel in
a valve.

SOLUTION

Change 33% to a decimal (0.33) and multiply by 19 lb:

```
       19
   x 0.33
       57
     5 7
     6.27
```

ANSWER

There are 6.27 lb of stainless steel in a valve.

Example 4–19

PROBLEM

How many pounds of welded metal are in the valve
described in Example 4-18?

SOLUTION

Change 2% to a decimal (0.02) and multiply by 19 lb:

```
      19
  x 0.02
    0.38
```

ANSWER

There is 0.38 lb of welded metal in the valve.

Example 4–20

PROBLEM

The gross income of a welding shop is $43,800. 24% is profit. Find the amount of profit.

SOLUTION

Change 24% to a decimal (0.24) and multiply the gross income by the decimal:

```
     43800
   x  0.24
    175200
    87600
   1051200
```

ANSWER

The profit is $10,512.

Applied Welding Problems

4-71. Basic stainless steel contains iron and chromium. What percent of the total volume is iron when 11% of the total volume is chromium?

4-72. A welder's pay is $422 per week. 6% of the pay is deducted for retirement; 20%, for federal tax; 12%, for state tax; and 6.65%, for social security.
a. How much is deducted for retirement?
b. How much is deducted for federal tax?
c. How much is deducted for state tax?
d. How much is deducted for social security?
e. What percent of the total amount does the welder receive?

4-73. A project required six hundred welds. Ninety-two percent are acceptable. How many welds are rejected?

4-74. A storage tank weighs 27,500 lb. Six thousand
      six hundred pounds of the total weight is stain-
      less steel. What percent of the total weight is
      mild steel? See Example 4-18.

4-75. A rail car hopper is welded with the flux cored
      process and the stick electrode process. Sixty-
      five percent of the five hundred inches of weld
      are done by the flux cored process. How many
      inches of weld are done by the stick electrode
      process?

4-76. A welder makes 147 welds on the job. If the job
      has 922 welds, what percent did the welder make?

4-77. A frame requires 922 welds. 147 welds are
      completed. What percent of the total welds are
      completed?

4-78. 25% of a 68,859 lb order is for aluminum. How
      many pounds of aluminum are in the order?

4-79. 29% of the order for Exercise 4-78 is for stain-
      less steel. How many pounds is this?

4-80. 36% of the order for Exercise 4-78 is for mild
      steel. How many pounds is this?

4-81. How many pounds of channel iron are in the order
      for Exercise 4-78 if 10.5% is channel iron?

4-82. A heavy-duty trailer fabricated in a welding shop
      required 6260 lb of mild steel. 2-1/2% of the
      steel was wasted. How many pounds were wasted?

4-83. In its first year of operation, a welding shop
      grossed $89,500. The profit for the first year of
      operation was $19,690; material costs were
      $41,170; building and warehouse costs were
      $13,425; and equipment payments were $15,215.
      a. What percent of the gross amount was profit?
      b. What percent of the gross amount was for
          material?
      c. What percent of the gross amount was for rent?
      d. What percent of the gross amount was for
          equipment payments?

4-84. A company employs three hundred fifty welders.
      What percent were at work on Monday if three
      hundred thirty-six welders were there?

4-85. A robot welder is operating for 5,747.5 hours
      at 95% efficiency. For how many hours would it
      operate at 100% efficiency?

4-86. A welder receives a ten percent raise after his
      probationary time period. He was earning eight
      dollars and eighty cents per hour. How much
      does he earn per hour after the raise?

4-87.   A cost-of-living increase was 7.8%. The present
        wage is $9.50 per hour. How much is the increase
        in pay per hour?

4-88.   An order of 130 hand trucks was completed by two
        welders. Welder A completed 52 trucks. Welder B
        completed 78 trucks.
        a.   What percent did welder A complete?
        b.   What percent did welder B complete?

4-89.   The time spent building a rail car hopper was
        divided as follows: welding, 126 hours; grinding,
        70 hours; layout and forming, 40 hours; fit-up
        and tack weld, 28 hours; and painting, 14 hours.
        a.   What percent of the time was used for welding?
        b.   What percent of the time was used for
             grinding?
        c.   What percent of the time was used for layout
             and forming?
        d.   What percent of the time was used for
             painting?

4-90.   A welding shop wants to increase business 12.5%
        next month. How much business should the welders
        do if they earned $4,275 this month?

Applied Welding Problems Survey Test

Name ____

Course/Sec. _______________   Date _______________________

4-1. During the month of July, a small welding shop had
      a gross income of $8540. 29% of this amount was
      used to pay shop and equipment rent; 24%, for
      materials; 6%, for insurance; 12%, for taxes; and
      14%, for utilities. The remainder of the money
      was used for miscellaneous spending.

   a. What percent of the income was used for   4-1a. _______________
       miscellaneous spending?

   b. How much did the shop pay for rent?   4-1b. _______________

   c. How much did it pay for materials?   4-1c. _______________

   d. How much did it pay for insurance?   4-1d. _______________

   e. How much did it pay for taxes?   4-1e. _______________

   f. How much did it pay for utilities?   4-1f. _______________

   g. How much did it pay for miscellaneous   4-1g. _______________
       spending?

4-2. Stainless steel electrodes have a list price of
      $128 for 50 lb. For prompt payments, a 21%
      discount is awarded.

   a. How much is the discount?   4-2a. _______________

   b. How much do the electrodes cost per pound (to 4-2b. _______________
       the nearest cent)?

4-3. A welding machine is rated at 250 amps. At the   4-3. _______________
      electrode holder, 150' from the machine, there are
      only 200 amps. What is the percent of power loss
      between the machine and the electrode holder?

4-4. A casting in need of repair weighs 540 lb. After 4-4. _______________
      the repair, it weighs 621 lb. What percent of
      metal is added for the repair?

4-5.   A double V joint like the one in Figure 4-6 is
       being prepared. The two plates weigh 35 lb each.
       What percent of the metal is removed if 1-3/4 lb
       are removed from each plate?

4-5. 

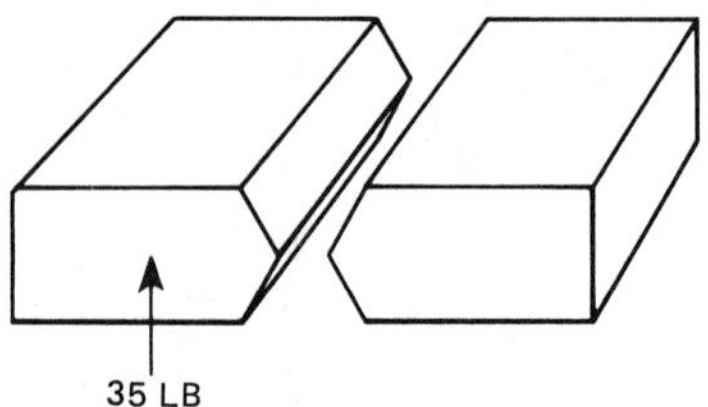

Figure 4—6

4-6.   The top 25-1/2' section of a 150' high exhaust
       stack as shown in Figure 4-7 needs replacing.
       What percent of the stack will be replaced?

4-6. 

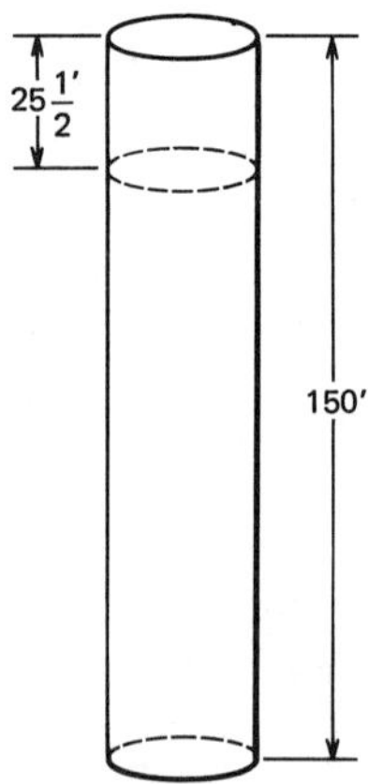

Figure 4—7

4-7.   There are 2" of waste for each 14" electrode, as
       shown in Figure 4-8. What percent of the
       electrode is wasted?

4-7. _______________

Figure 4—8

4-8.   After a repair, a pipe that had a rate of flow
       of 6200 gal per hour was reduced by 3% because of
       a restriction.

a. How many gallons per hour did the restriction represent? 4-8a. __________________

b. What was the rate per hour after the repair? 4-8b. __________________

4-9. Crusher plates in a rock crusher were worn. A welding shop built up the worn areas and hard-faced the plates. Before the work began, the plates weighed a total of 462 lb. After the repair work, the total weight was 483 lb. What percent of the weight was the repair metal? 4-9. __________________

4-10. The round end for a tank as shown in Figure 4-9 is cut from a blank that weighs 210 lb. The round end weighs 177 lb.

a. The end is what percent of the original blank? 4-10a. __________________

b. What percent of the original blank is waste? 4-10b. __________________

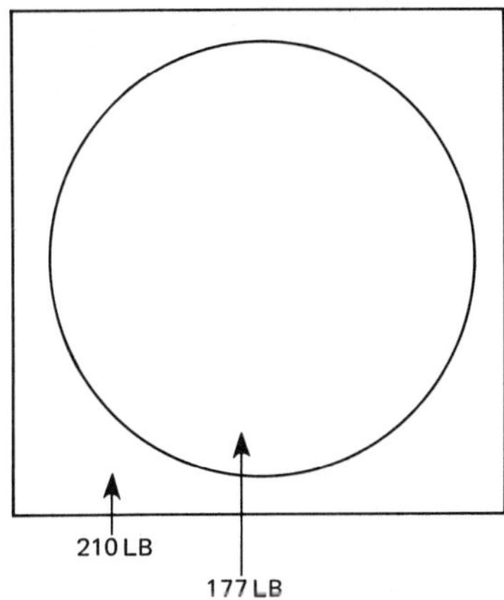

Figure 4—9

4-11. Of 80 completed welds, 20 welds are rejected. What percent are rejected? 4-11. __________________

Chapter 5: Basic Algebra
Preview and Self-Test

• •

Name ____

Course/Sec. _______________   Date _______________________

Chapter 5 covers the basic definitions of algebra and
the fundamental operations on signed numbers. Before
you turn to the chapter, complete the Self-Test to determine which sections in the chapter you need to study
carefully.

Self-Test

BASIC DEFINITIONS AND PROPERTIES OF ALGEBRA

5-1. $3x$ and $5x$ are called _________ terms.   5-1. _______________

5-2.   Letters that represent arithmetic numbers are   5-2. _______________
       called _________ numbers.

5-3. $bx + 4y - z$ is an _________ expression.   5-3. _______________

5-4.   The symbols < or > always point to the _________   5-4. _______________
       value.

5-5.   In the algebraic expression $3x^2 + 4y$, the 3 and   5-5. _______________
       the 4 are called _________ coefficients.

     Basic definitions and properties of algebra score   _______________

ADDITION AND SUBTRACTION OF SIGNED NUMBERS

Add the following algebraic numbers:

5-6. $+15 + (-4)$ 5-6. _______________

5-7. $-3286 + (+4384)$ 5-7. _______________

5-8. $+5280 + (-2890)$ 5-8. _______________

Subtract the following algebraic numbers:

5-9. $+58 - (-32)$ 5-9. _______________

5-10. $-3286 - (+4384)$ 5-10. _______________

5-11.   -567 - (-394)   5-11. ___________

            Addition and subtraction of signed numbers score ___________

MULTIPLICATION AND DIVISION OF SIGNED NUMBERS

Multiply the following algebraic numbers:

5-12.   (-3)(+6)(-2)   5-12. ___________

5-13.   (-7)(-12)(-56)   5-13. ___________

5-14.   (56)(-36)(0)(-54)   5-14. ___________

Divide the following algebraic numbers:

5-15.   -144 ÷ -12   5-15. ___________

5-16.   1980 ÷ -20   5-16. ___________

5-17.   -5600 ÷ 8000   5-17. ___________

            Multiplication and division of signed numbers score ___________

REMOVAL OF SIGNS OF GROUPING

Remove the parentheses and simplify:

5-18. $-5a(6 + 4 ÷ 2)$ 5-18. ___________

5-19. $17x(6 ÷ 3) - (39 ÷ 13) - 4$ 5-19. ___________

5-20. $[44 ÷ (11 \times 2)] + [36 ÷ (4 \times 2 - 8)]$ 5-20. ___________

                    Removal of signs of grouping score ___________

BASIC OPERATIONS WITH EXPONENTS

Perform the indicated operations:

5-21. x^{12}/x^5 5-21. ___________

5-22. $(s^4 j^3)^2$ 5-22. ___________

5-23. $5^3 - 5^2 + 5$ 5-23. ___________

5-24. $\dfrac{5280x^2 y^3 z^4}{660x^2 y^2 z^2}$ 5-24. ___________

5-25. $\dfrac{(468x^5)(9x^7)(27y^5)}{156x^6(9y^4)}$ 5-25. ___________

                    Basic operations with exponents score ___________

After you have checked your answers on the Self-Test,
transfer your scores to the Chapter 5 Objectives.

Chapter 5

....

Basic Algebra

Upon successful completion of this chapter, you will
know the basic definitions and properties of algebra.
You will be able to perform fundamental operations,
including addition, subtraction, multiplication, and
division on signed numbers; remove signs of grouping;
and perform basic operations with exponents.

Objectives

	Self-Test Scores	If you did poorly on the Self-Test, turn to:
• Basic definitions and properties of algebra	_________	Sections 5-1 through 5-5
• Addition and subtraction of signed numbers	_________	Section 5-6
• Multiplication and division of signed numbers	_________	Section 5-7
• Removal of signs of grouping	_________	Section 5-8
• Basic operations with exponents	_________	Section 5-9

5–1   Introduction to Basic Algebra

Algebra is a continuation of arithmetic and a branch of
mathematics that uses symbols (which often are letters)
to represent numbers. The rules of arithmetic apply also
to algebra.

Many problems that would be impossible to solve by
arithmetic can be solved by algebra. The algebra covered
in this text is intended to provide the background for
solving practical problems of industry. A knowledge of
basic algebra is essential in everyday work. Algebra is
also used to solve various problems in many trade and
industry handbooks.

5–2   Basic Definitions of Algebraic Terms

Algebra is a language in which symbols (letters) are
used to represent numbers and to perform operations with
numbers. A numeral is a symbol that names a number
according to a system of numeration. Arithmetic and
algebraic numerals are 0, 1, 2, 3, 4, 5, 6, 7, 8, and 9.

A definite number has an exact meaning or value.
For example, counting the people in a room would give a
definite, or counting, number. Definite numbers are
represented by the numerals 0, 1, 2, 3, . . . , 9.

A measured number is a number obtained by measuring
rather than by counting. Accuracy is determined by the
measuring instrument and the units of measurement (feet,
inches, centimeters, grams, and so on). Measured numbers
are represented by numerals and units of measure, such as
5", 3cm, 4.8km, 3-1/2'.

Arithmetic numbers include both definite and
measured numbers. Literal numbers are letters, such as
y, w, A, and so on, that represent arithmetic numbers.

An algebraic expression is a word statement
expressed in mathematical form by using literal numbers,
arithmetic numbers, and signs of operation.

A term is a part of an algebraic expression. Terms
of an algebraic expression are separated by plus or minus
signs. $5x^2$, $7y$, and 4 are terms in the algebraic
expression $5x^2 + 7y + 4$. In the term $8x^2y$, 8 is a
numerical coefficient, x^2 is a literal coefficient, and
y is a literal coefficient. The coefficients are all
considered factors of the term.

Exercises: Defining Algebraic Terms

Define the following terms according to the basic
definitions presented above:

5-1. $3x + 2y$ is an __________.

5-2.   3-1/2' is a __________ number.

5-3.   There are 11 members on a football team; 11 is a
__________ number.

5-4. $3xy$ is called a __________ number.

5-5.   A __________ is a symbol that names a number.

5–3   Algebraic Signs and Symbols

The <u>signs of operation</u> include the following:

    + , which indicates addition;
    − , which indicates subtraction;
    x and · , which indicate multiplication;
    ÷ and)‾ , which indicate division.

The signs of operation have the same meaning in algebra
as they do in arithmetic. They may be considered
"traffic signs" for basic operations.
    The <u>signs of grouping</u> include the following:

    ( ), which is called a set of parentheses;
    [ ], which is called a set of brackets;
    { }, which is called a set of braces.

    The <u>signs of quantity</u> indicate whether a number is
<u>positive</u> (+) or <u>negative</u> (−). For example, when you
earn $10, this quantity can be expressed as a positive
number: +$10. When you spend $10, this quantity can be
expressed as a negative number: −$10.
    The <u>signs of relation of order</u> include the fol-
lowing:

    = , which indicates equality;
    ≠ , which indicates inequality.

    The symbols > and < are used to indicate "greater
than" and "less than." The arrow points to the smaller
number or expression. Thus, 7 > 4 means 7 is greater
than 4, and 12 < 16 means 12 is less than 16.

Exercises: Identifying Algebraic Signs and Symbols

Determine whether > or < should go into the blanks:

5-6.   +10______+12   5-7.   −16______+11   5-8.   −3______−4

5-9.   14______−5   5-10.   −40______−76

Complete the following:

5-11.   Brackets are signs used in groupings. Draw a
        set of brackets.

5-12.   ( ) is a type of grouping sign. These signs of
        grouping are called __________.

5-13.   The sign of operation for adding is ______.

5-14.   The sign of operation for multiplying is x or
        _______.

5-15.   ÷ indicates that you should _________ the first
        quantity by the second.

5-16. $3x$ is called a _________, which is part of an
        algebraic expression.

5-17.   In the term $5xy$, the 5 is called the _________
        coefficient.

5-18.   In the term $12j$, the j is called the literal
        _________.

5–4   Like and Unlike Terms

Like terms are terms that have the same literal coefficient. $3x$ and $5x$, $3k^4j$ and $5k^4j$, and $14abc$ and $-32abc$
are examples of like terms. Like terms can be added and
subtracted.
     Unlike terms are terms that have different literal
coefficients. $3pq$ and $34s$, $8x^2y$ and $8yx^2$, and xy and
$2x^2y^2$ are examples of unlike terms. Unlike terms cannot
be added or subtracted.

Exercises: Identifying Like and Unlike Terms

Select the like terms and draw lines connecting them:

5-19. $3x$ $7pq$

         $23y^3$ $6z^2$

          $2pq$ $-66pg$

          $6pg$ $46y^3$

        $-27z^2$ $5x$

5-20. $12x^3y^2z^1$ and $44x^1y^2z^3$ are called _________ terms.

5–5   Positive and Negative Numbers

Many things in our daily lives have their opposites.
For example, on and off, up and down, and in and out are
words that indicate opposites. Temperature is measured
in opposites as above or below zero; for example, 22
degrees above zero Celsius is written as $+22°C$, and
10 degrees below zero Celsius is written as $-10°C$. In
algebra, we also have opposites, called positive and
negative numbers. A plus sign (+) is placed in front of
a positive number or quantity, and a negative sign (-)
precedes a negative number or quantity. If there is no
sign in front of a number, the number is understood to
be positive.

Exercises: Identifying Positive and Negative Numbers

Represent each of the following quantities with a + or
- sign. For example, a profit of $250 is +$250.

5-21.   A deposit of $150.

5-22.   A loss of $9.90.

5-23.   A drop in altitude of 1100m.

5-24.   A 9 yd gain.

5-25.   32° below zero.

5–6   Addition and Subtraction of Signed Numbers

The following number line helps to explain positive and
negative numbers:

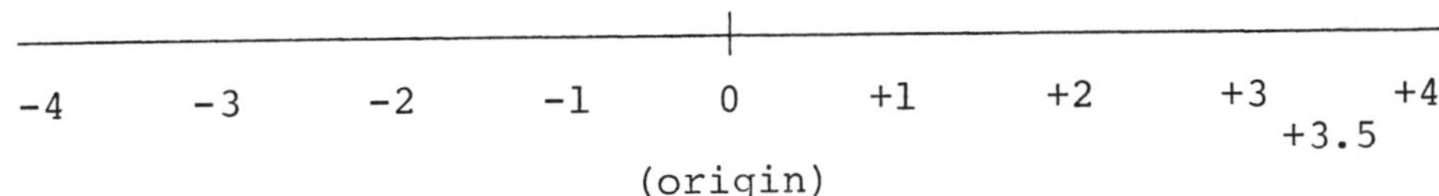

Numbers to the right of zero are positive, and numbers
to the left of zero are negative. Remember, if a number
does not have a sign, the number is understood to be
positive.

Procedure for Adding Signed Numbers

1. To add by using the number line, move to the right
    for positive numbers and move to the left for nega-
    tive numbers.
2. To add without using the number line, proceed as you
    normally would to add numbers if the numbers have
    the same sign and bring the sign down to the sum.
3. If one number is positive and the other is negative,
    subtract them and bring the sign of the larger number
    down to the difference.

Example 5–1

PROBLEM

Add +3, +1, and +4.

SOLUTION

1. If you use the number line, move to the right for
    each of the positive numbers:

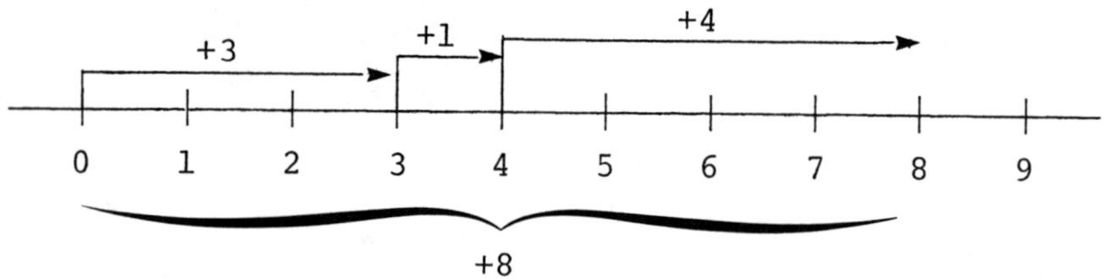

2. If you do not wish to use the number line, simply
    add the numbers and bring the sign down to the sum:

```
    3
    1
+   4
----
    8
```

Note: A number without a sign is understood to be
positive.

ANSWER

    +3 + 1 + 4 = 8

Example 5–2

PROBLEM

Add +3 and −5.

SOLUTION

1. If you use the number line, move 3 units to the
    right and 5 units to the left:

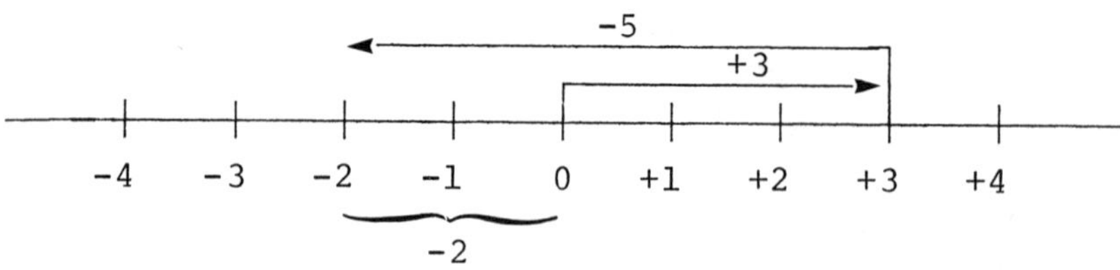

2. If you do not wish to use the number line, subtract
    the two numbers and bring the sign of the larger
    number down to the difference:

```
    +3
+ (−5)
------
   −2
```

ANSWER

    +3 + (−5) = −2

Exercises: Adding Signed Numbers

Add the following by using both the number line and the
regular procedure:

5-26. 5 + 12   5-27. -6 + 27   5-28. -6 + (-15)

5-29. (-23) + 16   5-30. $-3r + (-16r)$

Procedure for Subtracting Signed Numbers

●●

1. To subtract by using the number line, move to the
    right for positive numbers and move to the left for
    negative numbers. Then find the difference by
    finding the distance from the second quantity to the
    first quantity.
2. To subtract without using the number line, change
    the sign of the subtrahend (bottom number) and pro-
    ceed as in adding signed numbers.

●●

Example 5–3

PROBLEM

Subtract -5 from +15.

SOLUTION

1. If you use the number line, move 5 units to the left
    for -5 and 15 units to the right for +15:

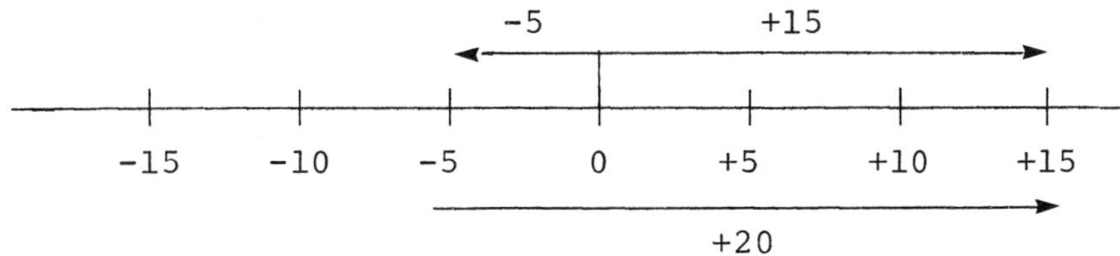

Note: Subtraction is finding the difference between the
numbers. To get from -5 to +15, you move +20 units to
the right.

2. If you do not wish to use the number line, change
    the sign of the subtrahend (bottom number) and add:

         +15
      + (-)5
         +20

ANSWER

      +15 - (-5) = +20

Example 5–4

PROBLEM

Subtract +4 from -8.

SOLUTION

1. If you use the number line, move 4 units to the
    right and 8 units to the left:

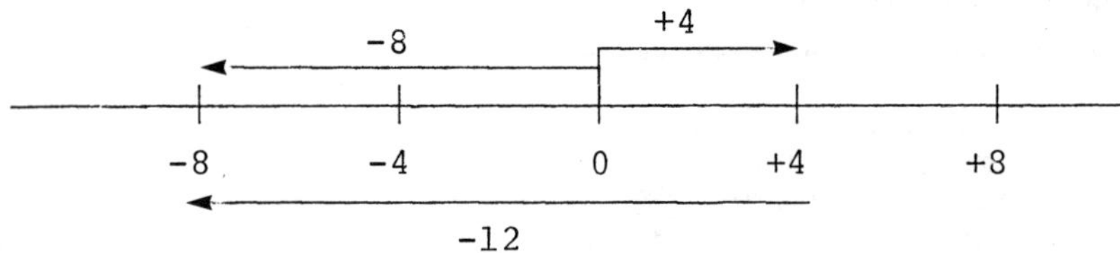

Note: To get from +4 to -8, you move 12 units to the
left, or -12.

2. If you do not wish to use the number line, change
    the sign of the subtrahend and add:

$$\begin{array}{r} -8 \\ -\ (+4) \\ \hline -12 \end{array}$$

ANSWER

$$-8 - (+4) = -12$$

Exercises: Subtracting Signed Numbers

Subtract the following by using both the number line
and the regular procedure:

5-31.   17 - (+7)   5-32.   -36 - (+12)   5-33.   -15 - (-10)

5-34.   37 - (-16)   5-35.   -39 - (-38)

5–7   Multiplication and Division of Signed Numbers

When multiplying and dividing signed numbers, like signs
give positive answers, and unlike signs give negative
answers. This fact is shown in the examples.

Procedure for Multiplying Signed Numbers

1. When multiplying numbers with like signs, the
    product is positive.
2. When multiplying two numbers with unlike signs, the
    product is negative.

3. When multiplying more than two numbers with unlike
    signs, count the negative signs. If the total
    number of negative signs is odd, the product is
    negative. If the total number of negative signs is
    even, the product is positive.

●●

Example 5–5

PROBLEM

Multiply -3 by 2.

SOLUTION

1. Note that -3 and (+)2 have unlike signs.

2. Multiply the two numbers; the product is negative.

ANSWER

    -3 x 2 = -6

Example 5–6

PROBLEM

Multiply (+6)(+2).

SOLUTION

1. Note that +6 and +2 have like signs.

2. Multiply the two numbers (sets of parentheses side
    by side indicate multiplication); the product is
    positive.

ANSWER

    (+6)(+2) = +12

Example 5–7

PROBLEM

Multiply (-4)(+12)(0).

SOLUTION

<u>Note</u>: Zero times any quantity is zero.

Because 0 is one of the numbers to be multiplied,
simply write 0 as the product.

ANSWER

    (-4)(+12)(0) = 0

<u>Note</u>: Zero does not have a sign.

Example 5–8

PROBLEM

Multiply $(+7)(-8)(-2)(3)(-1)$.

SOLUTION

<u>Note</u>: There is no sign in front of the 3; therefore, it <u>is</u> understood to be a positive number (+3).

1.   Multiply the numbers.

2.   Count the negative signs of the numbers you multiplied.

3.   The number of negative signs is odd, so the product is negative.

ANSWER

   $(+7)(-8)(-2)(3)(-1) = -336$

Example 5–9

PROBLEM

Multiply $(-2)(-3)(-4)(-5)$.

SOLUTION

1.   Multiply the numbers.

2.   Count the negative signs of the numbers you multiplied.

3.   The number of negative signs is even (4), so the product is positive.

ANSWER

   $(-2)(-3)(-4)(-5) = 120$

Exercises: Multiplying Signed Numbers

Multiply the following:

5-36.   9(-6)   5-37.   (19)(20)

5-38.   (-1)(-25)(-25)   5-39.   (-15)(-4)

5-40.   (-33)(14)(0)

Procedure for Dividing Signed Numbers

● ●

1.   When dividing numbers with like signs, the quotient is positive.

2. When dividing numbers with unlike signs, the quotient
    is negative.

•••

Example 5–10

PROBLEM

Divide -12 by -3.

SOLUTION

1. Note that the two numbers have like signs.

2. Divide the first number by the second; the quotient
    is positive.

ANSWER

$$-12 \div -3 = +4$$

Example 5–11

PROBLEM

Divide 125 by -5.

SOLUTION

1. Note that the two numbers have unlike signs.

2. Divide the first number by the second; the quotient
    is negative.

ANSWER

$$125 \div -5 = -25$$

Exercises: Dividing Signed Numbers

Divide the following:

5-41. $-55 \div 5$ 5-42. $48 \div (-8)$ 5-43. $-144 \div (-12)$

5-44. $-69 \div (-69)$ 5-45. $58 \div (-2)$

5–8   Removal of Signs of Grouping

The removal of signs of grouping is a necessary
operation in evaluating, or finding the value of, an
algebraic expression. This operation can also be called
simplifying or evaluating an expression.

Procedure for Removing Signs of Grouping

1. Perform all operations within the grouping (i.e., a set of parentheses).
2. If the sign in front of the grouping is negative, change all the signs (from positive to negative and from negative to positive) within the group when you remove the parentheses.
3. If the sign in front of the set of parentheses is positive, do not change any of the signs within the group when you remove the parentheses. (If there is no sign in front of the set of parentheses, a positive sign is understood.)
4. When removing several sets of grouping signs, start with the innermost set and work to the outermost, making the necessary changes according to whether a positive or a negative sign precedes the group.
5. After all the signs of grouping have been removed, perform the arithmetic operations.
6. Multiply and divide from left to right.
7. Add and subtract from left to right.
8. Combine all like quantities.

Example 5–12

PROBLEM

Evaluate $4(24 \div 2) - 8$.

SOLUTION

1. Perform the operation inside the parentheses:

$$(24 \div 2) = \left(\frac{24}{2}\right)$$

2. Remove the parentheses. The sign in front of the parentheses is understood to be positive:

$$4(24 \div 2) - 8 = 4\left(\frac{24}{2}\right) - 8$$

3. Perform the arithmetic operations. Working from left to right, multiply 4 by 12 and subtract 8:

$$4\left(\frac{24}{2}\right) - 8 = 4(12) - 8 = 48 - 8$$

ANSWER

$$4(24 \div 2) - 8 = 40$$

Example 5–13

PROBLEM

Evaluate $240 \div (6 \times 10 \times 2) \div 2$.

SOLUTION

1. Perform the multiplication within the parentheses from left to right:

    (6 x 10 x 2) = (120)

2. Remove the parentheses without changing any signs:

    240 ÷ (6 x 10 x 2) ÷ 2 = 240 ÷ 120 ÷ 2

3. Perform the arithmetic operations. Working from left to right, divide 240 by 120. Then divide the quotient by 2:

    $\frac{240}{120} \div 2 = 2 \div 2 = 1$

ANSWER

    240 ÷ (6 x 10 x 2) ÷ 2 = 1

Example 5–14

PROBLEM

Simplify $[24 \div (12 \times 2)] + (4x \times 5)$.

Note: In a complex set of groupings, an expression that includes a group in parentheses is enclosed by brackets (and an expression that includes the above is enclosed by braces).

SOLUTION

1. Perform the multiplication inside the parentheses:

    $[24 \div (12 \times 2)] + (4x \times 5) = [24 \div (24)] + (20x)$

2. Remove the parentheses without changing any signs:

    $[24 \div (12 \times 2)] + (4x \times 5) = [24 \div 24] + 20x$

3. Perform the division inside the brackets; remove the brackets and add:

    $\frac{24}{24} + 20x = 1 + 20x$

ANSWER

    $[24 \div (12 \times 2)] + (4x \times 5) = 1 + 20x$

Example 5–15

PROBLEM

Evaluate $8a - 5[6b - (a + 3b) + (4a - 3b) - 6a]$.

SOLUTION

1. Remove the innermost sets of parentheses first and change signs as necessary:

    $8a - 5[6b - (a + 3b) + (4a - 3b) - 6a] =$
    $8a - 5[6b - a - 3b + 4a - 3b - 6a]$

2. Combine like quantities inside the brackets:

    $8a - 5[6b - a - 3b + 4a - 3b - 6a] = 8a - 5[-3a + 0b]$

3. Remove the brackets by multiplying each term within the brackets by -5; then combine like terms:

    $8a - 5[-3a + 0b] = 8a + 15a = 23a$

<u>Note</u>: Remember that 0 times anything is 0, so $0b$ is 0.

ANSWER

   $8a - 5[6b - (a + 3b) + (4a - 3b) - 6a] = 23a$

Exercises: Removing Signs of Grouping

Solve the following problems:

5-46. $-8 + (-13)$

5-47. $42 - (-18) + 4$

5-48. $8a - (-24a) + 0a$

5-49. $4 \times (-12) + (-16) \times (-1)$

5-50. $(8m + 3n) + (4m - 5n)$

5-51. $(2x + 6y) - (7x - 16y)$

5-52. $35 - [18z - (7z + 2) + 3(-5z - 3)]$

5-53. $(102 - 63) \div 9 + 41(-1 + 2)$

5-54. $1/2(42 + 39.4) + 3.142(6)$

5-55. $4 \div 4 \times 4 \div 4 \times 4$

5-9   Basic Operations with Exponents

Before we can multiply algebraic expressions, we must know something about exponents and powers. An <u>exponent</u>, or <u>power</u>, is a number written above and to the <u>right of</u> a quantity. The exponent, or power, indicates how many times the <u>base</u>, or quantity, is used as a factor. Exponents are used to abbreviate algebraic expressions. For example:

$$j^4 = j \cdot j \cdot j \cdot j$$

(Exponent — 4; Base — j)

j^4 is read "j to the fourth power." x^3 is read "x to the third power."
  The following rules for multiplying and dividing with exponents are known as the <u>Laws of Exponents</u>. (There are no rules for adding and subtracting exponents; only like terms can be added or subtracted.)

  1. <u>When multiplying quantities with like bases, add the exponents</u>:

$$(k^3)(k^2) = k^{3\,+\,2} = k^5$$

because

$$(k \cdot k \cdot k)(k \cdot k) = k \cdot k \cdot k \cdot k \cdot k$$

Another example is:

$$(a^m)(a^n) = a^{m\,+\,n}$$

Follow the same procedure when like numbers are the bases:

$$(3^4)(3^2) = 3^{4\,+\,2} = 3^6$$

  2. <u>When dividing quantities with like bases, subtract the exponent of the denominator from the exponent of the numerator</u>:

$$\frac{n^5}{n^3} = n^{5\,-\,3} = n^2$$

because

$$\frac{\cancel{n} \cdot \cancel{n} \cdot \cancel{n} \cdot n \cdot n}{\cancel{n} \cdot \cancel{n} \cdot \cancel{n}} = n^2$$

<u>Note</u>: When moving an exponent from the denominator to the numerator, change the sign on the exponent, as in n^{5-3} in the example above. Do the same when moving a quantity from the numerator to the denominator.

Another example is:

$$\frac{a^m}{a^n} = a^{m\,-\,n}$$

Follow the same procedure when like numbers are the bases:

$$\frac{5^8}{5^5} = 5^{8-5} = 5^3 = 5 \times 5 \times 5 = 125$$

3.   <u>Any quantity with an exponent of zero is equal to 1.</u> This fact can best be illustrated by an example:

$$\frac{x^3}{x^3} = \frac{\cancel{x} \cdot \cancel{x} \cdot \cancel{x}}{\cancel{x} \cdot \cancel{x} \cdot \cancel{x}} = \frac{1}{1} = 1^0 = 1$$

or

$$\frac{x^3}{x^3} = x^{3-3} = x^0 = 1$$

Another example is:

$$a^0 = 1$$

Follow the same procedure when like numbers are the bases:

$$\frac{4^3}{4^3} = 4^{3-3} = 4^0 = 1$$

because

$$\frac{4^3}{4^3} = \frac{\cancel{4} \cdot \cancel{4} \cdot \cancel{4}}{\cancel{4} \cdot \cancel{4} \cdot \cancel{4}} = \frac{1}{1} = 1^0 = 1$$

Procedure for Multiplying and Dividing with Exponents

• •

1.   When multiplying a number and an algebraic term that has an exponent (an exponential term), multiply the numbers and carry the literal coefficient with the exponent into the product.
2.   When multiplying exponential terms, multiply the numbers (numerical coefficients) and add the exponents of the like bases.
3.   When dividing exponential terms, divide the numerical coefficient of the denominator into the numerical coefficient of the numerator and subtract the exponents of the like bases.

• •

Example 5–16

PROBLEM

Multiply $(2)(8k^2)$.

SOLUTION

Multiply the numbers and carry the literal coefficient into the product:

$$(2)(8k^2) = 16k^2$$

ANSWER

$$(2)(8k^2) = 16k^2$$

Example 5–17

PROBLEM

Multiply $(3x^2)(4x^3)$.

SOLUTION

Multiply the numbers (numerical coefficients) and add the exponents of the like bases (x):

$$(3x^2)(4x^3) = (3)(4)x^{2+3}$$
$$= 12x^5$$

ANSWER

$$(3x^2)(4x^3) = 12x^5$$

Example 5–18

PROBLEM

Divide $15p^4r^3$ by $5p^2r$.

<u>Note</u>: If there is no given exponent, the exponent is understood to be 1.

SOLUTION

Divide the denominator into the numerator and subtract the exponents of the like bases:

$$\frac{15p^4r^3}{5p^2r} = \frac{\overset{3}{\cancel{15}}p^{4-2}\,r^{3-1}}{\cancel{5}p^2r^1}$$
$$= 3p^2r^2$$

ANSWER

$$15p^4r^3 \div 5p^2r = 3p^2r^2$$

Example 5–19

PROBLEM

Divide $12^2a^2b^3c^4$ by $144a^2bc^4$.

SOLUTION

Divide the denominator into the numerator (remember, $12^2 = 144$) and subtract the exponents of the like bases (remember, $b = b^1$):

$$\frac{12^2 a^2 b^3 c^4}{144 a^2 bc^4} = \frac{(12 \times 12) a^{2-2} b^{3-1} c^{4-4}}{144 a^2 b^1 c^4}$$

$$= \frac{144 a^{2-2} b^{3-1} c^{4-4}}{144}$$

$$= a^0 b^2 c^0$$

$$= (1)(b)^2(1)$$

$$= b^2$$

ANSWER

$$12^2 a^2 b^3 c^4 \div 144 a^2 bc^4 = b^2$$

Exercises: Multiplying and Dividing with Exponents

Evaluate the following:

5-56. 4^3

5-57. $\dfrac{12^9}{12^8}$

5-58. t^3, if $t = 3$

5-59. $x^3 \cdot x^5$

5-60. $3(5x^6)$

5-61. $x^7 y^4 \cdot x^2 y^5$

5-62. $\dfrac{x^3 y^2}{x^2 y^2}$

5-63. $z^5(z^7)$

5-64. $(k^2 j^3)^2$

5-65. $\dfrac{P^{12}}{P^1}$

5-66. $\dfrac{168 A^3 B^2 C}{21 AB^2 C^3}$

5-67. $\dfrac{x^5 y}{x^5 y}$

5-68. $\dfrac{-10 C^4}{-2 C^3}$

5-69. $(X^0 Y^2)(X^2 Y^0)$

5-70. $-77 e^2 \div -11 e^2$

5-71. $(6^0)(6^2)$

5-72. $\dfrac{-15 Y^5}{Y^5}$

5-73. $\dfrac{225 X^1 YZ}{15^2 XY^2 Z^1}$

5-74. $\dfrac{5^6}{6^5}$

5-75. $(6X^3)(3^2 Y^6)$

Chapter 5

Basic Algebra

Solving Applied Welding Problems

Now that you have mastered the fundamental operations of
basic algebra, you are ready to apply these math skills
to typical problems that welders encounter on the job.
After studying the examples that follow, complete the
applied welding problems.

Example 5–20

PROBLEM

The length of the steel bar in Figure 5-1 is increased
by 4'. A represents the original dimension. Use an
algebraic expression to express the total length of the
steel bar.

SOLUTION

Add the increase to the original length.

ANSWER

Total length = A + 4.

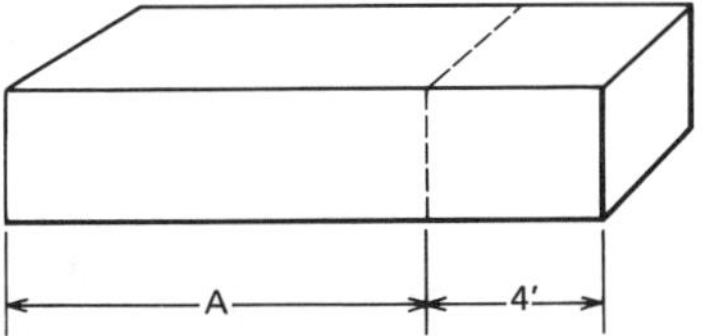

Figure 5—1

Example 5–21

PROBLEM

The height of a utility tower is increased by 30'. B
represents the original height. Use an algebraic
expression to express the total length of the tower.

SOLUTION

Add the increase to the original length.

ANSWER

Total length = B + 30'.

Example 5–22

PROBLEM

Sixty reinforcing rods are obtained for a tower base.
Some of the rods are too short. Let C equal the number
of rods that are too short. Use an algebraic expression
to indicate the number of rods that are the proper
length.

SOLUTION

Subtract the number of rods that are too short from the
total number of rods.

ANSWER

Number of rods that are the proper length = 60 - C.

Example 5–23

PROBLEM

Find the amount of material needed to make reinforcing
rods that are each 6" long. Let D equal the number of
reinforcing rods. Use an algebraic expression to indicate the total inches of material needed.

SOLUTION

Multiply the number needed (D) by the length of each rod.

ANSWER

Total amount of material = D(6").

Example 5–24

PROBLEM

The production rate of a wire feed welder is 4 times
greater than the rate of a stick electrode welder. The
stick electrode welder welds at a rate of E" per minute.
Use an algebraic expression to express the rate of the
wire feed welder.

SOLUTION

Multiply 4 ("times greater") by the E" welded per minute.

ANSWER

Rate of wire feed welder = 4(E).

Applied Welding Problems

5-76. The original length of a section of steel tubing
       is increased by 10' on one end. The other end
       of the tubing is increased by 11-1/2'. Use an

algebraic expression to express the overall length
of the tubing.

5-77.   The overall length of a metal plate is 16". The
plate is B" too long. Use an algebraic expression
to express the required length.

5-78.   A 20' length of pipe is J' too long. Use an
algebraic expression to express the required
length.

5-79.   A length of bar stock is C" too long. A 4'
length is cut from the bar. Use an algebraic
expression to express the remaining length.

5-80.   A 10' length of tubing has D" cut from it. It
also has E" cut from it. Use an algebraic expression to express the remaining length of
tubing.

5-81.   20 parts are cut from a length of round stock.
Each part is F" long. Use an algebraic expression
to express the length of the round stock before
cutting.

5-82.   A length of bar stock represented by G is cut
into 4 equal lengths. Use an algebraic expression
to express the length of the parts.

5-83.   20 welding rods weigh a total of 1 lb. Use an
algebraic expression to express the number of
welding rods in a pile weighing H lbs.

5-84.   An automatic welding machine welds 180" in T
minutes. Use an algebraic expression to express
the inches per minute.

5-85.   A plate is J" long. It has 3 equally spaced
holes cut in it. Use an algebraic expression to
express the distance between the centers of the
holes.

5-86.   A metal work table built by a welding student has
an overall height of 39". The table, however,
does not sit level. One leg is R" too long. Use
an algebraic expression to express the correct
length.

5-87.   A truck frame is increased by 4'. S represents
the original length. Use an algebraic expression
to express the overall length of the truck frame.

5-88.   The first shift at a welding shop has 3 times as
many employees as the second shift. The second
shift has T employees. Use an algebraic expression to express the number of employees on the
first shift.

5-89.   187 parts are welded in U hours. Use an algebraic expression to express the number of parts
that are welded per hour.

5-90.   450 parts are welded by V welders. Use an algebraic expression to express the number of welds that each welder produced.

Applied Welding Problems
Survey Test

Name ____

Course/Sec. ______________   Date ___________________________

5-1. K plates 12" long are required for metal cement   5-1. ________________
      form pads. Use an algebraic expression to express
      the total length of material needed.

5-2. An automatic welding setup can weld L" per minute.   5-2. ________________
      Use an algebraic expression to express the number
      of inches it can weld in 1 hour.

5-3. 3 sections are required to build a flagpole.   5-3. ________________
      Section 1 is M' long, section 2 is N' long, and
      section 3 is P' long. Use an algebraic expression
      to express the total length.

5-4. 5 pieces are cut from a length of tubing. Each   5-4. ________________
      piece is P' long. Use an algebraic expression to
      express the total length of the 5 parts.

5-5. A steel beam is cut into 8 equal sections. The   5-5. ________________
      beam is Q' long. Use an algebraic expression to
      express the length of each part.

Chapter 6: Evaluating Algebraic Expressions
Preview and Self-Test

Name ____

Course/Sec. ________________   Date ____________________

Chapter 6 covers the evaluation of algebraic expressions
and formulas. Before you turn to the chapter, complete
the Self-Test to determine which sections in the chapter
you need to study carefully.

Self-Test

EVALUATING ALGEBRAIC EXPRESSIONS

Given $a = 5$, $b = 3$, $c = 4$, and $e = 12$:

6-1. Find x when $x = a + b - c$.　　　　　　　　　　　6-1. ________________

6-2. Find y when $y = (ab - ce)/b$.　　　　　　　　　6-2. ________________

6-3. Find z when $z = abce$.　　　　　　　　　　　　6-3. ________________

6-4. Find p when $p = \sqrt{bce}$.　　　　　　　　　　6-4. ________________

6-5. Find k when $k = (20e/ace) + (ce/b)$.　　　　　6-5. ________________

Evaluating algebraic expressions score ________________

EVALUATING FORMULAS

6-6. Find A when $A = lw$, $l = 36$, and $w - 12.5$.　　6-6. ________________

6-7. Find A when $A = 0.7854d^2$ and $d = 42$.　　　　　6-7. ________________

6-8. Find A when $A = 0.5bh$, $b \doteq 29.5$, and $h = 15.8$.　6-8. ________________

6-9. Find P when $P = 2(l + w)$, $l = 36$, and $w = 12.5$.　6-9. ________________

6-10. Find R when $r_1 = 2$, $r_2 = 4$, and $r_3 = 8$, and　6-10. ________________

$$R = \frac{1}{\frac{1}{r_1} + \frac{1}{r_2} + \frac{1}{r_3}}$$

Evaluating formulas score ________________

EVALUATING COMPLEX FORMULAS

Evaluate the following formulas and round the answer
according to the indicated accuracy:

6-11.   Evaluate for x when $x = \sqrt{a^2 + b^2 - c^2}$ and $a = 100$,   6-11. _________________
        $b = 50$, and $c = 25$.   Round the answer to the
        nearest hundredth.

6-12.   Find V when $V = \pi r^2 h/3$ and $\pi = 22/7$, $r = 28$, and   6-12. _________________
        $h = 1/7$.

6-13.   Find c when $c = \sqrt{a^2 + b^2}$, $a = 32.4$, and $b = 42.3$.   6-13. _________________
        Round the answer to the nearest tenth.

6-14.   Find C when $C = (a^2 + b^2 - c^2)/2ab$ and $a = 19$,   6-14. _________________
        $b = 32$, and $c = 47$.

6-15.   Find V when $V = h(B_1 + 4M + B_2)/6$ and $h = 36$,   6-15. _________________
        $B_1 = 10.5$, $M = 6.8$, and $B_2 = 5.8$.

                    Evaluating complex formulas score   _________________

After you have checked your answers on the Self-Test,
transfer your scores to the Chapter 6 Objectives.

••

Evaluating Algebraic Expressions

Upon successful completion of this chapter, you will be
able to evaluate algebraic expressions, formulas, and
complex formulas and to round the results to the given
level of accuracy.

Objectives

	Self-Test Scores	If you did poorly on the Self-Test, turn to:
• Evaluating algebraic expressions	__________	Section 6-2
• Evaluating formulas	__________	Section 6-3
• Evaluating complex formulas	__________	Section 6-4

6–1 Introduction to Evaluating Algebraic Expressions

The evaluation of algebraic expressions can be a vital
part of work in many trade areas. We usually do not
consider finding the amount of interest on savings or
loans, figuring the area of a room, or determining miles
per gallon as evaluating algebraic expressions. But
we actually do evaluate expressions when we find the
answers to these problems. Although the algebraic
expressions we use in everyday living and those used in
technology vary greatly, the procedures for evaluation
are the same. Thus, learning to evaluate expressions is
very important because this knowledge will carry over to
evaluating more complex expressions.

6–2 Evaluating Algebraic Expressions

You remember that an <u>algebraic expression</u> consists of
numbers, letters, and mathematical symbols. Examples of
algebraic expressions are:

$$p = 2l + 2w, \quad X = w - l, \quad c = \sqrt{a^2 + b^2}$$

You also recall that the regularly used mathematical
symbols are +, -, x, and ÷.
   Multiplication and division signs are seldom used
in algebra because these operations are included when
the expression is written. For example, the expression
$I = prt$ indicates that the value of I is to be found by
multiplying p by r by t. The expression $x = yz$ indicates that the value of x is to be found by multiplying
y by z. The expression $j = kl/2$ indicates that the
value of j is to be found by multiplying k by l and
dividing the product by 2.
   The value of any algebraic expression can be determined if the numerical values of the literal quantities
are given. This process is called <u>evaluating an algebraic expression.</u>
   Evaluating an algebraic expression is not
difficult. Most students who have difficulty do so
because they have not read the expression correctly.

Procedure for Evaluating Algebraic Expressions

• •

1. Copy the expression to be evaluated.
2. Substitute the given numerical values for the literal
    quantities.
3. Perform the indicated operations by using the correct
    order of operation.

• •

   The examples that follow illustrate how to evaluate
algebraic expressions.

Example 6–1

PROBLEM

Evaluate $a = b + c - d$ when $b = 10$, $c = 4$, and $d = 6$.

SOLUTION

1. Copy the expression:

$$a = b + c - d$$

2. Substitute the given numerical values for the literal quantities:

$$a = 10 + 4 - 6$$

3. Perform the indicated operations:

$$a = 10 + 4 - 6 = 14 - 6 = 8$$

ANSWER

$$a = 8$$

Example 6–2

PROBLEM

Evaluate $x = (3k + 2q)/j$, given $k = 12$, $q = 22$, and $j = 6$.

SOLUTION

1. Copy the expression:

$$x = \frac{3k + 2q}{j}$$

2. Substitute the given numerical values for the literal quantities:

$$x = \frac{3(12) + 2(22)}{6}$$

3. Perform the indicated operations:

$$x = \frac{36 + 44}{6} = \frac{80}{6} = 13.\overline{3}$$

ANSWER

$$x = 13.\overline{3}$$

Example 6–3

PROBLEM

Evaluate $n = \sqrt{abc}$, given $a = 2$, $b = 8$, and $c = 9$.

SOLUTION

1. Copy the expression:

   $n = \sqrt{abc}$

2. Substitute the given numerical values for the literal quantities:

   $n = \sqrt{(2)(8)(9)}$

3. Perform the indicated operations:

   $n = \sqrt{(16)(9)} = \sqrt{144} = 12$

Note: The symbol $\sqrt{}$, called the radical, indicates that you are to find the square root, or the factor that, when multiplied by itself, equals the quantity under the radical. Thus, $\sqrt{9} = 3$, $\sqrt{25} = 5$, and $\sqrt{144} = 12$. Similarly, the symbol $\sqrt[3]{}$ indicates that you are to find the cube root, or the factor that, when multiplied by itself three times, equals the quantity under the radical. Thus, $\sqrt[3]{27} = 3$, $\sqrt[3]{125} = 5$, and $\sqrt[3]{343} = 7$. Use a calculator to find square roots and cube roots of numerical values.

ANSWER

   $n = 12$

Exercises: Evaluating Algebraic Expressions

Evaluate the following:

6-1.   If $x = 2$, find $5x$.

6-2.   If $p = -5$, find $-5p$.

6-3.   If $y = -6$, find $8y$.

6-4.   If $j = 1/2$, find $16j$.

6-5.   Find y if $y = ab$, $a = 4$, and $b = 7$.

6-6.   Find a if $a = pq$ when $p = -3$ and $q = 14$.

6-7.   Given $K = fgh$, find K when $f = 4$, $g = 6$, and $h = 2$.

6-8.   Given $M = jsm$, find M when $j = -6$, $s = -8$, and $m = 1/2$.

6-9.   Given $x = 3p - 4g$, find x when $p = 9$ and $g = -3$.

6-10.   Given $z = a/3 + b/4$, find z when $a = 9$ and $b = -12$.

6-11.   Given $j = x^2 + y^2$, find j when $x = -3$ and $y = 5$.

6-12.   If $p = (st)^2$, find p when $s = 6$ and $t = 3$.

6-13.   If $d = (pk)^2$, find d when $p = 1/16$ and $k = 64$.

6-14.   If $N = \left(\dfrac{B + b}{2}\right)^2$, find N when $B = 12$ and $b = -2$.

6-15.   If $Z = (n - m - 2)^2$, find Z when $n = 39$ and $m = 27$.

6-16.   If $E = 10a^2b^2$, find E when $a = 4$ and $b = 1/2$.

6-17.   Given $J = \sqrt{12a}$, find J when $a = 3$.

6-18.   Given $R = \sqrt{1350b}$, find R when $b = 1/6$.

6-19.   Given $W = \sqrt{19r + 3t}$, find W when $r = 10$ and $t = 2$.

6-20.   Given $h = \sqrt{10z^2 - 5a}$, find h when $z = 5$ and $a = 5$.

6–3   Evaluating Formulas

A <u>formula</u> is a rule or law used in mathematics, science,
or industry to solve specific problems. A formula can
be considered an algebraic expression. Formulas are
evaluated in the same manner as algebraic expressions.
      Formula evaluation may involve all the arithmetic
operations. The operations should be performed in the
proper order, and the proper dimensions and labels
should be used.
      Finding the distance traveled when going at a
certain rate is a simple application of a formula. The
formula is written $d = rt$, where d = distance, r = rate,
and t = time. The distance is equal to the rate
traveled times the time traveled. For example, if the
rate is 55 miles per hour and the time is 4 hours, the
distance equals rate times time. The problem is solved
by formula as follows:

$$d \;=\; rt$$
$$\;=\; \frac{55 \text{ miles}}{\text{hour}} \times 4 \text{ hours}$$
$$\;=\; 220 \text{ miles}$$

Procedure for Evaluating Formulas

The procedure for evaluating formulas is the same as
that for evaluating algebraic expressions.

1. Copy the expression to be evaluated.
2. Recopy the expression but substitute the given
    numerical values for the literal quantities.
3. Perform the indicated operations by using the
    correct order of operation.

Example 6–4

PROBLEM

Given $A = lw$, find A when $l = 45'$ and $w = 32'$.

SOLUTION

1. Copy the given formula:

$$A = lw$$

2. Substitute the known values:

$$A = (45)(32)$$

3. Perform the indicated operations:

$$A = (45)(32) = 1440$$

Note: When you multiply a unit by the same kind of unit, the result is square units, such as square feet, sq ft or ft^2; square yards, sq yd or yd^2; square meters, or m^2; and so on.

ANSWER

$$A = 1440 \text{ ft}^2$$

Example 6–5

PROBLEM

Given $A = bh/2$, find A when $b = 38$ yd and $h = 28$ yd.

SOLUTION

1. Copy the given formula:

$$A = \frac{bh}{2}$$

2. Substitute the known values:

$$A = \frac{(38)(28)}{2}$$

3. Perform the indicated operations:

$$A = \frac{(38)(28)}{2} = 532 \text{ sq yd}$$

ANSWER

$$A = 532 \text{ sq yd}$$

Example 6–6

PROBLEM

Given $I = prt$, find I when $p = \$4500$, $r = 9.5\%$, and $t = 3$ years.

SOLUTION

1. Copy the given formula:

$$I = prt$$

2.   Substitute the known values:

$$I = (4500)(0.095)(3)$$

<u>Note</u>: Remember, 9.5% = 0.095.

3.   Perform the indicated operations:

$$I = (427.50)(3) = 1282.50$$

ANSWER

$$I = 1282.50$$

Example 6–7

PROBLEM

Given the right triangle (a triangle with a 90° angle)
shown in Figure 6-1, find c.

SOLUTION

<u>Note</u>: The Pythagorean Theorem, named after the Greek
mathematician Pythagoras, is a formula stating that the
sum of the squares of the sides of a right triangle
equals the hypotenuse (the side opposite the 90° angle)
squared (multiplied by itself). Written as a formula,
the Pythagorean formula is $c^2 = a^2 + b^2$. Solved for c,
the formula is

$$c = \sqrt{a^2 + b^2}$$

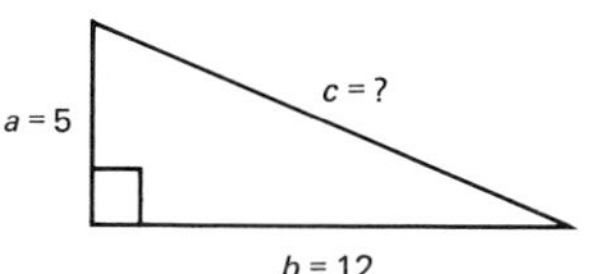

Figure 6—1

1.   To solve for c, copy the given formula:

$$c = \sqrt{a^2 + b^2}$$

2.   Substitute the known values:

$$c = \sqrt{5^2 + 12^2}$$

3.   Perform the indicated operations:

$$c = \sqrt{25 + 144} = \sqrt{169} = 13$$

ANSWER

$$c = 13$$

Exercises: Evaluating Formulas

Solve the following to the nearest hundredth, unless
other directions are given:

6-21.   Given $A = lw$, find A when $l = 4'$ and $w = 3\text{-}1/2'$.

6-22.   Given $P = 4s$, find P when $s = 16.5\text{m}$.

6-23.   Given $P = 2(l + 2)$, find P when $l = 28$ and
        $w = 10.5$.

6-24. Given $P = a + b + c$, find P when $a = 11.3$, $b = 16.6$, and $c = 21.9$.

6-25. Given $P = 2l + 2w$, find P when $l = 39.8$m and $w = 19.9$m.

6-26. If $A = bh/2$, find A when $b = 17$ and $h = 42$.

6-27. If $S = (a + b + c)/2$, find S when $a = 12$, $b = 20$, and $c = 28$.

6-28. If $A = [(B + b)h]/2$, find A when $B = 18$, $b = 26$, and $h = 14$.

6-29. If $Q = 0.26LWT$, find Q when $L = 130$, $W = 0.01$, and $T = 2000$.

6-30. If $A = 0.5W/(P + 10)$, find A when $W = 125$ and $P = 2.5$.

6-31. If $A = D/[0.5(D + d)]$, find A when $D = 148$ and $d = 2$.

6-32. If $A = W/0.0012V^2$, find A when $W = 1200$ and $V = 100$.

6-33. If $V = 0.7854d^2h$, find V when $d = 18$ and $h = 31.4$.

6-34. If $C = 5(F - 32)/9$, find C when $F = 72$.

6-35. Given $E = A - 0.6\sqrt{A}$, find E when $A = 1024$.

6-36. Given $D = \sqrt{s/5500}$, find D when $s = 198{,}000$.

6-37. Given $C = \sqrt{a^2 + b^2}$, find C when $a = 15$ and $b = 8$.

6-38. Given $a = \sqrt{c^2 - b^2}$, find a when $c = 26$ and $b - 10$.

6-39. Given $b = \sqrt{c^2 - a^2}$, find b when $c = 148$ and $a = 37$.

6-4 Evaluating Complex Formulas

Some formulas are more complex than the ones given above. Complex formulas involve several operations. Some complex formulas may involve subscripts. A subscript is a symbol--generally a number--written below and to the right of a quantity. For example, when finding the area of a trapezoid (a four-sided figure with only two parallel sides), we use the two bases (the parallel sides); one base may be labeled b_1 and the other base, b_2. They are read as "b sub one" and "b sub two." We use subscripts when it makes sense to use the same letter more than once in a formula.

Procedure for Evaluating Complex Formulas

The same procedure used to solve simple formulas can be used to solve complex formulas.

1. Copy the expression to be evaluated.
2. Substitute the given numerical values for the literal quantities.
3. Perform the indicated operations by using the correct order of operation.

Example 6–8

PROBLEM

Given $A = \pi r^2$, find A when $\pi = 3.142$ and $r = 3.5$m.

SOLUTION

1. Copy the given formula:

$$A = \pi r^2$$

2. Substitute the known values:

$$A = (3.142)(3.5)^2$$

3. Perform the indicated operations:

$$A = (3.142)(3.5)(3.5) = (3.142)(12.25) = 38.489$$

ANSWER

$$A = 38.4895\text{m}^2$$

Example 6–9

PROBLEM

Given $r_1 = 4$ and $r_2 = 5$, find R in the following formula:

$$R = \frac{1}{\dfrac{1}{r_1} + \dfrac{1}{r_2}}$$

SOLUTION

1. Copy the given formula:

$$R = \frac{1}{\dfrac{1}{r_1} + \dfrac{1}{r_2}}$$

2. Substitute the known values:

$$R = \cfrac{1}{\cfrac{1}{4} + \cfrac{1}{5}}$$

<u>Note</u>: Remember, 1/4 = 0.25 and 1/5 = 0.20.

3. Perform the indicated operations:

$$R = \frac{1}{0.25 + 0.20} = \frac{1}{0.45} = 2.2\overline{2}$$

ANSWER

$$R = 2.2\overline{2}$$

Example 6–10

PROBLEM

Given $A = \pi(R + r)(R - r)$, find A when $\pi = 3.142$, $R = 3/4$, and $r = 1/2$.

SOLUTION

1. Copy the given formula:

$$A = \pi(R + r)(R - r)$$

2. Substitute the known values:

$$A = 3.142 \, (0.75 + 0.50)(0.75 - 0.50)$$

<u>Note</u>: Remember, 3/4 = 0.75 and 1/2 = 0.50.

3. Perform the indicated operations:

$$A = 3.142(1.25)(0.25) = 3.142(0.3125) = 0.981875$$

ANSWER

$$A = 0.982 \text{ (to the nearest thousandth)}$$

Example 6–11

PROBLEM

Given $h = \sqrt{a^2 - (b^2/4)}$, find h when $a = 26'$ and $b = 20'$.

SOLUTION

1. Copy the given formula:

$$h = \sqrt{a^2 - \frac{b^2}{4}}$$

2. Substitute the known values:

$$h = \sqrt{26^2 - \frac{20^2}{4}}$$

3. Perform the indicated operations:

$$h = \sqrt{676 - \frac{400}{4}} = \sqrt{676 - 100} = \sqrt{576} = 24$$

ANSWER

$$h = 24'$$

Exercises: Evaluating Complex Formulas

6-40.   Given $C = Kab/(b - a)$, find C when $K = 10$, $a = 12.5$, and $b = 14.5$.

6-41.   Given $A = m(p + t)/t$, find A when $m = 55$, $t = 5$, and $p = 11.8$.

6-42.   Given $I = prt$, find I when $p = \$4500$, $r = 0.095$, and $t = 3\text{-}1/2$.

6-43.   Given $W = 2PR/(R - r)$, find W when $P = 25.8$, $R = 11.9$, and $r = 6.4$.

6-44.   Given $K = L(D - d)/V$, find K when $L = 35$, $D = 28$, $d = 21$, and $V = 14$.

6-45.   Given $V = 0.7854d^2h$, find V when $d = 38.2$ centimeters and $h = 16$ centimeters.

6-46.   Given $I = (E - e)/R$, find I when $E = 14.754$, $e = 9.731$, and $R = 4.862$.

6-47.   Given $E/e = (R + r)/r$, find E when $R = 78$, $r = 31$, and $e = 10.5$.

6-48.   Given $V = \pi r^2 h/3$, find V when $\pi = 3.142$, $r = 8\text{-}1/2$, and $h = 12\text{-}1/2$.

6-49.   Given $W = 2PR/(R - r)$, find W when $P = 32$, $R = 27.4$, and $r = 23.9$.

6-50.   Given $M = (D - 1.732P) + 3W$, find M when $D = 16.754$, $P = 3.01$, and $W = 2$.

6-51.   Given $A = 0.7854nd^2/360$, find A when $n = 60$ and $d = 12.8$m.

6-52.   Given $R = \dfrac{1}{\dfrac{1}{r_1} + \dfrac{1}{r_2} + \dfrac{1}{r_3}}$

find R (to the nearest thousandth) when $r_1 = 10.5$, $r_2 = 11.6$, and $r_3 = 12.8$.

6-53.   Given $R = r_1 r_2/(r_1 + r_2)$, find R when $r_1 = 50$ and $r_2 = 55$.

6-54.   Given $G = 0.7854 d^2 h/231$, find G when $d = 55.2$ and $h = 148$.

6-55.   Given $R = \sqrt{r^2 + (s^2/4)}$, find R when $r = 15.5$ and $s = 24$.

6-56.   The formula for the volume of a sphere is $V = 4\pi r^3/3$. Find the volume of a sphere when $r = 12$m.

6-57.   The formula for the perimeter of an ellipse is $P = \pi(a + b)$. Find P when $\pi = 3.142$, $a = 22$, and $b = 32$.

6-58.   The formula for the area of a sector of a circle is $A = d^2/1440$. Find A when $\pi = 3.142$ and $d = 15$cm.

6-59.   One of the formulas for horsepower is $H = D^2 N/2.5$. Find H when $D = 4\text{-}3/4"$ and $N = 8$.

6-60.   The formula for the area of a triangle, given all sides, is $A = \sqrt{s(s - a)(s - b)(s - c)}$ where $s = (a + b + c)/2$. Find A given $a = 33$, $b = 67$ and $c = 82$.

Chapter 6

Evaluating Algebraic Expressions

Solving Applied Welding Problems

Now that you have mastered the fundamental operations
of evaluating algebraic expressions, you are ready to
apply these math skills to typical problems that welders
encounter on the job. After studying the examples that
follow, complete the applied welding problems.

Example 6–12

PROBLEM

Given a length of 17" and a tolerance of ±1/4", what is
the shortest acceptable length for side A in Figure 6-2?

SOLUTION

Note: The dimensions given on a drawing indicate how
accurate the finished product must be. The acceptable
range above or below the given dimensions is called the
tolerance. This range is denoted by a plus-or-minus
sign (±), which is an application of positive and
negative numbers.

Subtract the tolerance from the length of the side:

$$A = 17 - \frac{1}{4} = 16\frac{3}{4}$$

ANSWER

    A = 16-3/4"

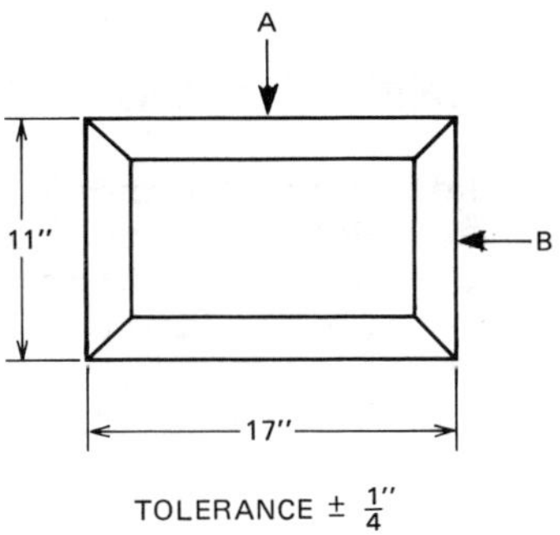

Figure 6—2

Example 6–13

PROBLEM

What is the longest acceptable length for side A in
Figure 6-2?

SOLUTION

Add the tolerance to the length of the side:

$$A = 17 + \frac{1}{4} = 17\frac{1}{4}$$

ANSWER

A = 17-1/4"

Example 6–14

PROBLEM

Given a length of 11" and a tolerance of ±1/4", what is the longest acceptable length for side B in Figure 6-2?

SOLUTION

Add the tolerance to the length of the side:

$$B = 11 + \frac{1}{4} = 11\frac{1}{4}$$

ANSWER

B = 11-1/4"

Applied Welding Problems

6-61.   Refer to Figure 6-3.
    a.   How wide is the opening in part A?
    b.   Given a tolerance of ±1/16", what is the lower limit of the width of the opening in part A?
    c.   How wide is part B if it fits inside part A?

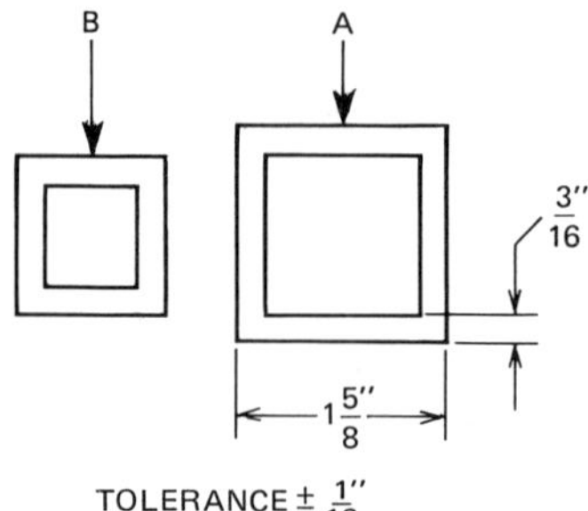

Figure 6–3

6-62.   Refer to the dimensions shown in Figure 6-4 and a tolerance of ±3/32".
    a.   What is the smallest acceptable opening?
    b.   What is the largest acceptable opening?

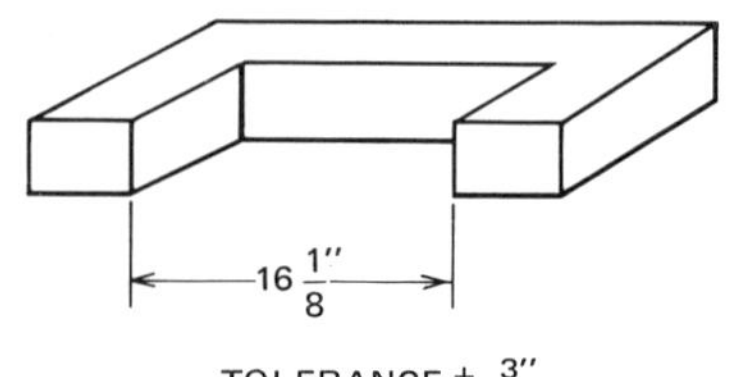

Figure 6–4

6-63. Refer to the dimension shown in Figure 6-5 and a
       tolerance of ±3/8". (a) What is the lower limit
       of the diameter in the figure? (b) What is the
       upper limit of the diameter?

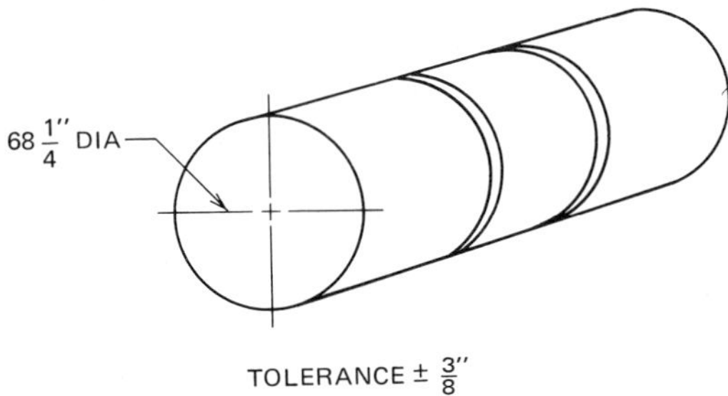

Figure 6—5

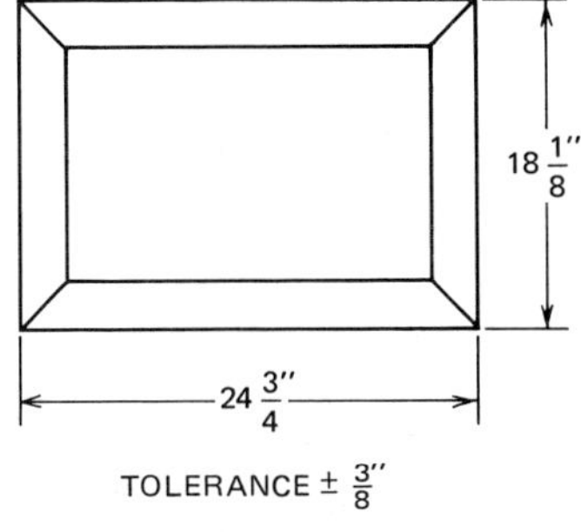

Figure 6—6

6-64. The completed frame in Figure 6-6 measures
       24-7/8" long. The original length is 24-3/4" and
       the tolerance is ±3/8". (a) Is the dimension of
       24-7/8" within tolerance? (b) Is a 17-15/16"
       width within tolerance?

6-65. For the pipe in Figure 6-7, the original length
       is 97-1/8" and the tolerance is ±1/8". (a) Is
       a length of 96-7/8" within tolerance? (b) What
       is the longest acceptable length for the pipe?
       (c) What is the lower limit of the pipe?

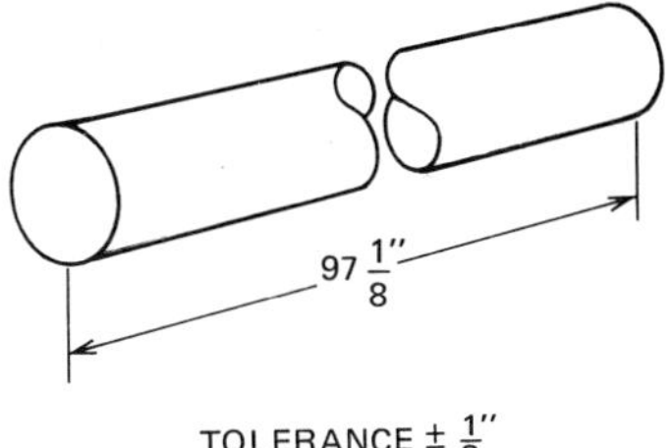

Figure 6—7

6-66. Figure 6-8 shows a 90° angle and a tolerance of
       ±3°20' (3 degrees, 20 minutes).
       a. What is the largest acceptable angle? (There
           are 60 minutes in a degree.)
       b. What is the smallest acceptable angle?
       c. Is 87°40' acceptable?
       d. Is 92°20' acceptable?

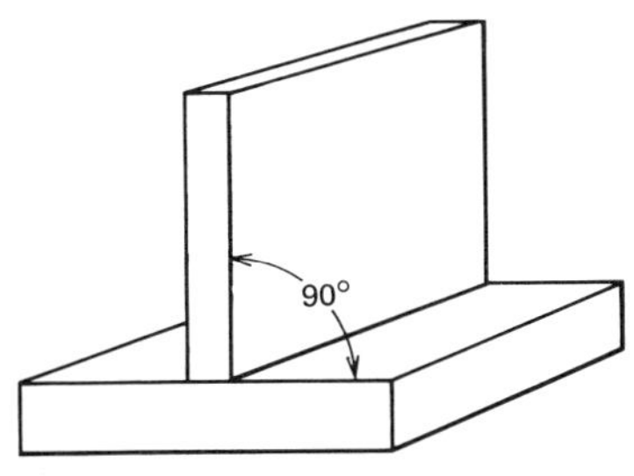

Figure 6—8

6-67. Dimension A in Figure 6-9 is 11-7/16" when com-
       pleted. The tolerance is ±1/8".
       a. Is the 11-7/16" width within tolerance?
       b. If A is 11-1/2" when completed, is this
           width within tolerance?

    c.   If dimension B is 12-9/16" when completed,
        is this width acceptable?
    d.   If dimension B is 12-7/8" when completed, is
        this width acceptable?
    e.   What is the upper limit of A?
    f.   What is the lower limit of A?
    g.   What is the upper limit of B?
    h.   What is the lower limit of B?

6-68. Refer to the dimensions shown in Figure 6-10 and
      a tolerance of ±0.005".
    a.   What is the upper limit of side C?
    b.   What is the lower limit of side D?
    c.   What is the upper limit of the hole?
    d.   What is the lower limit of the hole?
    e.   What is the lower limit of side C?
    f.   What is the upper limit of side D?

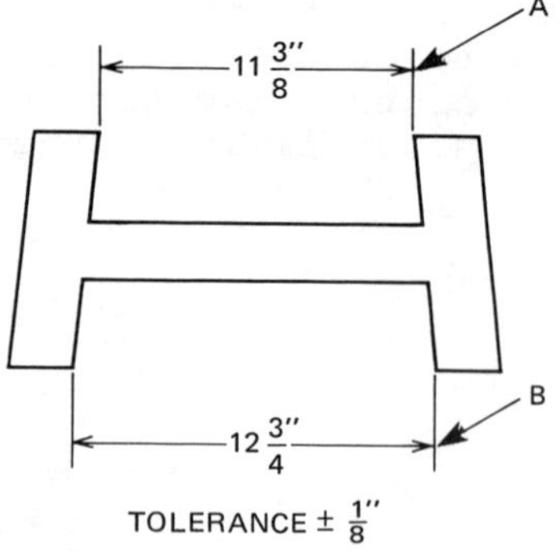

Figure 6—9

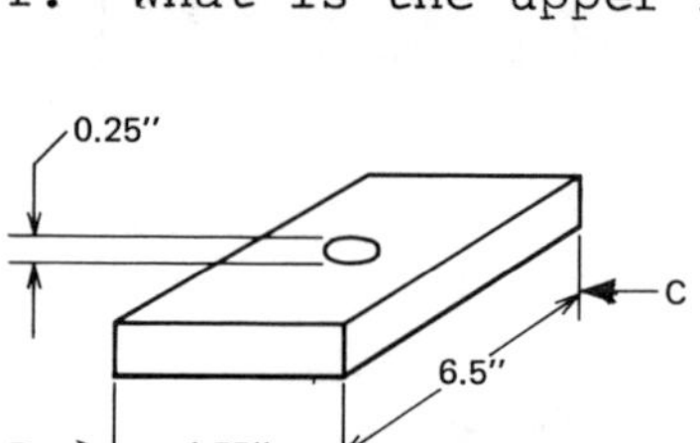

Figure 6—10

Chapter 6: Evaluating Algebraic Expressions

Applied Welding Problems Survey Test

Name ______________________________

Course/Sec. ______________ Date ______________

6-1. The return line in Figure 6-11 will be installed in an oil refinery. The assembly can be 3/16" longer or shorter than the measurements given.

a. What is the maximum total length the piping assembly can be? 6-1a. ____________

b. What is the minimum total length the piping assembly can be? 6-1b. ____________

c. What is the maximum total width the assembly can be? 6-1c. ____________

d. What is the minimum total width the assembly can be? 6-1d. ____________

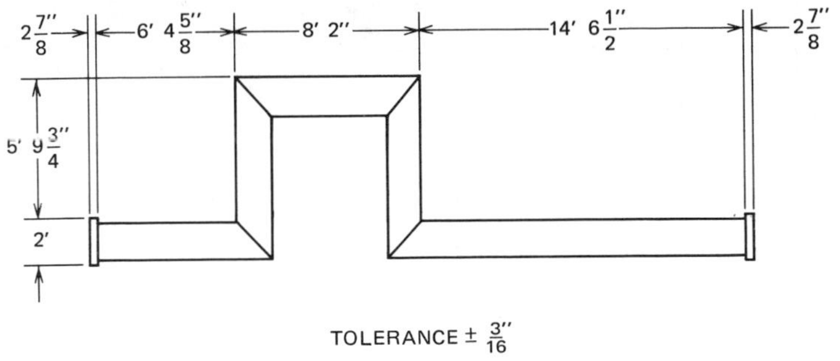

Figure 6–11

6-2. The acid storage tank in Figure 6-12 is designed to fit in a predetermined space in a plating shop. The dimensions can vary 5/16" larger or smaller than the given measurements.

a. What is the maximum height the assembly can be? 6-2a. ____________

b. What is the minimum height the assembly can be? 6-2b. ____________

c. What is the maximum length the assembly can
    be?

6-2c. _________________

d. What is the minimum length the assembly can
    be?

6-2d. _________________

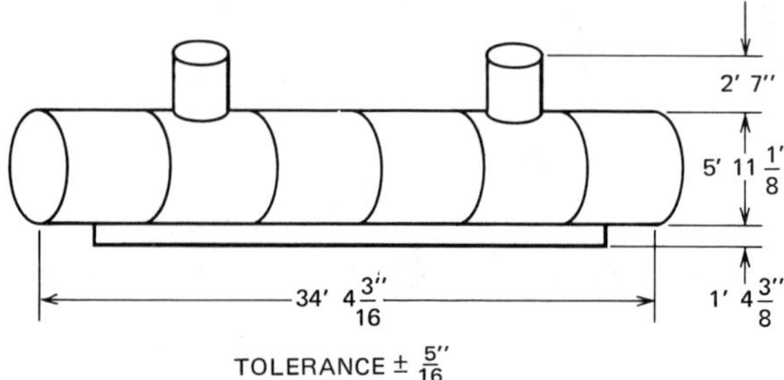

Figure 6—12

6-3. The tolerance of the set of steps in Figure 6-13
      is 3/4" larger or smaller than the given dimen-
      sions.

a. If the 3 steps are a total of 36-5/16" high,
    is this dimension within tolerance?

6-3a. _________________

b. If the 3 steps are a total of 49" deep, is
    this dimension within tolerance?

6-3b. _________________

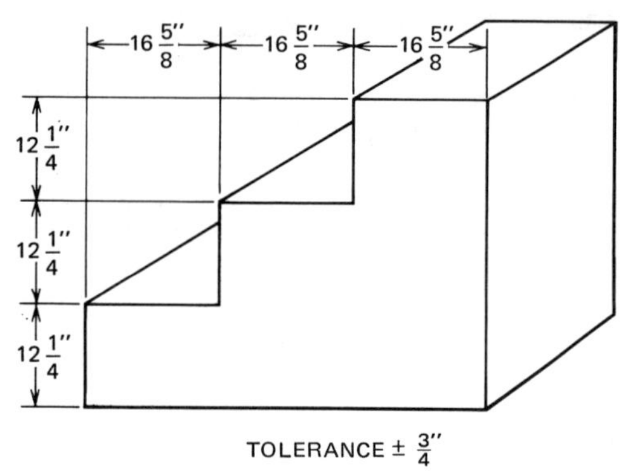

Figure 6—13

Preview and Self-Test

Name ____

Course/Sec. _______________ Date _____________________

Chapter 7 covers solving simple equations by various operations. Before you turn to the chapter, complete the Self-Test to determine which sections in the chapter you need to study carefully.

Self-Test

SOLVING SIMPLE EQUATIONS BY ADDITION AND SUBTRACTION

7-1. Solve for x when $x - 4 = 7(4)$.   7-1. _______________

7-2. Solve for a when $7 - 11 = a - 3$.   7-2. _______________

7-3. Find y when $-y + 32 = 56 - 12$.   7-3. _______________

7-4. Find j when $3j + 7 = 2j - 29$.   7-4. _______________

7-5. Solve for k when $2k + 3k + 5k = 39 + 19 - 8$.   7-5. _______________

Solving simple equations
by addition and subtraction score

____

SOLVING SIMPLE EQUATIONS BY MULTIPLICATION

7-6. Find b when $b/2 = 17$.   7-6. _______________

7-7. Find C when $C/3 = -61.4$.   7-7. _______________

7-8. Find D when $7D/8 = 42.7$.   7-8. _______________

Solving simple equations by multiplication score

____

SOLVING SIMPLE EQUATIONS BY DIVISION

7-9.   Find E when $12E = 96$.   7-9. _______________

7-10. Solve for F when $32F = 17.92$.   7-10. _______________

7-11. Find G when $16.7G = 23.7G - 202$.   7-11. _______________

7-12. Solve for GM when $386.74 - 56 - 36.74 = 12GM$.   7-12. _______________

Solving simple equations by division score

____

SOLVING SIMPLE EQUATIONS BY SPECIAL OPERATIONS

7-13.   Solve for H when $H^2 = 6.25$.

7-14.   Find J when $\sqrt{J} = 7.054$.

7-15.   Find K when $\sqrt{32K - 30K + K} = 6$.

Solving simple equations by special operations score

SOLVING SIMPLE EQUATIONS BY SEVERAL OPERATIONS

7-16.   Solve for x when $3 - 5x - 38 = 2x - 5 - 4x$.

7-17.   Solve for y when $2(5y - 6) - 15 = 33$.

7-18.   Find W when $15W - 5(6W - 1) - 50 = 0$.

7-19.   Solve for Z when $9 - (3 + 2Z) = 12 - 4(3 - Z)$.

7-20.   Find a when $16 - 3(a - 2) = 5a - 2(2a + 1)$.

Solving simple equations by several operations score

After you have checked your answers on the Self-Test,
transfer your scores to the Chapter 7 Objectives.

···

Solving Simple Equations

Upon successful completion of this chapter, you will
know the properties of simple equations. You will be
able to solve simple equations by addition, subtraction,
multiplication, division, and special operations.

Objectives

	Self-Test Scores	If you did poorly on the Self-Test, turn to:
• Solving simple equations by addition and subtraction	__________	Section 7-3
• Solving simple equations by multiplication	__________	Section 7-4
• Solving simple equations by division	__________	Section 7-5
• Solving simple equations by special operations	__________	Section 7-6
• Solving simple equations by several operations	__________	Section 7-7

7–1   Introduction to Solving Simple Equations

Mathematics is a problem-solving tool. An important part of mathematics is solving simple equations. Many routine problems, and many industrial problems as well, can be solved by simple applications of procedures for solving equations. These procedures are similar for nearly all types of problems. By learning the suggested procedures, you will be able to solve simple equations with ease.

7–2   Properties of Equations

An equation in mathematics is a statement declaring that two expressions are equal. For example, $5x = 15$ is an equation (a statement). By substituting 3 for the x in $5x = 15$, we get $5(3) = 15$. The two expressions are now equal.

The process of finding the value that makes the expressions equal is called solving the equation. The correct value is called the root, the solution, or the solution set of the equation.

The quantities on the left side of the equal sign are called the left member of the equation. The quantities on the right side of the equal sign are called the right member of the equation. The equal sign in the center is the sign of equality, or the balance point of the equation.

The three important parts of an equation are the left member, the equal sign, and the right member. Many students find it helpful to think of an equation as a balance scale, such as the one in Figure 7-1. The equal sign is vital. It tells you that the left and right members are equal. Whatever you do to one side of the equal sign you must also do to the other side. Remembering this rule is essential in solving equations.

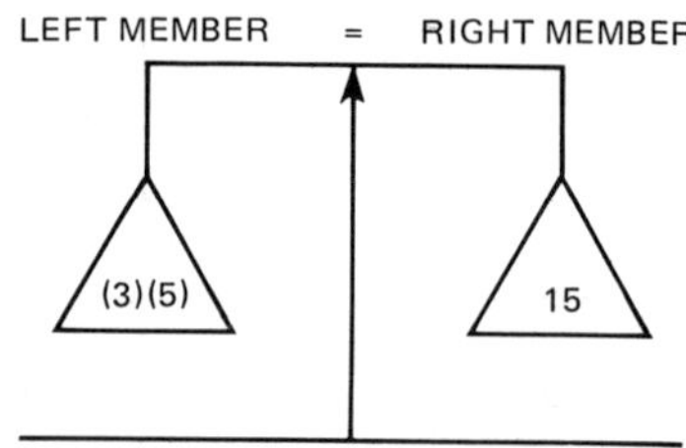

Figure 7-1

Exercises: Identifying the Properties of Equations

Given the equation $4x + 6 = 18$:

7-1. Name the right member.

7-2. Name the left member.

Given the equation $5y + 7 - 1y + 8 = 39(21)$:

7-3. Name the left member.

7-4. Simplify the left member.

7-5. Simplify and name the right member.

7–3   Solving Simple Equations by Addition and Subtraction

The four basic operations used to solve equations are
the fundamental arithmetic operations: addition,
subtraction, multiplication, and division. The procedures for performing these operations are called the
axioms, or rules, of algebra. The rules require that
whatever operation you perform on one member of an
equation you must also perform on the other. The use
of these rules is logical. You can visualize their
application by thinking of the balance scale principle.

In solving equations, isolate the unknown
literal factor (letter) on one side of the equation.
Try to write the answer with the unknown factor on the
left side of the equation and the other quantities on
the right side. The value of the unknown should be
positive. If the unknown is negative, you may need to
multiply both sides by -1.

When a quantity is moved from one side of the
equal sign to the other side, its sign of quantity is
changed. Moving a quantity from one side of an equal
sign to the other side is called <u>transposition</u>. This
procedure is shown in the examples.

Always check your answer (the root of the
equation). This step will eliminate careless errors
and ensure that you have the correct solution. The
procedure for checking or proving your answers is also
shown in the examples.

Procedure for Solving Simple Equations by Addition and Subtraction

••

Add or subtract the same number or quantity from one
side of an equation that is added or subtracted from
the other side.

••

Example 7–1

PROBLEM

Solve $k - 5 = 2$ for k.

SOLUTION 1

To find what value of k will make the left member
equal to the right member, recall that $-5 + 5 = 0$.
Add $+5$ to both sides of the equation:

$$
\begin{array}{rcr}
k - 5 &=& 2 \\
+\,5 &=& +\,5 \\
\hline
k &=& 7
\end{array}
$$

ANSWER

$$k = 7$$

SOLUTION 2

Move -5 across the equal sign to the right member:

$$k - 5 = 2$$
$$k = 2 + 5$$
$$k = 7$$

<u>Note</u>: Remember that when you move a quantity from one side of the equal sign to the other side, its sign of quantity changes.

ANSWER

$$k = 7$$

CHECK

Substitute 7 for k in the original equation and evaluate:

$$k - 5 = 2$$
$$7 - 5 = 2$$
$$2 = 2$$

Example 7–2

PROBLEM

Solve $13 - 31 = a + 9$ for a.

SOLUTION 1

1. To find what value of a will make the left member equal to the right member, recall that $+9 - 9 = 0$.

2. Subtract +9 from or add -9 to each side of the equation:

$$
\begin{array}{rcr}
13 - 31 &=& a + 9 \\
- 9 &=& - 9 \\
\hline
13 - 40 &=& a
\end{array}
$$

3. Combine the 13 and the -40:

$$13 - 40 = a$$
$$-27 = a$$

ANSWER

$$a = -27$$

SOLUTION 2

1. Move +9 to the left member:

$$13 - 31 = a + 9$$
$$13 - 31 - 9 = a$$

2. Combine the numbers in the left member:

$$\underbrace{13 - 31 - 9}_{-27} = a$$

ANSWER

$$a = -27$$

CHECK

To check the root (the answer), substitute -27 for a in the original equation and evaluate:

$$13 - 31 = a + 9$$
$$\underbrace{13 - 31}_{-18} = \underbrace{-27 + 9}_{-18}$$

Example 7–3

PROBLEM

Solve $4x + 11 = 3x + 28$ for x.

SOLUTION 1

1. To find what value of x will make the left member equal to the right member, recall that $+11 - 11 = 0$ and $+3x - 3x = 0$.

2. Subtract $3x$ from each side of the equation:

$$
\begin{array}{rcl}
4x + 11 &=& 3x + 28 \\
-\ 3x &=& -\ 3x \\
\hline
1x + 11 &=& 28
\end{array}
$$

3. Subtract 11 from each side of the equation:

$$
\begin{array}{rcl}
x + 11 &=& 28 \\
-\ 11 &=& -\ 11 \\
\hline
x &=& 17
\end{array}
$$

ANSWER

$$x = 17$$

SOLUTION 2

1. Move $+3x$ from right to left and $+11$ from left to right (remember to change the signs):

$$4x + 11 = 3x + 28$$
$$4x - 3x = 28 - 11$$

2. Combine like quantities on each side of the equal sign:

$$\underbrace{4x - 3x}_{x} = \underbrace{28 - 11}_{17}$$

ANSWER

$x = 17$

CHECK

Substitute 17 for x in the original equation and evaluate:

$$4x + 11 = 3x + 28$$
$$4(17) + 11 = 3(17) + 28$$
$$68 + 11 = 51 + 28$$
$$79 = 79$$

Exercises: Solving Simple Equations by Addition and Subtraction

Solve for the unknown and prove:

7-6. $x + 6 = 17$ 7-7. $k - 7 = 24$

7-8. $r - 1 = +1$ 7-9. $y - 11 = -11$

7-10. $2x + 6 = x + 7$ 7-11. $3x + 4 = 2x - 14$

7-12. $2y - 8 = 4 + y$ 7-13. $5x + 7 = 4x - 4$

7-14. $3 + 2x = x - 5$

7-15. $14x + 14 = 10x + 3x + 21$

7–4 Solving Simple Equations by Multiplication

Procedure for Solving Simple Equations by Multiplication

•••

Multiply the same number or quantity on one side of an equation that is multiplied on the other side.

•••

Example 7–4

PROBLEM

Solve $x/3 = 12$ for x.

SOLUTION

1. To find what value of x will make the left member equal to the right member, recall that $(3/1)(1/3) = 1$.

2. Multiply both sides of the equation by 3/1, or 3:

$$\frac{x}{3} = 12$$

$$\left(\frac{3}{1}\right)\left(\frac{x}{3}\right) = \left(\frac{12}{1}\right)\left(\frac{3}{1}\right)$$

$$\frac{x}{1} = \frac{36}{1} \quad \text{or} \quad x = 36$$

ANSWER

$$x = 36$$

CHECK

Substitute 36 for x in the original equation and evaluate:

$$\frac{x}{3} = 12$$

$$\frac{36}{3} = 12$$

$$12 = 12$$

Example 7–5

PROBLEM

Solve $3y/4 = 24$ for y.

SOLUTION

1. To find what value of y will make the left member
    equal to the right member, recall that $(3/4)(4/3)$
    $= 1$.

2. Multiply both sides of the equation by 4/3:

$$\frac{3}{4}y = 24$$

$$\left(\frac{4}{3}\right)\left(\frac{3}{4}y\right) = \left(\frac{24}{1}\right)\left(\frac{4}{3}\right)$$

$$y = 8(4) = 32$$

ANSWER

$$y = 32$$

CHECK

Substitute 32 for y in the original equation and evaluate:

$$\frac{3}{4}y = 24$$

$$\frac{3}{4}(32) = 24$$

$$24 = 24$$

Example 7–6

PROBLEM

Solve $(+1.51k)/(-0.5) = 30.2$ for k.

SOLUTION

1. To find what value of k will make the left side equal to the right, recall that $(+1.51/-0.5)(-0.5/+1.51) = 1$.

2. Multiply both sides of the equation by $(-0.5)/(+1.51)$:

$$\frac{+1.51k}{-0.5} = 30.2$$

$$\left(\frac{-0.5}{+1.51}\right)\left(\frac{+1.51k}{-0.5}\right) = (30.2)\left(\frac{-0.5}{+1.51}\right)$$

$$k = 20(-0.5)$$
$$= -10$$

ANSWER

$k = -10$

CHECK

Substitute $k = -10$ into the original equation and evaluate:

$$\frac{1.51k}{-0.5} = 30.2$$

$$\frac{1.51(-10)}{-0.5} = 30.2$$

$$30.2 = 30.2$$

Exercises: Solving Simple Equations by Multiplication

Solve for the unknown and prove:

7-16. $x/4 = 16$ 7-17. $K/5 = 15$

7-18. $x/6 = -4$ 7-19. $-z/2 = 11$

7-20. $j/5 = 24$ 7-21. $x/-3 = -27$

7-22. $k/0.5 = 46.7$ 7-23. $2y/3 = 30$

7-24. $5x/6 = -35$

7-25. $7x/8 = -15.4$

7–5 Solving Simple Equations by Division
Procedure for Solving Simple Equations by Division

Divide the same number or quantity into one side of
an equation that is divided into the other side.

Example 7–7

PROBLEM

Solve $7x = 28$ for x.

SOLUTION

1. To find what value of x will make the left member
    equal to the right member, recall that $7/7 = 1$.

2. Divide both sides of the equation by 7:

$$7x = 28$$

$$\frac{\cancel{7}x}{\cancel{7}} = \frac{\cancel{28}^{4}}{\cancel{7}}$$

$$x = 4$$

ANSWER

$$x = 4$$

CHECK

Substitute $x = 4$ into the original equation and evaluate:

$$\begin{aligned} 7x &= 28 \\ 7(4) &= 28 \\ 28 &= 28 \end{aligned}$$

Example 7–8

PROBLEM

Solve $-5x = 3(7 - 32)$ for x.

SOLUTION

To find what value of x will make the left member equal
to the right member, first perform the operations inside
the parentheses and then divide both sides by -5:

$$-5x = 3(7 - 32)$$

$$-5x = 3\ (-25)$$

$$\frac{\cancel{-5}x}{\cancel{-5}} = \frac{3(-\cancel{25}^{5})}{\cancel{-5}}$$

$$x = 3(5) = 15$$

ANSWER

$x = 15$

CHECK

Substitute $x = 15$ into the original equation and evaluate:

$$-5x = 3(7 - 32)$$
$$-5(15) = 3(-25)$$
$$-75 = -75$$

Exercises: Solving Simple Equations by Division

Solve for the unknown and prove:

7-26. $3x = 24$ | 7-27. $5k = 45$

7-28. $3j = 39$ | 7-29. $5x = 1$

7-30. $-2x = -0.8$ | 7-31. $-14x = +15.4$

7-32. $-16x = 6.4$ | 7-33. $-0.5x = -12.5$

7-34. $-1.2r = 0.4$ | 7-35. $-y = -17$

7-6   Solving Simple Equations by Special Operations

The special operations rule presented below should be used to solve any equation that has not been covered by previous procedures. Squaring and taking the square root are examples of special operations. To solve equations for the unknown after the unknown is isolated on one side, always perform the operation that is the inverse (opposite) of the one indicated on the unknown. For example, division is the inverse of multiplication, and subtraction is the inverse of addition. You perform the inverse operation when you divide $2x$ by 2, multiply $x/5$ by 5, add $+4$ to -4, and subtract -5.5 from $+5.5$. To solve $\sqrt{x} = 4$ for x, square both sides of the equation:

$$\sqrt{x} = 4$$
$$(\sqrt{x})^2 = 4^2$$
$$(\sqrt{x})(\sqrt{x}) = (4)(4)$$
$$x = 16$$

To solve $x^2 = 144$ for x, take the square root of 144. The square root of x^2 is x:

$$x^2 = 144$$
$$\sqrt{x^2} = \sqrt{144}$$
$$x = 12$$

The square root of any number can be positive or negative. For example, if $x^2 = 36$, $x = \pm 6$. However, we accept the + value as the root. Similar procedures apply to cubing and taking the cube root.

Procedure for Solving Simple Equations by Special Operations

Perform the same special operation (squaring, taking the square root, and so on) on one side of an equation that is performed on the other side.

Example 7–9

PROBLEM

Solve $x^2 = 225$ for x.

SOLUTION

1. To find what value of x will make the left member equal to the right member, recall that $\sqrt{x^2} = x$. (The inverse operation of squaring x is taking the square root of x^2).

2. Take the square root of each side of the equation:

$$x^2 = 225$$
$$\sqrt{x^2} = \sqrt{225}$$
$$x = 15$$

ANSWER

$$x = 15$$

CHECK

Substitute 15 for x in the original equation and evaluate:

$$x^2 = 225$$
$$15^2 = 225$$
$$225 = 225$$

Example 7–10

PROBLEM

Solve $\sqrt{x} = 7$ for x.

SOLUTION

1. To find the value of x that will make the left member equal to the right member, recall that the

inverse operation of $\sqrt{x}$ (taking the square root) is $(\sqrt{x})^2$ (squaring).

2. Square both sides of the equation:

$$\sqrt{x} = 7$$
$$(\sqrt{x})^2 = 7^2$$
$$x = 49$$

ANSWER

$$x = 49$$

CHECK

Substitute 49 for x in the original equation and evaluate:

$$\sqrt{x} = 7$$
$$\sqrt{49} = 7$$
$$7 = 7$$

Exercises: Solving Simple Equations by Special Operations

Solve for the unknown and prove:

7-36. $x^2 = 81$ 7-37. $x^2 = 121$

7-38. $x^2 = 19.36$ 7-39. $\sqrt{x} = 16$

7-40. $\sqrt{x} = 5.42$

7-7   Solving Simple Equations by Several Operations

Several operations may be necessary to solve certain equations. What look like complex equations can be solved by following the proper procedure.

Procedure for Solving Simple Equations by Several Operations

1. Remove parentheses when necessary by starting with the innermost set of parentheses.
2. Combine like quantities on each side of the equal sign. Starting from the left of each side, perform multiplication, division, subtraction, and addition as necessary, in this order.
3. Transpose quantities to isolate the unknown on one side of the equal sign.
4. Combine like quantities.
5. Solve for the unknown by using the addition, sub-

traction, multiplication, division, or special operations rule.
6. Check your solution by substituting the value obtained for the unknown in the original equation.

•••

By following this procedure carefully, you will be able to solve any simple equation. Application of the procedure is illustrated in the examples.

Example 7–11

PROBLEM

Solve $3y - 1 = 2(y - 5)$ for y.

SOLUTION

1. To find the value of y that will make the right side equal to the left side, first remove the parentheses:

$$3y - 1 = 2(y - 5)$$
$$= 2y - 10$$

2. Then transpose quantities to isolate the unknown on one side (remember to change the sign when moving quantities from one side of the equal sign to the other side):

$$3y - 1 = 2y - 10$$
$$3y - 2y = +1 - 10$$

3. Combine like quantities:

$$3y - 2y = +1 - 10$$
$$y = -9$$

ANSWER

$$y = -9$$

CHECK

Substitute -9 for y in the original equation and evaluate:

$$3y - 1 = 2(y - 5)$$
$$3(-9) - 1 = 2(-9 - 5)$$
$$-27 - 1 = 2(-14)$$
$$-28 = -28$$

Example 7–12

PROBLEM

Solve $5x/3 + 2 = 16 - x/3$ for x.

SOLUTION

1. To find what value of x will make the left member
    equal to the right member, transpose $-x/3$ to the
    left and $+2$ to the right:

$$\frac{5x}{3} + 2 = 16 - \frac{x}{3}$$

$$\frac{5x}{3} + \frac{x}{3} = 16 - 2$$

2. Combine like quantities:

$$\frac{5x}{3} + \frac{x}{3} = 16 - 2$$

$$\frac{6x}{3} = 14$$

3. Reduce the fraction:

$$\frac{\overset{2}{\cancel{6}}x}{\underset{1}{\cancel{3}}} = 14$$

4. Divide both sides of the equation by 2:

$$\frac{\overset{1}{\cancel{2}}x}{\underset{1}{\cancel{2}}} = \frac{\overset{7}{\cancel{14}}}{\underset{1}{\cancel{2}}}$$

ANSWER

$$x = 7$$

CHECK

Substitute 7 for x in the original equation and
evaluate:

$$\frac{5x}{3} + 2 = 16 - \frac{x}{3}$$

$$\frac{5(7)}{3} + 2 = 16 - \frac{7}{3}$$

$$\frac{35}{3} + \frac{6}{3} = \frac{48}{3} - \frac{7}{3}$$

$$\frac{41}{3} = \frac{41}{3}$$

Example 7–13

PROBLEM

Solve $14 + 3(k - 7) = 26 + 2(k + 1) + 8k$ for k.

SOLUTION

1. To find what value of k will make the left member
   equal to the right member, first remove the
   parentheses:

   $$14 + 3(k - 7) = 26 + 2(k + 1) + 8k$$
   $$14 + 3k - 21 = 26 + 2k + 2 + 8k$$

2. Combine like quantities:

   $$14 + 3k - 21 = 26 + 2k + 2 + 8k$$
   $$3k - 7 = 28 + 10k$$

3. Transpose $3k$ to the right and 28 to the left:

   $$3k - 7 = 28 + 10k$$
   $$-7 - 28 = -3k + 10k$$

4. Combine like quantities:

   $$\underbrace{-7 - 28}_{-35} = \underbrace{-3k + 10k}_{7k}$$

5. Finally, divide both sides by 7:

   $$\frac{\overset{-5}{\cancel{-35}}}{\cancel{7}} = \frac{\overset{1}{\cancel{7}}k}{\cancel{7}}$$

   $$-5 = k$$

ANSWER

   $$k = -5$$

CHECK

Substitute -5 for k in the original equation and
evaluate:

$$14 + 3(k - 7) = 26 + 2(k + 1) + 8k$$
$$14 + 3(-5 - 7) = 26 + 2(-5 + 1) + 8(-5)$$
$$14 + 3(-12) = 26 + 2(-4) - 40$$
$$14 - 36 = 26 - 8 - 40$$
$$-22 = 26 - 48$$
$$-22 = -22$$

Example 7–14

PROBLEM

Solve $\sqrt{x - 3} = 8$ for x.

SOLUTION

1. To find the value of x that will make the left
    member equal to the right member, first square both
    sides of the equation:

$$\sqrt{x - 3} = 8$$
$$(\sqrt{x - 3})^2 = (8)^2$$
$$x - 3 = 64$$

2. Transpose the -3 to the right side:

$$x - 3 = 64$$
$$x = 64 + 3$$

3. Finally, combine like quantities:

$$x = 64 + 3$$
$$= 67$$

ANSWER

$$x = 67$$

CHECK

Substitute 67 for x in the original equation and
evaluate:

$$\sqrt{x - 3} = 8$$
$$\sqrt{67 - 3} = 8$$
$$\sqrt{64} = 8$$
$$8 = 8$$

Exercises: Solving Simple Equations by Several Operations

Solve for the unknown and prove:

7-41. $3n + 7 = 42 - 4n$ 7-42. $7y - 3 = y + 9$

7-43. $3k - 18 = 18 - 3k$ 7-44. $14 + 17j = 9j - 18$

7-45. $17y - 5.3 = 28.7$ 7-46. $4a + 13 = 9a - 52$

7-47. $3b - 3 = 2(b + 2)$ 7-48. $8(c - 2) = 2 + 5c$

7-49. $38 = 2(3 - 4b)$ 7-50. $8(p - 4) = 8 + 3p$

7-51. $6b/5 - 5 = 13$ 7-52. $1 + 12x/4 = 9 - x$

7-53. $6x/7 - 2 = 16$ 7-54. $x/4 + 8 = 2(5.8 - 0.3)$

7-55. $5x/3 + 2/12 = 7(x - 3)/14$

7-56. $3(a - 2) + 1 = 11 - (2 - a)$

7-57. $2a - 3(a - 2) = 2/4(a + 1)$

7-58. $k - k/5 + 1 = 1/3(k - 5)$

7-59. $\sqrt{3x - 2} = \sqrt{x + 4}$

7-60. $3(x^2 + 4) = 2(x^2 + 16) + 0.5(10)$

Chapter 7

Solving Simple Equations

Solving Applied Welding Problems

Now that you have mastered the fundamental operations
of solving simple equations, you are ready to apply
these math skills to typical problems that welders
encounter on the job. After studying the examples that
follow, complete the applied welding problems.

Example 7–15

PROBLEM

Welding machine number one welds 5.57 lb of metal in
an hour. Machine number two welds 3.89 lb of metal per
hour. How many more pounds of metal must machine
number two weld to equal machine number one?

SOLUTION

1. Let A = the additional pounds of metal that machine
    number two must weld to equal machine number one.

2. Express the problem as an equation:

    3.89 + A = 5.57

3. To find the value of A that will make the left
    member of the equation equal to the right member,
    transpose 3.89:

    A = 5.57 - 3.89

4. Combine the quantities in the right member:

    A = 5.57 - 3.89
      =   1.68

ANSWER

Machine number two must weld 1.68 more pounds to equal
machine number one.

Example 7–16

PROBLEM

Metal bar B is 21.6" long. Metal bar C is 7.96" long.
How many inches of metal must be added to bar C to make
it equal in length to bar B?

SOLUTION

1.   Let X = the amount of metal that must be added to
     bar C to make it equal in length to bar B.

2.   Express the problem as an equation:

     $7.96 + X = 21.6$

3.   To find the value of X that will make the left
     member of the equation equal to the right member,
     transpose 7.96:

     $X = 21.6 - 7.96$

4.   Combine the quantities in the right member:

     $$X = \underbrace{21.6 - 7.96}$$
     $$= 13.64$$

ANSWER

13.64 additional inches of metal must be added to bar C
to make it equal in length to bar B.

Example 7–17

PROBLEM

Trailer frame F weighs 140.85 lb. Trailer frame K
weighs 704.25 lb. How many times heavier is frame K
than frame F?

SOLUTION

1.   Let T = the number of times K is heavier than F.

2.   Express the problem as an equation:

     $140.85(T) = 704.25$

3.   To find the value of T that will make the left
     member of the equation equal to the right member,
     transpose 140.85:

     $$T = \frac{704.25}{140.85} = 5$$

ANSWER

Frame K is 5 times heavier than frame F.

Applied Welding Problems

7-61. The angle iron in Figure 7-2 is 24.875" long.
The required length is 18.9375". How many
inches should be cut from the original length?
Let X = the number of inches that should be cut.
Express the problem as an equation and solve
for X.

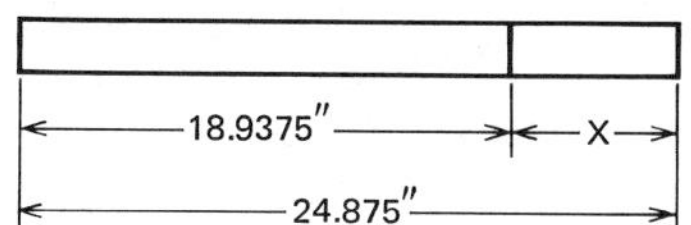

Figure 7-2

7-62. The trailer frame in Figure 7-3 is 108.5" wide.
The legal width is 96". How many inches should
be cut from the trailer to bring it to the legal
width? Let Y = the number of inches that should
be cut. Express the problem as an equation and
solve for Y.

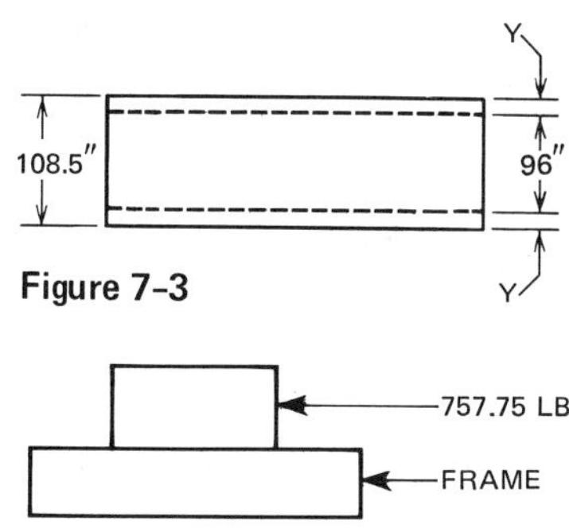

Figure 7-3

Figure 7-4

7-63. The channel iron frame in Figure 7-4 is designed
to support 450 lb. A 757.75 lb load is applied
to the frame. How many pounds of extra weight
are placed on the frame? Let X = the pounds of
extra weight. Express the problem as an equation
and solve for X.

7-64. A welder is paid $11.47 per hour. After a pay
increase, she is paid $12.76 per hour. How much
was the increase in pay? Let Y = the amount of
increase. Express the problem as an equation and
solve for Y.

7-65. The required amperage when using an E6012 5/32"
electrode is 145 amps. A welder is using 180
amps. By how many amps will he need to lower
the setting to be at the required setting? Let
X = the number of amps. Express the problem as
an equation and solve for X.

7-66. A welder used 16 lb of welding rod on a rail car
hitch. 2.5 times as much rod are still needed.
How many more pounds of rod are needed? Let Y =
the number of additional rods needed. Express
the problem as an equation and solve for Y.

7-67. A welder uses 2 spacers to level a fixture. The
total height of the spacers is 6.125". The top
spacer is 3.5 times as high as the bottom
spacer. How high is the bottom spacer? Let X
= the height of the bottom spacer. Express the
problem as an equation and solve for X.

7-68. Five equal sections are cut from a 92.5" length
of pipe. How long is each section? Let Y = the
length of each section. Express the problem as
an equation and solve for Y.

7-69. A 12' length of tubing is cut into 2 unequal
lengths. One piece is 4 times longer than the
other. How long is each piece? Let X = the
length of one piece and 4X = the length of the
other. Express the problem as an equation and
solve for both X and 4X.

7-70.   Two sections of pipe are welded together. Their
        total length is 86.375". One section is 36.75"
        long. How long is the other section? Let Y =
        the length of the second section. Express the
        problem as an equation and solve for Y.

7-71.   The pipe in problem 7-70 is cut into 5 equal
        pieces. What is the length of each piece? Let
        X = the length of each of the 5 pieces. Express
        the problem as an equation and solve for X.

7-72.   A length of angle iron is 172.75" long. It is
        cut into 5 equal parts. What is the length of
        each part? Let Y = the length of each part.
        Express the problem as an equation and solve
        for Y.

7-73.   There are 15 welding rods in a pound. How many
        rods are in 20 lb? Let X = the total number of
        rods. Express the problem as an equation and
        solve for X.

7-74.   A 30.25' length of channel iron is cut into 2
        unequal lengths. One piece is 4.5 times longer
        than the other. How long is each piece? Let
        Y = the length of one piece and 4.5Y = the length
        of the other. Express the problem as an equation
        and solve for both Y and 4.5Y.

7-75.   Five pieces of flat stock, each twice as long as
        the previous piece, are cut from a 232.5" length,
        as shown in Figure 7-5. Let X = the length of
        the first piece. Express the problem as an
        equation and solve for X. Then find the lengths
        of the other 4 pieces.

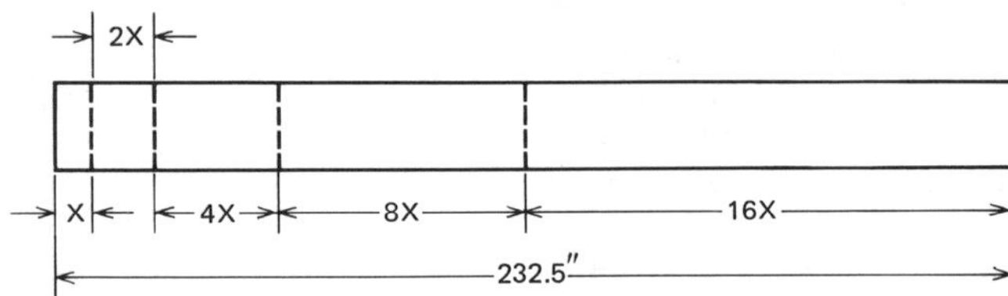

Figure 7-5

Applied Welding Problems Survey Test

• •

Name ____

Course/Sec. _______________   Date ___________________

7-1.   A steel beam is extended by 8'. The overall
       length is 149'. What was the original length?
       Let X = the original length. Express the problem
       as an equation and solve for X.

7-1. _________________

7-2.   A steel beam 16.5' long is required for a bridge
       support. How many feet must be cut from a 20.1'
       beam to obtain the required length? Let Y = the
       number of feet that must be cut. Express the
       problem as an equation and solve for Y.

7-2. _________________

7-3.   How long is dimension X of the backing plate in
       Figure 7-6? Express the problem as an equation
       and solve for X.

7-3. _________________

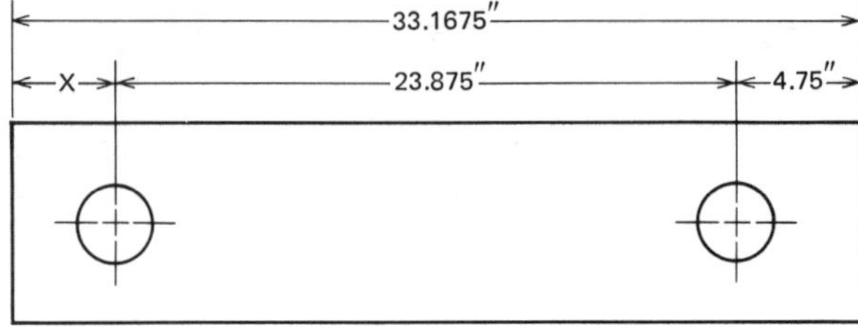

Figure 7-6

7-4.   How long is dimension Y of the formed channel in
       Figure 7-7? Express the problem as an equation
       and solve for Y.

7-4. _________________

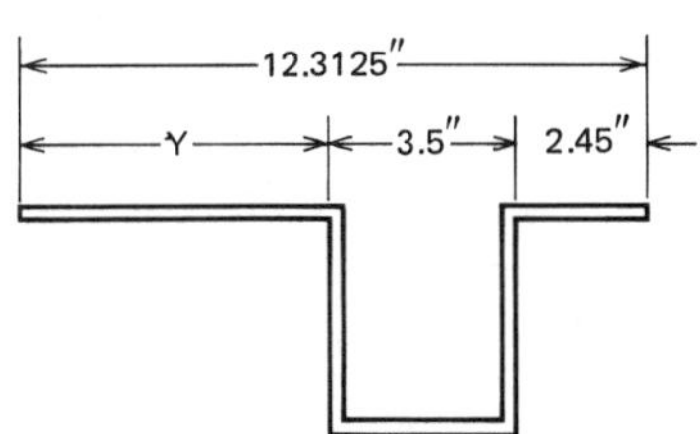

Figure 7-7

7-5. A 1.5" stop is welded onto the steel plate shown
      in Figure 7-8. What is dimension X? Express the
      problem as an equation and solve for X.

7-5. __________________

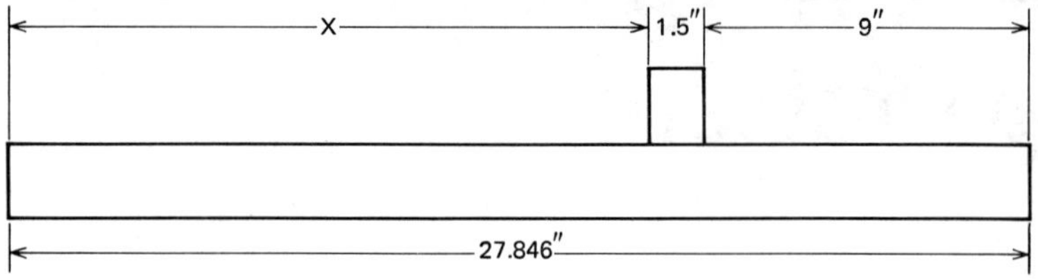

Figure 7-8

Chapter 8: Solving Simple Formulas
Preview and Self-Test

• •

Name ____

Course/Sec. _______________   Date _______________________

Chapter 8 covers the solving of simple formulas by
various operations. Before you turn to the chapter,
complete the Self-Test to determine which sections in
the chapter you need to study carefully.

Self-Test

SOLVING SIMPLE FORMULAS BY ADDITION AND SUBTRACTION

8-1. Solve $c + d = y$ for d.   8-1. _______________

8-2. Solve $P = a + b + c$ for c.   8-2. _______________

8-3. Solve $P = s + B + b + s$ for B.   8-3. _______________

Solving simple formulas by
addition and subtraction score

____

SOLVING SIMPLE FORMULAS BY MULTIPLICATION AND DIVISION

8-4. Find r when $I = prt$.   8-4. _______________

8-5. Find s when $P = 3s$.   8-5. _______________

8-6. Find b when $A = 0.5bh$.   8-6. _______________

8-7. Find a when $A = \pi ab/4$.   8-7. _______________

Solving simple formulas by
multiplication and division score

____

SOLVING SIMPLE FORMULAS BY SPECIAL OPERATIONS

8-8.   Solve $A = s^2$ for s.   8-8. _______________

8-9.   Solve $A = 0.7854d^2$ for d.   8-9. _______________

8-10. Solve $V = \pi r^2 h$ for r.   8-10. _______________

Solving simple formulas by special operations score

____

SOLVING SIMPLE FORMULAS BY SEVERAL OPERATIONS

8-11.   Solve $P = \pi(a + b)$ for b.   8-11. ________________

8-12.   Given $A = \theta\pi r^2/360$, solve for r.   8-12. ________________

8-13.   Solve $s = 2h/\sqrt{3}$ for h.   8-13. ________________

8-14.   Solve $A = 3s^2\sqrt{3}/2$ for s.   8-14. ________________

8-15.   Solve $T = ph + aA$ for h.   8-15. ________________

8-16.   Solve $D_1 = D - 1.732/N$ for N.   8-16. ________________

8-17.   Solve $V = \pi d^3/6$ for d.   8-17. ________________

8-18.   Solve $A = 4\pi^2 Rr$ for r.   8-18. ________________

8-19.   Solve $P = \pi\sqrt{s(a^2 + b^2)}$ for a.   8-19. ________________

8-20.   Solve $V = h(B_1 + 4M + B_2)/6$ for M.   8-20. ________________

Solving simple formulas by several operations score   ________________

After you have checked your answers on the Self-Test,
transfer your scores to the Chapter 8 Objectives.

Solving Simple Formulas

Upon successful completion of this chapter, you will know the properties of formulas. You will be able to solve simple formulas by addition, subtraction, multiplication, division, and special operations.

Objectives

	Self-Test Scores	If you did poorly on the Self-Test, turn to:
• Solving simple formulas by addition and subtraction	__________	Section 8-2
• Solving simple formulas by multiplication and division	__________	Section 8-3
• Solving simple formulas by special operations	__________	Section 8-4
• Solving simple formulas by several operations	__________	Section 8-5

8–1   Introduction to Solving Simple Formulas

For many people, formulas play a vital part in everyday work. For example, a technician's ability to work with formulas is often related to good work performance and job advancement.

Handbooks for many vocations contain the various formulas used in a given profession. However, the formulas are not always in the desired form. Thus, learning to manipulate and solve formulas is important. The procedures that follow are applications of those you learned in the previous chapter on equations.

8–2   Solving Simple Formulas by Addition and Subtraction

A _formula_ is a rule or law used in mathematics, science, business, or industry to solve a problem. A _literal equation_ is an equation in which some or all of the quantities are represented by letters rather than by numbers. Equations, literal equations, and formulas are all solved by the same set of rules. The same definitions also apply to equations, literal equations, and formulas.

To solve a formula, first read the formula carefully. Be sure you know what is given and what you are solving for. Many formulas can be solved by performing the operation that is the inverse (opposite) of the one written in the formula. For example, a term to be multiplied can be transposed and divided into the other member. A term to be subtracted can be transposed and added to the other member, and so on. This procedure is illustrated in the examples.

The checking procedure is sometimes more difficult than actually working the problem. Reworking the formula is often easier than substituting and evaluating.

Procedure for Solving Simple Formulas by Addition and Subtraction

• •

Add or subtract the same number or quantity from one side of an equation that is added to or subtracted from the other side.

• •

Example 8–1

PROBLEM

Solve $P = a + b + c$ for a.

SOLUTION

To find what a equals, transpose (move) b and c to the other side of the equal sign:

$$P = a + b + c$$
$$P - b - c = a$$

Note: Remember to express the answer by placing the unknown on the left side of the equal sign and the other quantities on the right side.

ANSWER

$$a = P - b - c$$

CHECK

Substitute $P - b - c$ for a in the original formula and evaluate:

$$P = a + b + c$$
$$= P - b - c + b + c$$
$$= P$$

Example 8–2

PROBLEM

Solve $A = 180 - (B + C)$ for C.

SOLUTION

1. To find what C equals, first remove the parentheses:

    $$A = 180 - B - C$$

2. Then transpose $- C$ and A:

    $$A = 180 - B - C$$
    $$+C = 180 - B - A$$

ANSWER

$$C = 180 - A - B$$

CHECK

Substitute $180 - A - B$ for C in the original formula and evaluate:

$$A = 180 - (B + C)$$
$$= 180 - [B + (180 - A - B)]$$
$$= 180 - (B + 180 - A - B)$$
$$= 180 - B - 180 + A + B$$
$$= A$$

**Exercises: Solving Simple Formulas by Addition
and Subtraction**

Solve for the indicated unknown:

8-1.   Solve $R = C + P$ for P.

8-2.   Solve $180 = A + B + C$ for B.

8-3.   Solve $P = B + 25 + b$ for b.

8-4.   Solve $NP = SP - D$ for D.

8-5.   Solve $D = A - L$ for L.

8-6.   Solve $D = H + T + S$ for S.

8-7.   Solve $C = d + ab + w$ for ab.

8-8.   Solve $P = B + S_1 + b + S_2$ for B.

8-9.   Solve $F_t = F_1 + F_2 + F_3$ for $F_1 + F_2$.

8-10. Solve $T = ps/2 + a$ for a.

8–3   Solving Simple Formulas by Multiplication and Division

Procedure for Solving Simple Formulas by Multiplication and Division

• •

Multiply or divide the same number or quantity into
one side of an equation that is multiplied or divided
into the other side.

• •

Example 8–3

PROBLEM

Solve $d = rt$ for r.

SOLUTION

To find what r equals in terms of d and t, divide both
sides by t:

$$\frac{d}{t} = \frac{r\cancel{t}}{\cancel{t}} = r$$

ANSWER

$$r = \frac{d}{t}$$

Example 8–4

PROBLEM

Solve $y = mx + b$ for x.

SOLUTION

1. To find what x equals in terms of y, m, and b,
    first transpose b:

    $$y = mx + b$$
    $$y - b = mx$$

2. Then divide both sides by m:

    $$\frac{y - b}{m} = \frac{mx}{m} = x$$

ANSWER

$$x = \frac{y - b}{m}$$

Example 8–5

PROBLEM

Solve the following formula for B:

$$A = \frac{1}{2}(B + b)h$$

SOLUTION

1. To find what B equals in terms of the other literal
    factors, first multiply both sides by 2:

    $$(2)A = 2\frac{1}{2}(B + b)h$$
    $$2A = (B + b)h$$

2. Remove the parentheses:

    $$2A = Bh + bh$$

3. Transpose bh:

    $$-bh + 2A = Bh$$

4. Finally, divide both sides by h:

    $$\frac{-bh + 2A}{h} = \frac{Bh}{h}$$

Note: When you express the answer, remember to write
the positive term ($2A$) before the negative term ($-bh$).

ANSWER

$$B = \frac{2A - bh}{h}$$

Exercises: Solving Simple Formulas by Multiplication and Division

Solve for the indicated unknown:

8-11. Solve $ax = b$ for x.

8-12. Solve $D = B/N$ for B.

8-13. Solve $I = prt$ for r.

8-14. Solve $R = (p + q)/2$ for q.

8-15. Solve $V = 4st/a$ for t.

8-16. Solve $C = fgt$ for g.

8-17. Solve $ax = bc$ for x.

8-18. Solve $V = 4st/a$ for a.

8-19. Solve $b(y + 1) = d$ for y.

8-20. Solve $Z = (X - M)/s$ for M.

8-21. Given $P = Fs/t$, find F.

8-22. Given $F = m_1 m_2/rd^2$, find m_2.

8-23. Given $E = P(Z + R)$, find Z.

8-24. Solve $K = mt(x - r)$ for r.

8-25. Given $8y - 12j = 2y + 9j - y$, find y.

8-26. Solve $A = [(B + b)a]/2$ for a.

8-27. Solve $S = (a + b + c)/2$ for c.

8-28. Solve $N = P(L - Y)$ for Y.

8-29. Solve $3mt = c(x - r)$ for r.

8-30. Given $m = (y_2 - y_1)/(x_2 - x_1)$, find x_1.

8–4 Solving Simple Formulas by Special Operations

The special operations rule presented below should be
used to solve any formulas that have not been covered
by previous rules. To solve formulas for the unknown,
always perform the necessary operations to isolate the

unknown on one side of the equal sign.   Then perform
the opposite (inverse) operation on the unknown as
needed.   For example, if $A = e^2$, solve for e by taking
the square root of both sides:

$$A = e^2$$
$$\sqrt{A} = \sqrt{e^2}$$
$$\sqrt{A} = e$$
$$e = \sqrt{A}$$

The examples show other special operations.

Procedure for Solving Simple Formulas by Special Operations

••

Perform the same special operation (squaring, taking
the square root, cubing, taking the cube root, and so
on) on one side of a formula that is performed on the
other side.

••

Example 8–6

PROBLEM

Solve the following formula for t:

$$S = \frac{g t^2}{2}$$

SOLUTION

1. To find what t equals in terms of S and g, first
    multiply both sides by 2:

$$(2)S = \left(\frac{g t^2}{2}\right) 2$$
$$2S = g t^2$$

2. Divide both sides by g:

$$\frac{2S}{g} = \frac{\cancel{g} t^2}{\cancel{g}}$$

Note: The first two steps could have been combined by
multiplying both sides by $2/g$.

3. Finally, take the square root of both sides:

$$\sqrt{\frac{2S}{g}} = \sqrt{t^2}$$
$$\sqrt{\frac{2S}{g}} = t$$

ANSWER

$$t = \sqrt{\frac{2S}{g}}$$

Example 8–7

PROBLEM

Solve the following formula for A:

$$r = \sqrt{\frac{A}{\pi}}$$

SOLUTION

1. To find what A equals in terms of r and π, first square both sides:

$$r^2 = \left(\sqrt{\frac{A}{\pi}}\right)^2 = \frac{A}{\pi}$$

2. Then multiply both sides by π and reduce:

$$(\pi)\,r^2 = \frac{A}{\not{\pi}}(\not{\pi})$$
$$\pi r^2 = A$$

ANSWER

$$A = \pi r^2$$

Example 8–8

PROBLEM

Solve $d = \sqrt{b^2 - 4ac}$ for b.

SOLUTION

1. To find what b equals in terms of a, c, and d, first square both sides:

$$(d)^2 = (\sqrt{b^2 - 4ac})^2$$
$$d^2 = b^2 - 4ac$$

2. Transpose $-4ac$ to the other side:

$$+4ac + d^2 = b^2$$

3. Finally, take the square root of both sides:

$$\sqrt{4ac + d^2} = \sqrt{b^2}$$
$$\sqrt{4ac + d^2} = b$$

ANSWER

$$b = \sqrt{4ac + d^2}$$

Exercises: Solving Simple Formulas by Special Operations

Solve for the indicated unknown:

8-31.   Solve $T = 6s^2$ for s.

8-32.   Solve $A = 3.142r^2$ for r.

8-33.   Given $P = I^2R$, find R.

8-34.   Solve $T = 4\pi r^2$ for r.

8-35.   Given $V = \pi r^2 h$, find r.

8-36.   Given $A = 0.7854d^2$, find d.

8-37.   Given $V = Fcd/r^2$, find r.

8-38.   Solve $HP = D^2N/2.5$ for D.

8-39.   Given $P = 0.44d^2 Kn$, find d.

8-40.   Given $A = \Theta/360\pi r^2$, find r.

8–5   Solving Simple Formulas by Several Operations

Several operations may be necessary to solve certain
formulas. What look like complex formulas can be
solved by following the proper procedure.

Procedure for Solving Simple Formulas by Several Operations

●●●

1. Remove the parentheses when necessary by starting
    with the innermost set of parentheses.
2. Combine like quantities on each side of the equal
    sign. Starting from the left of each side, perform
    multiplication, division, subtraction, and addition
    as necessary, in this order.
3. Transpose quantities to isolate the unknown on one
    side of the equal sign.
4. Combine like quantities.
5. Solve for the unknown by using the addition,
    subtraction, multiplication, division, or special
    operations rule. (Sometimes certain operations
    must be repeated.)
6. When possible, check your solution by substituting
    the value obtained for the unknown in the original
    equation.

●●●

By following this procedure carefully, you will be able
to solve any simple formula. Application of the
procedure is illustrated in the examples.

Example 8–9

PROBLEM

Solve the following formula for r_1:

$$R = \frac{r_1 r_2}{r_2 - r_1}$$

SOLUTION

1. To find what r_1 equals in terms of r_2 and R, first
    multiply both sides by $(r_2 - r_1)$:

$$R = \frac{r_1 r_2}{r_2 - r_1}$$

$$(r_2 - r_1)R = \frac{r_1 r_2}{r_2 - r_1}(r_2 - r_1)$$

$$(r_2 - r_1)R = r_1 r_2$$

2. Remove the parentheses:

$$Rr_2 - Rr_1 = r_1 r_2$$

3. Transpose Rr_1:

$$Rr_2 = r_1 r_2 + Rr_1$$

Note: r_1 divides evenly into $r_1 r_2 - Rr_1$. Thus we can
simplify and write this member of the equation as
$r_1(r_2 + R)$:

$$Rr_2 = r_1(r_2 + R)$$

4. Finally, divide both sides by $(r_2 + R)$:

$$\frac{Rr_2}{r_2 + R} = \frac{r_1(r_2 + R)}{(r_2 + R)}$$

$$\frac{Rr_2}{r_2 + R} = r_1$$

ANSWER

$$r_1 = \frac{Rr_2}{r_2 + R}$$

Example 8–10

PROBLEM

Solve the following formula for F:

$$C = \frac{5}{9}(F - 32)$$

SOLUTION

1. To find what F equals in terms of C and the
    numerical factors, first multiply both sides by 9:

$$C = \frac{5}{9}(F - 32)$$

$$(9)\,C = \left(\frac{9}{1}\right)\frac{5}{9}(F - 32)$$

$$(9)\,C = 5(F - 32)$$

2. Remove the parentheses:

$$9C = 5F - 160$$

3. Transpose -160:

$$+160 + 9C = 5F$$

4. Finally, divide each term on both sides by 5:

$$\frac{\overset{32}{\cancel{160}}}{\cancel{5}} + \frac{9C}{5} = \frac{\overset{}{\cancel{5}}F}{\cancel{5}}$$

$$32 + \frac{9}{5}C = F$$

ANSWER

$$F = \frac{9}{5}C + 32$$

Exercises: Solving Simple Formulas by Several Operations

Solve for the indicated unknown:

8-41.   Solve $A = bh/2$ for h.

8-42.   Solve $S = at^2/2$ for t.

8-43.   Solve $V = Bh/3$ for B.

8-44.   Given $S = PT/(P + T)$, solve for P.

8-45.   Solve $H = DV/385$ for V.

8-46.   Solve $a = (V_2 - V_1)/t$ for V_1.

8-47.   Solve $Q = 62.5h$ for h.

8-48.   Given $P = 3.142ab$, find b.

8-49.   Given $A = \pi d^2/4$, find d.

8-50.   Solve $P = I^2R$ for I.

8-51.   Given $A = (B + b)h/2$, find h.

8-52. Solve $V = \pi r^2 h$ for r.

8-53. Solve $P = \pi(a + b)$ for b.

8-54. Solve $T = ps/2 + A$ for p.

8-55. Solve $Z^2 = X^2 + R^2$ for R.

8-56. Given $K = L(D - d)/j$, find D.

8-57. Solve $P = \pi\sqrt{2(a^2 + b^2)}$ for b.

8-58. Solve $SA = \pi r^2 h + 2\pi r^2$ for h.

8-59. Solve $V = 4\pi r^3/3$ for r.

8-60. Solve $S = 2tw + 2ht + 2hw$ for h.

Chapter 8

Solving Simple Formulas

•••

Solving Applied Welding Problems

Now that you have mastered the fundamental operations
of solving simple formulas, you are ready to apply
these math skills to typical problems that welders
encounter on the job. After studying the examples that
follow, complete the applied welding problems.

To solve the following Applied Welding Problems,
you need to know that forming metal changes the overall
size of the material. As a rule of thumb, add 50% of
the material thickness for each <u>inside bend</u> and subtract
150% of the material thickness for each <u>outside bend</u> of
the pieces shown in these problems. The formulas for
finding the size of formed metal are given here.

For example, Figure 8-1 shows the inner dimensions
of a bracket. To find the total length of angle iron
needed to make the bracket, add lengths A and B and 50%
of the material thickness T for the <u>inside bend</u>:

    length = A + B + 0.50T

Figure 8-2 shows the outer dimensions of a bracket.
To find the total length of angle iron needed to make
the bracket, add lengths C and D and subtract 150% of
the material thickness T for the <u>outside bend</u>:

    length = C + D - 1.50T

Example 8–11

PROBLEM

What length of material is needed to build the spacer
bracket in Figure 8-3?

SOLUTION

<u>Note</u>: The formula, or rule, for finding the length of
an object is simply to add the lengths of the component
parts. For each of the two inside bends of the bracket,
add 50% of the material thickness.

1. Let X = the total length of material needed.

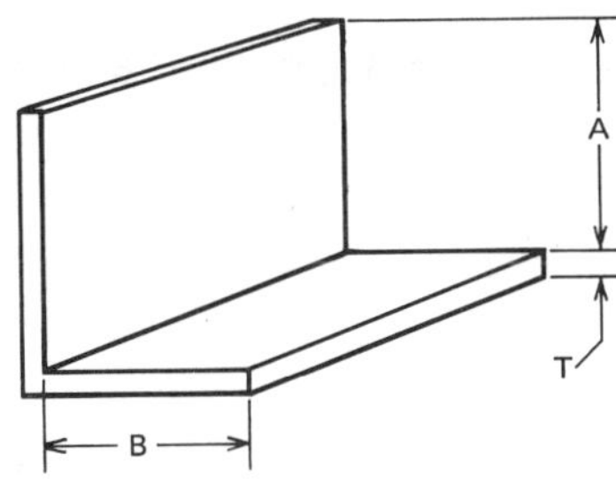

Figure 8-1

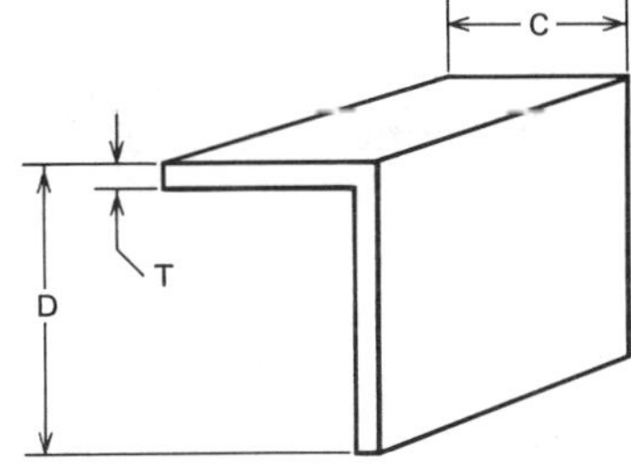

Figure 8-2

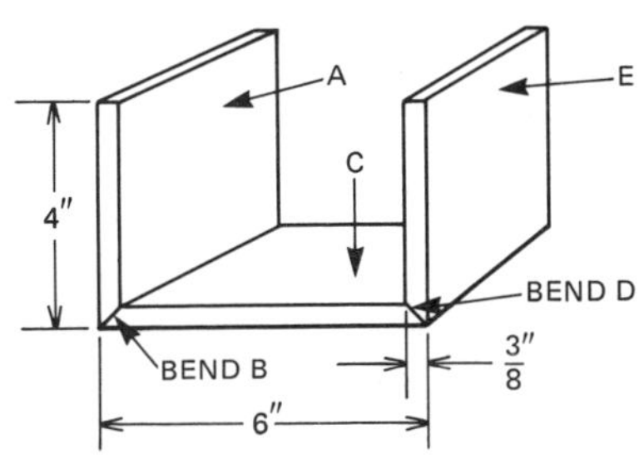

Figure 8-3

2. Identify the given values of the dimensions needed
    to find the total length:

    side A = 4"
    bend B = 50% x 3/8" = 0.50(0.375")
    side C = 6"
    bend D = 50% x 3/8" = 0.50(0.375")
    side E = 4"

3. Express the problem as a formula to be solved by
    addition:

    X = A + B + C + D + E

4. Substitute the given values into the formula:

    X = 4 + 0.50(0.375) + 6 + 9.50(0.375) + 4

5. Remove the parentheses:

    X = 4 + 0.1875 + 6 + 0.1875 + 4

6. Combine the quantities in the right member:

    X = 14.375

ANSWER

14.375" of material are needed to build the spacer
bracket.

Example 8–12

PROBLEM

How many inches of welding material are needed to weld
the baffle to the pipe in Figure 8-4?

SOLUTION

1. Let X = the number of inches of welding material
    needed.

2. Observe that the baffle must be welded to <u>half</u>
    the circumference (distance around) of the circle
    formed by the opening of the pipe. Note also that
    the diameter of the circle is 6".

3. Express the problem as a formula to be solved by
    multiplication, given that the formula for finding
    the circumference of a circle is π (3.142) times
    the diameter (C = πD):

    X = 0.5πD

4. Substitute the given values into the formula:

    X = 0.5(3.142)(6)

5. Multiply the terms in the right member:

    X = 9.426

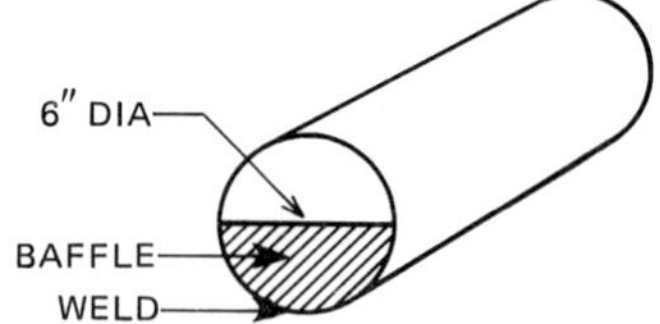

Figure 8-4

ANSWER

9.426" of welding material are needed.

Example 8–13

PROBLEM

What length of material is needed to form the cylinder in Figure 8-5?

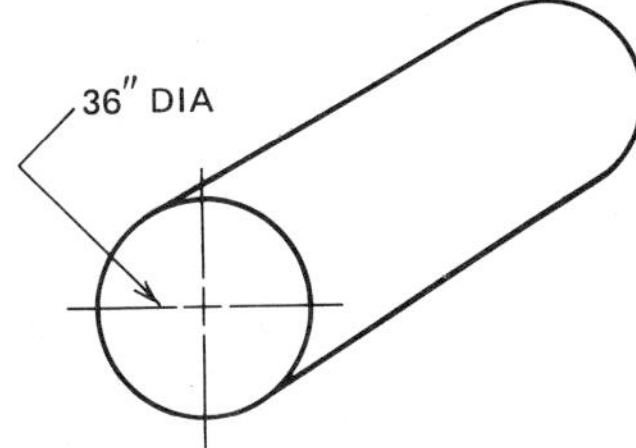

Figure 8-5

SOLUTION

1. Let X = the length of material needed.

2. Observe that the length of material needed is equal to the circumference of the circle formed by the cylinder opening. Note also that the diameter of the circle is 36".

3. Express the problem as a formula to be solved by multiplication, given that the formula for finding the circumference is π (3.142) times the diameter (C = πD):

    X = πD

4. Substitute the given values into the formula:

    X = 3.142(36)

5. Multiply the terms in the right member:

    X = 113.112

ANSWER

113.112" length of material is needed.

Applied Welding Problems

8-61. What length of material is needed for the angle iron in Figure 8-6?

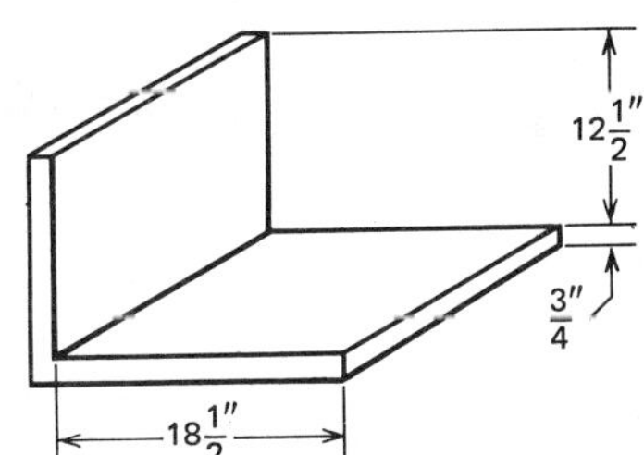

Figure 8-6

8-62. What length of material is needed for the inspection cover in Figure 8-7?

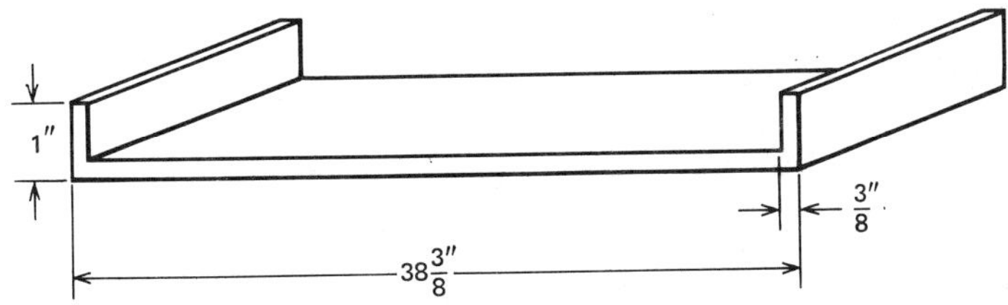

Figure 8-7

8-63. What length of material is needed to build the hanger bracket in Figure 8-8 with one inside and one outside bend?

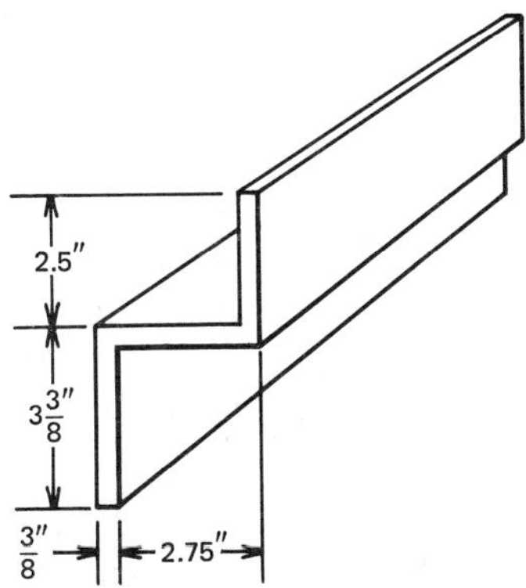

Figure 8-8

8-64. What length of material is needed for the special sill channel in Figure 8-9?

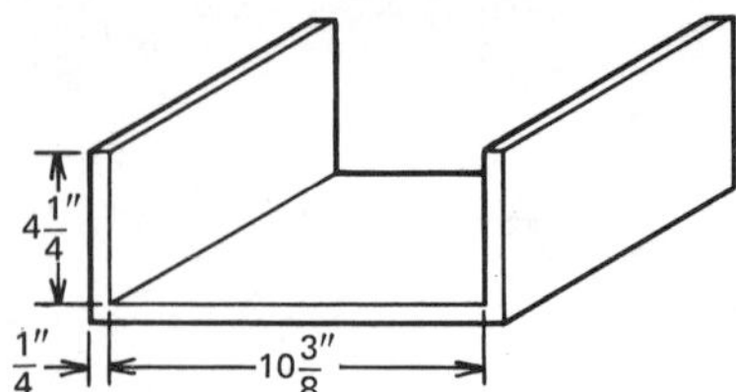

Figure 8-9

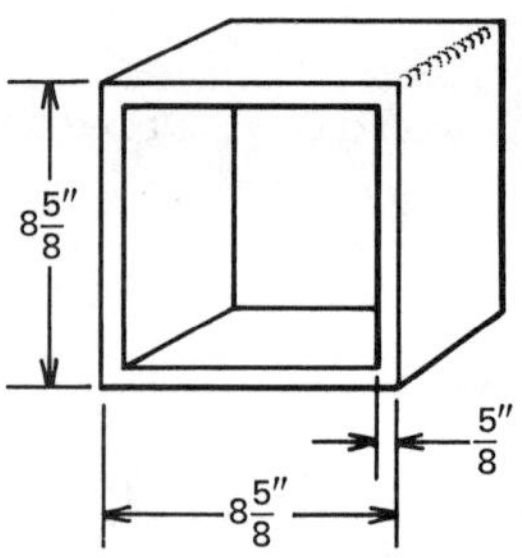

Figure 8-10

8-65. The light pole base section in Figure 8-10 has three outside bends and one welded corner. What length of material is needed to make the pole base?

8-66. What length of material is needed for the stiffener channel in Figure 8-11?

8-67. How many inches of weld are required to weld an end cap on the square opening in Figure 8-12?

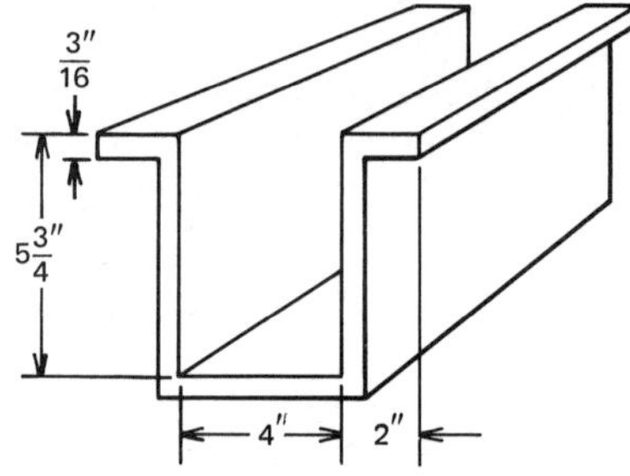

Figure 8-11

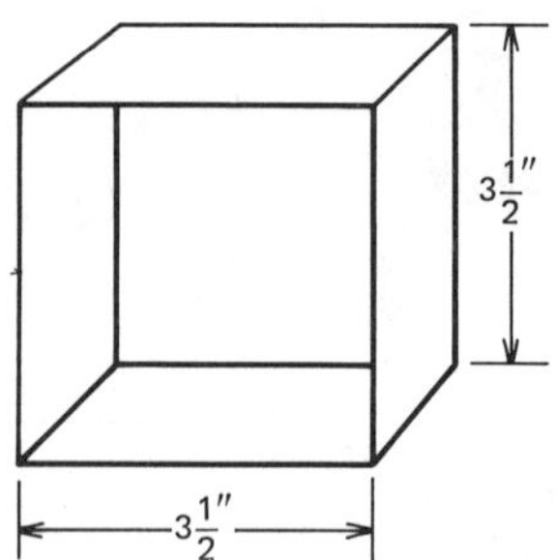

Figure 8-12

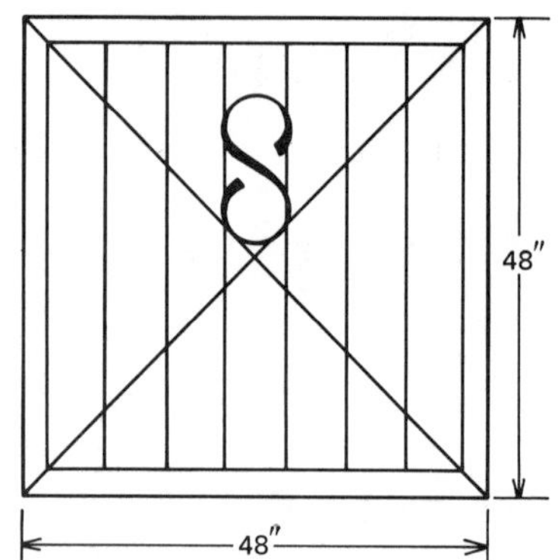

Figure 8-13

8-68. A welding shop borrowed $15,000 to buy two new machines. The loan was for 4 years with a simple interest rate of 11.5% per year. Simple interest is equal to principal ($15,000) times time (4 years).
   a. How much interest did the shop pay per year?
   b. How much interest did the shop pay in 8 months?
   c. How much interest did the shop pay in 18 months?

8-69. Find the length of the two diagonal cross pieces in the gate in Figure 8-13. Use the formula stating that the square of the longest side C of a right triangle is equal to the sum of the squares of the other two sides A and B ($C^2 = A^2 + B^2$).

8-70. What is the outside diameter (distance across the center) of the pipe in Figure 8-14?

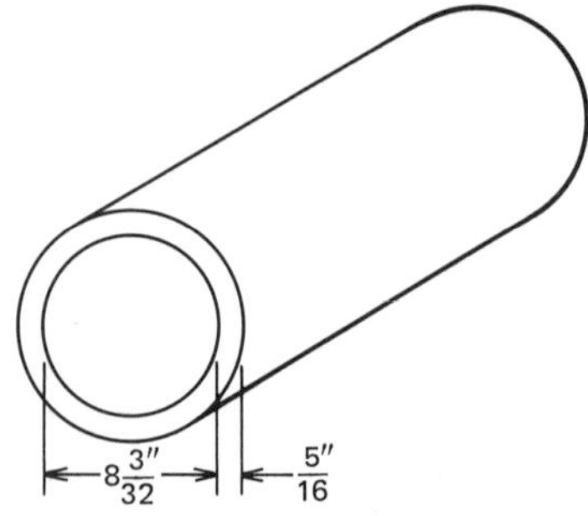

Figure 8-14

8-71. Refer to the tool box in Figure 8-15.
    a. What length of material is needed for the
       cover? (Length = sides + 25% of the material
       thickness for each 122-1/2° bend.)
    b. What length of material is needed for the
       bottom of the tool box?

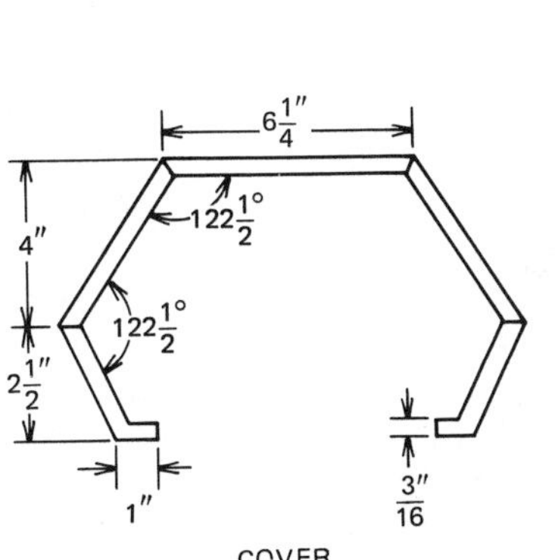
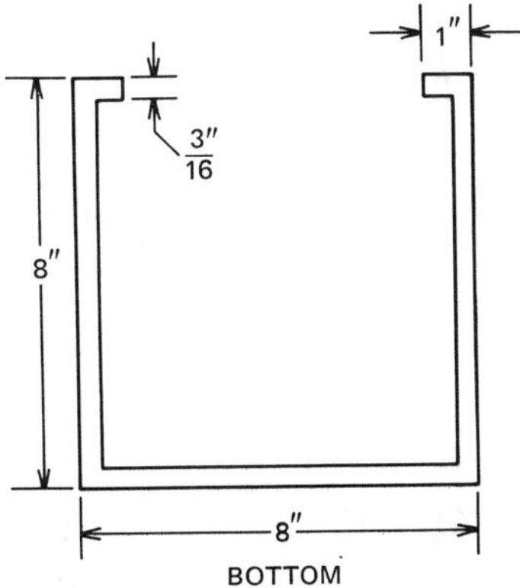

Figure 8-15

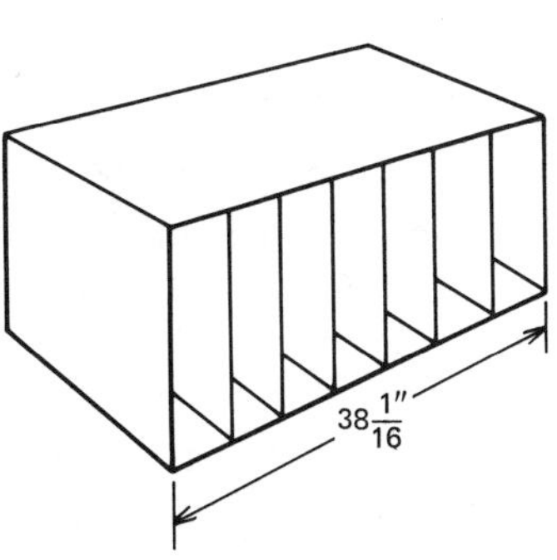

Figure 8-16

8-72. What is the distance from center to center
between each pair of the 6 support parts in
Figure 8-16?

8-73. What length of material is needed for the
stainless steel cement form in Figure 8-17?
(Length = sides - 75% of material thickness per
bend.)

8-74. What length of material is needed for the welded
metal reinforcement in Figure 8-18?

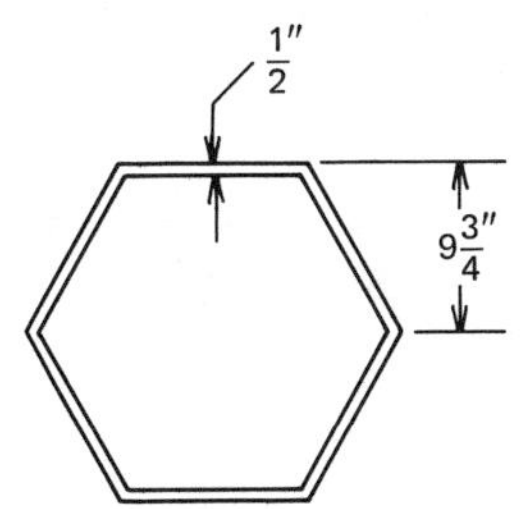

Figure 8-17

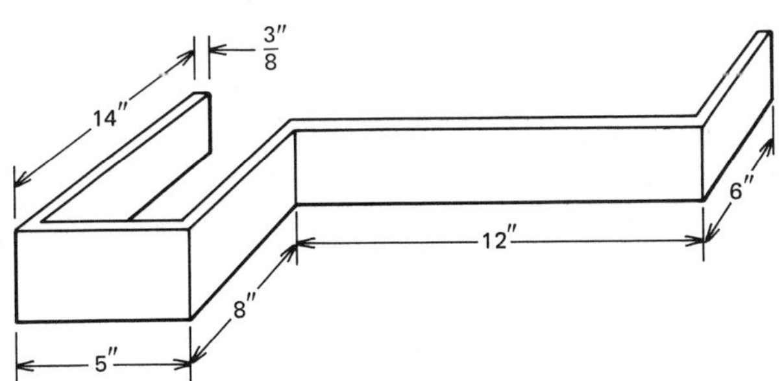

Figure 8-18

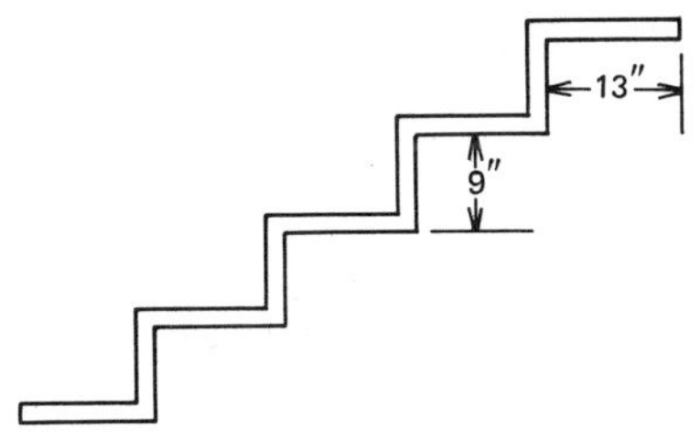

Figure 8-19

8-75. What length of material is needed for the 5
welded stair steps in Figure 8-19?

8-76. What length of material is needed for the welded
shelf bracket in Figure 8-20?

8-77. What is distance X between each pair of evenly
spaced support rings on the tank in Figure 8-21?

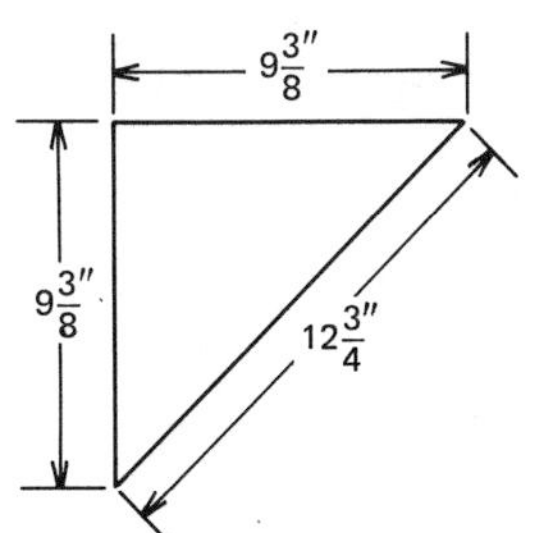

Figure 8-20

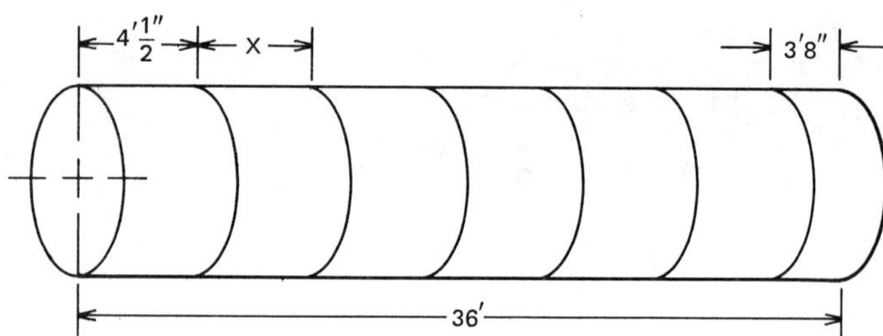

Figure 8-21

8-78. What is distance Y between the holes in the sill plate in Figure 8-22?

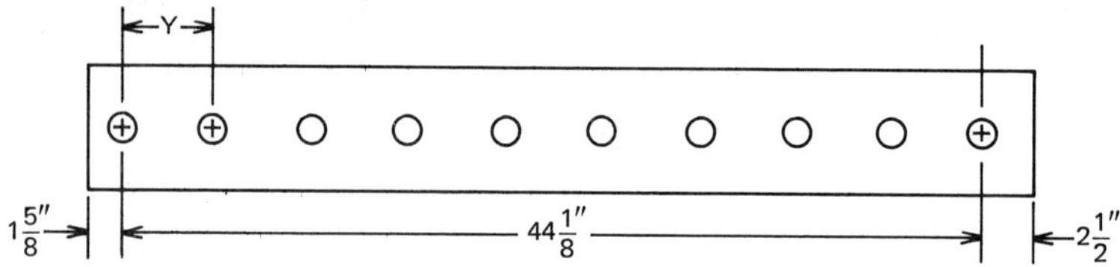

Figure 8-22

8-79. What is the taper per foot of the light pole in Figure 8-23? (Taper per foot = 12(A − B)/L, where A = 46", B = 8", and L = 480".)

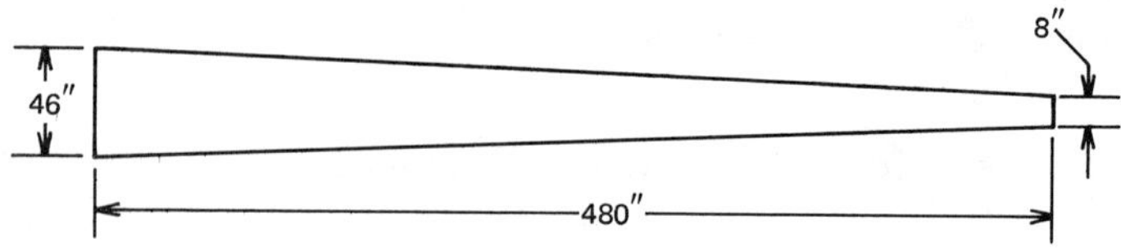

Figure 8-23

8-80. What is the radius of the center hole of the spacer in Figure 8-24? (Radius = 1/2 diameter, or R = D/2.)

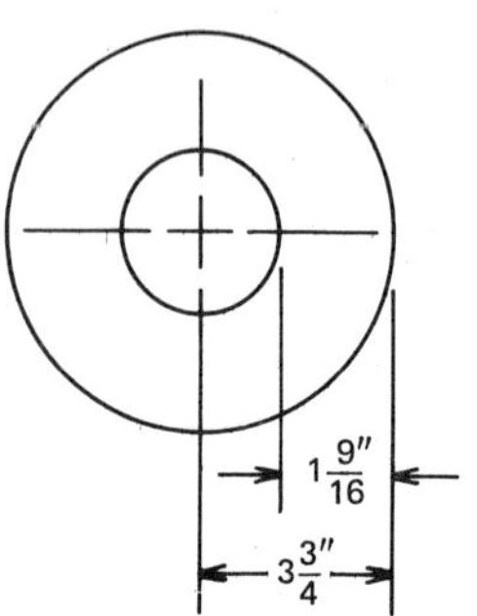

Figure 8-24

Chapter 8: Solving Simple Formulas
Applied Welding Problems
Survey Test

Name ____

Course/Sec. _______________   Date _______________________

8-1.   The length of steel in Figure 8-25 is cut into
        4 equal sections.

    a.   How long is dimension A?   8-1a. _________________

    b.   How long is dimension B?   8-1b. _________________

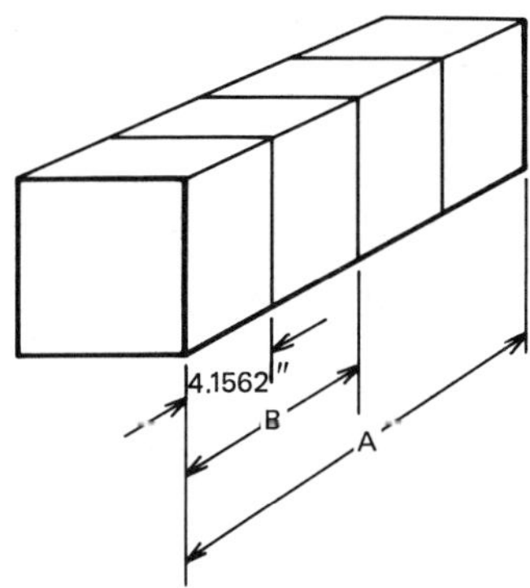

Figure 8-25

8-2.   An iron fence is shown in Figure 8-26.

    a.   How wide is dimension C?   8-2a. _________________

    b.   How wide is dimension D?   8-2b. _________________

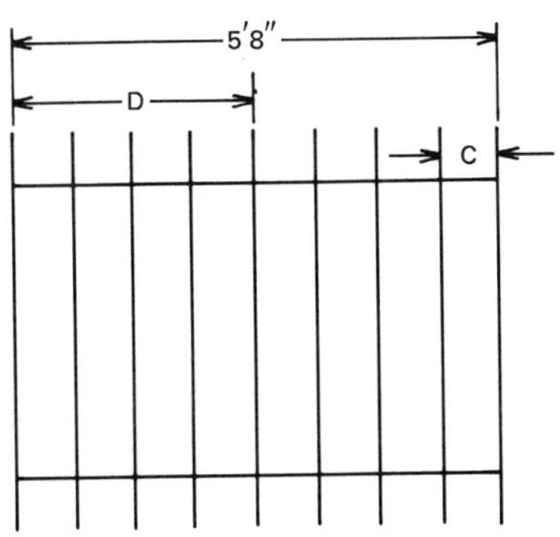

Figure 8-26

8-3.   The stationary engine-mounting frame shown in
       Figure 8-27 is to be made with 1" square tubing.

    a.   What length of material is needed for the
            top?

    b.   What length of material is needed for the
            4 legs?

8-3a. ______________

8-3b. ______________

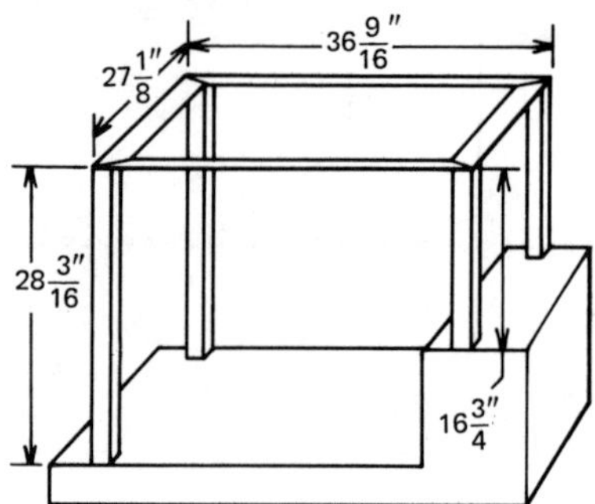

Figure 8-27

8-4.   The pipes in Figure 8-28 are evenly spaced on a
       steel header attached to a water reservoir.

    a.   How long is dimension E?

    b.   How long is dimension F?

8-4a. ______________

8-4b. ______________

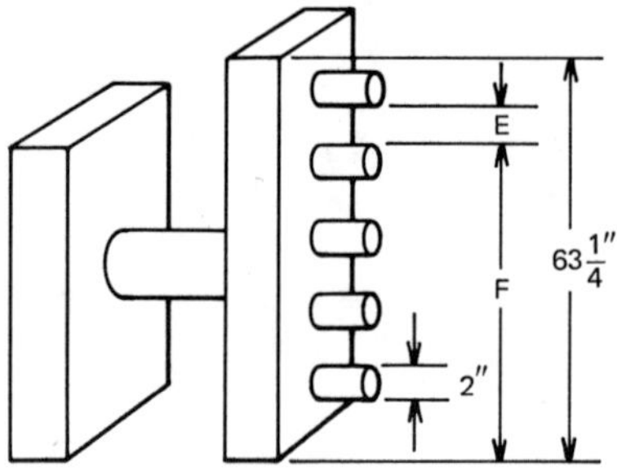

Figure 8-28

8-5.   Reinforcing gussets are needed for the top rail
       of a barge. Gussets are cut from the 4' x 8'
       sheet of steel shown in Figure 8-29. How many
       12" x 12" gussets can be cut from the sheet?

8-5. ______________

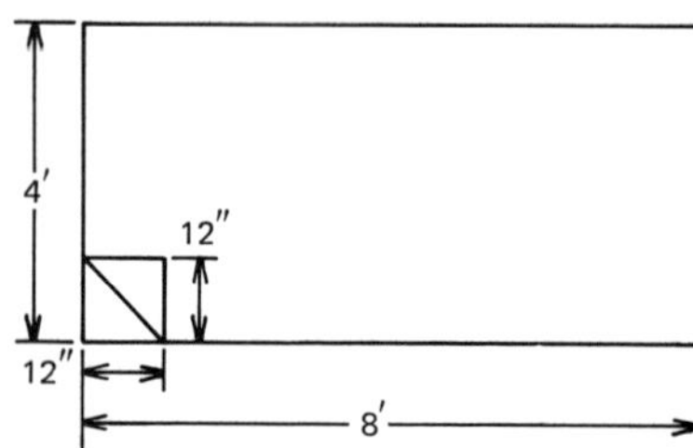

Figure 8-29

8-6.   The gussets in Figure 8-30 are evenly spaced on
       a barge side rail. How wide is dimension X?

8-6. ______________

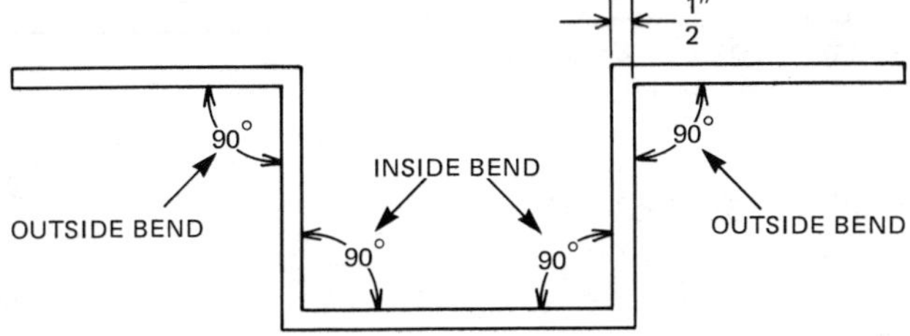

Figure 8-30

8-7.   What length of 1/2" thick material is needed
       to make the dock hanger bracket in Figure 8-31?
       For each inside bend of 90°, add 25% of the
       material thickness. For this type of material,
       subtract 75% of the material thickness for each
       outside bend of 90°.

8-7. ______________________

Figure 8-31

8-8.   Refer to Figure 8-32.

       a.   What are the measurements required to center
            the hole in part A of the figure?

       8-8a. ______________________

       b.   What is distance X between the two evenly
            centered holes in part B?

       8-8b. ______________________

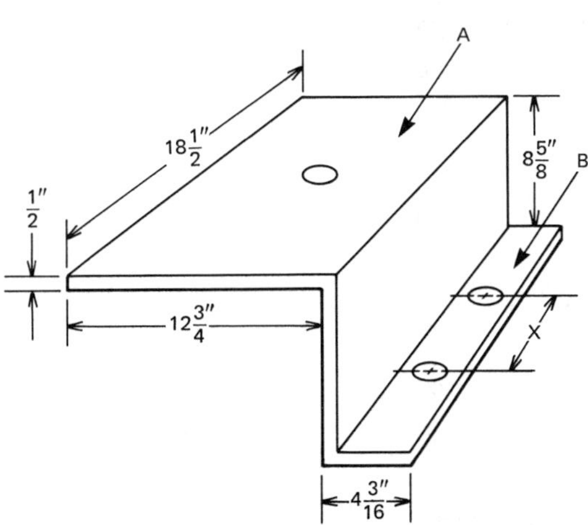

Figure 8-32

8-9.   What length of 1/4" thick metal is needed to build the edge protector in Figure 8-33? The protector has one inside bend of 90° and one outside bend of 90°. See Problem 8-7 for amount to add and subtract for the bends.

8-9. ____________________

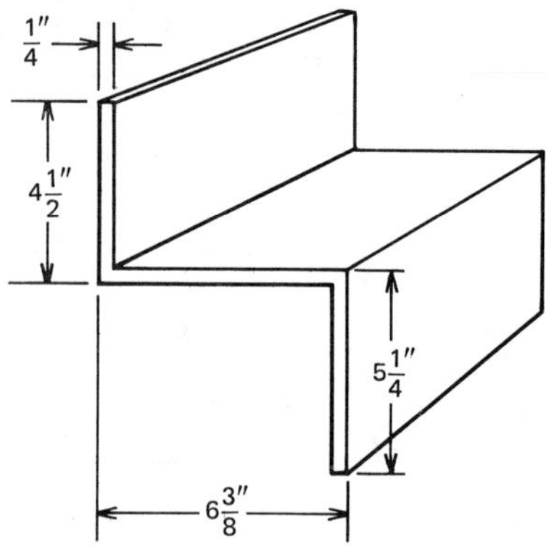

Figure 8-33

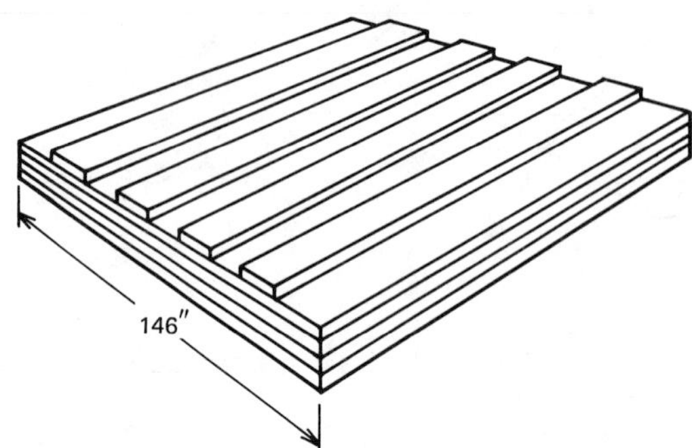

Figure 8-34

8-10. Four reinforcing channels are equally spaced on the welded stack of steel sheets in Figure 8-34. How wide is the spacing between the channels?

8-10. ____________________

8-11. The welding fixture shown in Figure 8-35 holds pieces of metal in place while they are being welded. For each bend in the fixture, add 25% of the material thickness.

   a. What length of 3/8" thick material is required to build the fixture?

8-11a. ____________________

   b. How long is dimension A?

8-11b. ____________________

   c. How long is dimension B?

8-11c. ____________________

   d. How long is dimension C?

8-11d. ____________________

   e. Hole X is centered in the metal. How long is dimension D?

8-11e. ____________________

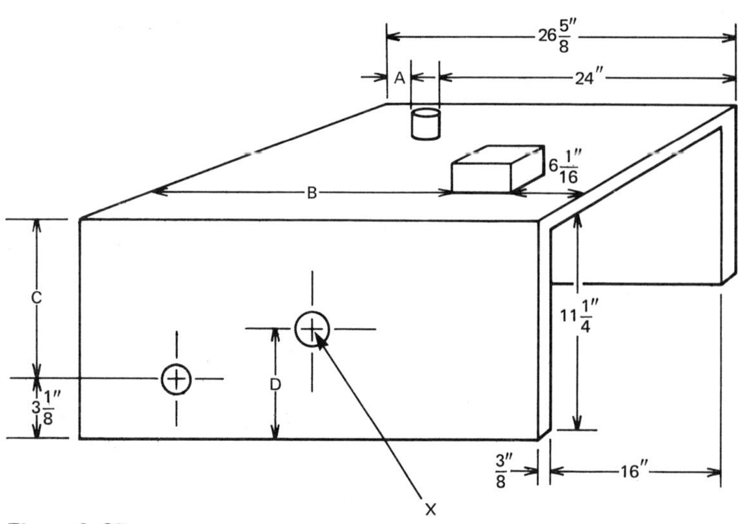

Figure 8-35

••

Name ____

Course/Sec. ______________   Date ___________________________

Chapter 9 covers ratios, proportions, and applied
problems involving variation. Before you turn to the
chapter, complete the Self-Test to determine which
sections in the chapter you need to study carefully.

Self-Test

SIMPLIFYING RATIOS

Simplify the following ratios:

9-1. $\dfrac{3\frac{5}{8}}{1\frac{5}{8}}$

9-2. $\dfrac{4 \text{ months}}{7\frac{1}{2} \text{ days}}$

9-1. ___________________________

9-2. ___________________________

Simplifying ratios score ___________________________

SOLVING PROPORTIONS

Solve the following proportions:

9-3. $\dfrac{a}{16} = \dfrac{9}{48}$

9-4. $\dfrac{24}{36} = \dfrac{y}{8}$

9-5. $\dfrac{6.5}{r} = \dfrac{13}{10}$

9-6. $\dfrac{8}{0.07} = \dfrac{188}{F}$

9-7. $\dfrac{4\frac{1}{2}}{6\frac{7}{8}} = \dfrac{144}{A}$

9-8. $56:64 = 21:X$

9-9. $35:7 :: 10:X$

9-10. $0.007:0.200 = B:0.04$

9-11. $\dfrac{c + 5}{c} = \dfrac{115}{140}$

9-12. $\dfrac{D - 4.2}{D} = \dfrac{3}{10}$

9-3. ___________________________

9-4. ___________________________

9-5. ___________________________

9-6. ___________________________

9-7. ___________________________

9-8. ___________________________

9-9. ___________________________

9-10. ___________________________

9-11. ___________________________

9-12. ___________________________

9-13. $\dfrac{E}{E - 4\frac{3}{4}} = \dfrac{4}{2\frac{1}{2}}$ 9-14. $\dfrac{18.6}{6.2} = \dfrac{G - 4}{G}$ 9-13. _______________

9-14. _______________

Solving proportions score   _______________

PRACTICAL APPLICATIONS OF VARIATION

9-15.   The weight of 150 board feet of red oak is
        360 lb. Find the weight of 2400 board feet of
        red oak.   9-15. _______________

9-16.   A lot containing 15,840 sq ft costs $7500. How
        much is the cost per acre if an acre contains
        43,560 sq ft?   9-16. _______________

9-17.   The scale on a map indicates that 1" equals 12
        mi. What is the distance between two cities that
        are 15-1/4" apart on the map?   9-17. _______________

9-18.   A carpet can be installed in a commercial
        building by 3 carpet layers in 12 hours. How
        long would the job take 7 carpet layers?   9-18. _______________

9-19.   Ice weighs .0324 lb per cu in. Find the weight
        of 9.5 cu ft of ice. (1 cu ft = 1728 cu in.)   9-19. _______________

9-20.   The largest gear with a diameter of 11-1/4"
        rotates at 1500 revolutions per minute. How fast
        does the smallest gear rotate if its diameter is
        6-1/2"?   9-20. _______________

Practical applications of variation score   _______________

After you have checked your answers on the Self-Test,
transfer your scores to the Chapter 9 Objectives.

Chapter 9

··

Ratio, Proportion, and Variation

Upon successful completion of this chapter, you will be
able to simplify ratios, solve proportions, and solve
applied problems involving variation.

Objectives

	Self-Test Scores	If you did poorly on the Self-Test, turn to:
• Simplifying ratios	_________	Section 9-2
• Solving proportions	_________	Section 9-3
• Practical applications of variation	_________	Section 9-4

9–1   Introduction to Ratio, Proportion, and Variation

We make comparisons between things or quantities almost daily. We do so to help answer the questions how much, how long, or how many. For example, if we average 50 miles per hour when driving, how long will a 350-mile trip take? By comparing 50 miles to 350 miles, we see that the trip will take 7 hours. This problem is an application of the concepts of ratio and proportion.

Ratio and proportion are powerful mathematical tools. Knowing how to apply them will make your work easier and more rewarding.

9–2   Properties of Ratio and Simplifying Ratios

A <u>ratio</u> is the comparison of one like quantity to another. A fraction is actually a ratio that compares the quantity in the numerator to the quantity in the denominator. Comparisons are sometimes expressed in "ratio language." For example, if a family has 4 girls and 1 boy, we could say the ratio of girls to boys is 4 girls to 1 boy, or 4 to 1. This ratio can be written in the following ways:

   4 to 1,   4/1 (fractional form),   4:1

The colon form is an older way to write ratios. 4:1 is read as "4 to 1." Although we may write ratios in any of the above ways, the fractional form is preferred.

Like quantities must be compared in the same units of measure. For example, to compare 50¢ and $4.00, change the $4.00 to 400¢:

$$\frac{50¢}{400¢} = \frac{1}{8}$$

Always reduce ratios when possible in order to simplify them. If the fractional form reduces to a whole number, write 1 as the denominator. For example, the ratio of 60 to 12 should be reduced as follows:

$$\frac{60}{12} = \frac{5}{1}\quad \text{(not 5)}$$

Procedure for Simplifying Ratios

●●●

1.   Express the comparison as a fraction in equivalent units of measure and reduce when possible.
2.   If the fraction reduces to a whole number, write 1 as the denominator.

●●●

Example 9–1

PROBLEM

What is the ratio of 16 to 144?

SOLUTION

To compare the first quantity to the second, express
the ratio as a fraction and reduce:

$$\frac{16}{144} = \frac{1}{9}$$

ANSWER

The ratio of 16 to 144 is 1 to 9.

Example 9–2

PROBLEM

What is the ratio of 3" to 4'?

SOLUTION

To compare the first quantity to the second, express
the ratio as a fraction in equivalent units of measure
(remember, 1' = 12") and reduce:

$$\frac{3"}{4'} = \frac{3"}{48"} = \frac{1}{16}$$

ANSWER

The ratio of 3" to 4' is 1 to 16.

Example 9–3

PROBLEM

Simplify the ratio of 6-3/4 to 3/8.

SOLUTION

1. To compare the first quantity to the second,
    express the ratio as a fraction:

$$\frac{6\frac{3}{4}}{\frac{3}{8}}$$

2. Change 6-3/4 to an improper fraction and divide by
    3/8:

$$\frac{27}{4} \div \frac{3}{8}$$

3. Invert, reduce, and multiply:

$$\frac{\overset{9}{\cancel{27}}}{\underset{1}{\cancel{4}}} \times \frac{\overset{2}{\cancel{8}}}{\underset{1}{\cancel{3}}} = \frac{18}{1}$$

ANSWER

$$\frac{6\frac{3}{4}}{\frac{3}{8}} = \frac{18}{1}$$

Exercises: Simplifying Ratios

Simplify the following by reducing to lowest terms:

9-1. $\dfrac{14}{21}$ 9-2. 260:520   9-3. 96 to 12

9-4. $\dfrac{2.4}{0.3}$ 9-5. $\dfrac{8\frac{1}{2}}{\frac{1}{2}}$ 9-6. $\dfrac{40}{\frac{1}{4}}$

9-7. 15 yd to 5'   9-8. $\dfrac{3\frac{1}{2}\text{ minutes}}{5\text{ seconds}}$

9-9. 9 months to 6 weeks   9-10. 6 mi to 1320'

9–3   Properties of Proportion and Solving Proportions

A proportion is an equation stating that two ratios
are equal. An example of a proportion is:

$$\frac{5}{2} = \frac{25}{10} \quad \text{(fractional form)}$$

$$5:2::25:10 \quad \text{(colon form)}$$

This proportion is read: "5 is to 2 as 25 is to 10."
    A proportion has 4 terms. In the example just
given, 5 is the first term; 2, the second term; 25,
the third term; and 10, the fourth term. The first and
fourth terms (5 and 10) are called the extremes (the
outside terms). The second and third terms (2 and 25)
are called the means (the inside terms).
    The product of the extremes equals the product of
the means. These products are sometimes called the
cross products or loop products. They are obtained by
cross multiplication:

$$\frac{5}{2}\ \frac{\text{(first term)}}{\text{(second term)}} = \frac{25}{10}\ \frac{\text{(third term)}}{\text{(fourth term)}}$$

$$(5)(10)\quad \text{(extremes)} = (2)(25)\quad \text{(means)}$$

When three out of the four terms of a proportion are given, we can solve for the unknown term. The unknown term is called the <u>variable</u>.

Procedure for Solving Proportions

• •

1. Write the proportion in fractional form.
2. Find the cross products of the extremes and the means by cross multiplication.
3. Solve for the variable by dividing the numerical coefficient of the variable into both sides of the equation.

• •

Example 9–4

PROBLEM

Solve the proportion x is to 4 as 16 is to 2 for x.

SOLUTION

1. To find the value of x that will make the ratios equal, first write the proportion in fractional form:

$$\frac{x}{4} = \frac{16}{2}$$

2. Then find the cross products:

$$2x = (4)(16)$$

3. Finally, divide both sides by 2:

$$\frac{2x}{2} = \frac{(4)(16)}{2}$$

ANSWER

$$x = 32$$

CHECK

Substitute 32 for x in the original proportion and evaluate:

$$\frac{x}{4} = \frac{16}{2}$$

$$\frac{32}{4} = \frac{16}{2}$$

$$\frac{32}{4} = \frac{16}{2}$$

$$8 = 8$$

Example 9–5

PROBLEM

Solve the proportion 0.6 is to 48 as 0.4 is to b for b.

SOLUTION

1. To find the value of b that will make the ratios
    equal, first write the proportion in fractional
    form:

    $$\frac{0.6}{48} = \frac{0.4}{b}$$

2. Then find the cross products:

    $$\frac{0.6}{48} = \frac{0.4}{b}$$

    $(0.6)b = (0.4)(48)$

3. Finally, divide both sides by 0.6:

    $$\frac{(0.6)b}{0.6} = \frac{(0.4)(48)}{0.6}$$

ANSWER

$b = 32$

CHECK

Substitute 32 for b in the original proportion and
evaluate:

$$\frac{0.6}{48} = \frac{0.4}{b}$$

$$\frac{0.6}{48} = \frac{0.4}{32}$$

$0.0125 = 0.0125$

Example 9–6

PROBLEM

Solve the proportion $(j + 2)$ is to 1 as 35 is to 7 for j.

SOLUTION

1. To find the value of j that will make the ratios
    equal, first write the proportion in fractional
    form:

    $$\frac{j + 2}{1} = \frac{35}{7}$$

2. Find the cross products:

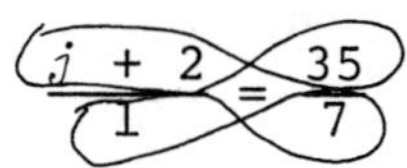

$$7(j + 2) = (35)(1)$$

3. Remove the parentheses:

$$7j + 14 = 35$$

4. Transpose +14:

$$7j = 35 - 14$$

5. Combine like quantities:

$$7j = 21$$

6. Divide both sides by 7:

$$\frac{\cancel{7}j}{\cancel{7}} = \frac{\overset{3}{\cancel{21}}}{\cancel{7}}$$

ANSWER

$$j = 3$$

CHECK

Substitute 3 for j in the original proportion and evaluate:

$$\frac{j + 2}{1} = \frac{35}{7}$$

$$\frac{3 + 2}{1} = \frac{35}{7}$$

$$5 = 5$$

Example 9–7

PROBLEM

Solve the proportion $x:4 :: 5:10$ for x.

SOLUTION

1. To find the value of x that will make the proportion true, recall that the product of the extremes (outside terms) must equal the product of the means (inside terms).

    <u>Note</u>: :: means =.

2. Find the products of the means and the extremes:

$$x : 4 = 5 : 10$$

$$10(x) = (4)(5)$$

3. Divide both sides by 10:

$$\frac{10(x)}{10} = \frac{(4)(5)}{10}$$

ANSWER

$$x = 2$$

CHECK

Substitute 2 for x in the original proportion and evaluate:

$$x:4 :: 5:10$$

$$2 : 4 :: 5 : 10$$

$$\frac{2}{4} = \frac{5}{10}$$

$$\frac{1}{2} = \frac{1}{2}$$

Exercises: Solving Proportions

Solve the following for the unknown and check:

9-11. $\dfrac{x}{12} = \dfrac{3}{4}$

9-12. $\dfrac{x}{8} = \dfrac{32}{4}$

9-13. $\dfrac{7}{x} = \dfrac{21}{63}$

9-14. $\dfrac{x}{96} = \dfrac{10}{24}$

9-15. $\dfrac{16}{14} = \dfrac{2}{x}$

9-16. $\dfrac{24}{18} = \dfrac{x}{14}$

9-17. $60:360 = k:15$

9-18. $\dfrac{20}{y} = \dfrac{8}{18}$

9-19. $\dfrac{11}{x} = \dfrac{4}{14.2}$

9-20. $32:j :: 8:9$

9-21. $\dfrac{3.5}{10\frac{1}{2}} = \dfrac{8}{k}$

9-22. $\dfrac{0.200}{0.007} = \dfrac{0.04}{c}$

9-23. $\dfrac{x}{8.6} = \dfrac{39.4}{42.355}$

9-24. $\dfrac{700}{10x} = \dfrac{2}{1}$

9-25. $\dfrac{10}{y} = \dfrac{3.4}{1.2}$

9-26. $\dfrac{x - 4}{20} = \dfrac{14}{70}$

9-27. $\dfrac{2}{1} = \dfrac{10 - a}{7}$ 9-28. $\dfrac{x}{x + 1} = \dfrac{3}{1}$

9-29. $\dfrac{j + 8}{j} = \dfrac{0.75}{1}$ 9-30. $\dfrac{y - 5}{28} = \dfrac{1.9}{13.3}$

9–4 Practical Applications of Variation

Numerous applications of ratio and proportion occur in our daily activities: cost (price per gallon), map scale (1" = 40 mi), yield (bushels per acre), batting average (hits per time at bat), and so on.

Many physical laws state that one quantity varies as another or as a certain combination of other quantities. These laws are frequently translated into formulas. The expression of these formulas is called the language of variation. We often accept and use certain formulas without considering the underlying physical laws. The laws, or formulas of variation, are actually an application of the rules of proportion.

The four types of variation are direct, inverse, joint, and combined. The following list gives a common example of each type of variation:

Direct variation: The circumference of a circle varies according to the diameter multiplied by π.

Inverse variation: The number of days needed to complete a job decreases if there are more workers.

Joint variation: The area of a rectangle varies according to the length and the width.

Combined variation: The weight of a load varies according to length, width, height, and weight per unit.

The examples that follow illustrate various applications of ratio and proportion that could be considered applications of variation. Remember, when working the proportion, always compare equal units of measure (ft/sec) to equal units (ft/sec) and use the correct physical law, fact, or formula.

Example 9–8

PROBLEM

How many inches are in 7.5 ft, given 12 in = 1 ft?

SOLUTION

1. Set up a proportion to determine the number of inches in 7.5 ft. Let x = the inches in 7.5 ft:

$$\frac{x}{7.5 \text{ ft}} = \frac{12 \text{ in}}{1 \text{ ft}}$$

2. Multiply both sides by 7.5 ft:

$$(7.5 \text{ ft}) \frac{x}{7.5 \text{ ft}} = \frac{12 \text{ in}}{1 \text{ ft}} (7.5 \text{ ft})$$

3. Cancel the feet and multiply 12 in by 7.5:

$$x = \frac{12 \text{ in} (7.5 \text{ ft})}{1 \text{ ft}}$$

$$= 90 \text{ in}$$

ANSWER

There are 90 in in 7.5 ft.

Example 9–9

PROBLEM

How much will 21.5 gal of gas cost (to the nearest cent) if 2.82 gal costs $4.00?

SOLUTION

1. Set up a proportion to find the cost of 21.5 gal.
    Let c = the cost of 21.5 gal:

$$\frac{c}{21.5 \text{ gal}} = \frac{\$4.00}{2.82 \text{ gal}}$$

2. Multiply both sides by 21.5 gal:

$$(21.5 \text{ gal}) \frac{c}{21.5 \text{ gal}} = \frac{\$4.00}{2.82 \text{ gal}} (21.5 \text{ gal})$$

3. Cancel the gallons, multiply $4.00 by 21.5, and
    reduce:

$$c = \frac{\$4.00}{2.82 \text{ gal}} (21.5 \text{ gal})$$

$$= \frac{\$86.00}{2.82}$$

$$= \$30.496$$

ANSWER

21.5 gallons will cost $30.50 (rounded to the nearest cent).

Example 9–10

PROBLEM

The scale on a blueprint indicates that 1/4 in equals 1 ft. How long is a room whose length measures 3-3/4 in on the blueprint?

SOLUTION

1. Set up a proportion to find the length of the room.
    Let l = the length of the room:

$$\frac{l}{3\text{-}3/4 \text{ in}} = \frac{1 \text{ ft}}{1/4 \text{ in}}$$

2. Change the fractions to decimals and multiply both
    sides by 3.75 in:

$$(3.75 \text{ in})\frac{l}{3.75 \text{ in}} = \frac{1 \text{ ft}}{.25 \text{ in}}(3.75 \text{ in})$$

3. Cancel the inches and evaluate:

$$l = \frac{1 \text{ ft } (3.75 \text{ in})}{.25 \text{ in}}$$

$$= 15 \text{ ft}$$

ANSWER

A room whose length measures 3-3/4 in on the blueprint
is 15 ft long.

Example 9–11

PROBLEM

Two pulleys are connected by a V belt. The larger
pulley is 16" in diameter and rotates 800 revolutions
per minute (RPM). The smaller pulley is 4" in diameter.
How fast does the smaller pulley rotate?

SOLUTION

1. Draw a sketch like the one in Figure 9-1.

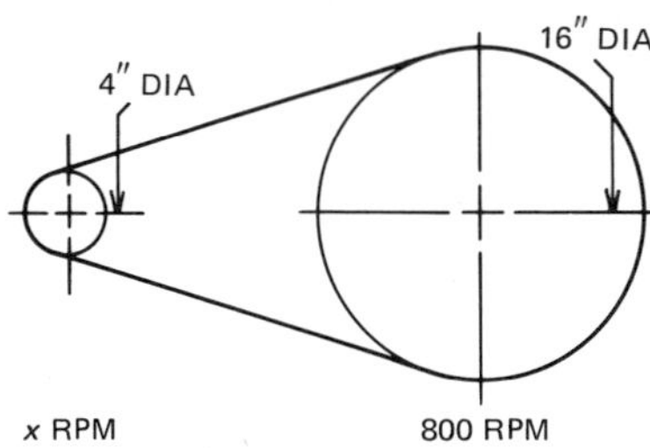

Figure 9–1

Note: This problem is an application of inverse
variation. A wheel with a smaller diameter needs to
turn more times to cover the same distance as a wheel
with a larger diameter. The number of revolutions
turned varies inversely with the diameter of the
wheel. The relationship between the two pulleys and
their RPMs can be described as follows: The product
of the diameter of the smaller pulley times its RPMs
equals the product of the diameter of the larger pulley
times its RPMs.

2.   Set up an equation (formula) expressing this
     relationship. Use a lowercase letter for the
     diameter of the smaller pulley and a capital letter
     for the diameter of the larger pulley:

$$(d)(RPM) = (D)(RPM)$$

3.   Substitute the given values:

$$(4")(RPM) = (16")(RPM)$$

4.   Let x = the RPMs of the smaller pulley and divide
     both sides by 4":

$$\frac{(\cancel{4}")(x)}{\cancel{4}"} = \frac{(\cancel{16}")(800\ RPM)}{\cancel{4}"}$$

$$x = 3200\ RPM$$

ANSWER

The smaller pulley rotates at a rate of 3200 RPM.

CHECK

Substitute 3200 for x in the equation and evaluate:

$$(4)(3200) = (16)(800)$$
$$12800 = 12800$$

Example 9–12

PROBLEM

If 12 men take 10.5 days to do a job, how long will
18 men take to do the same job?

SOLUTION

Note: This problem is an application of inverse
variation. As the number of men increases, the amount
of time needed to complete the same amount of work
decreases. The relationship between the number of men
and the time needed to do the job can be described as
follows: The product of the original number of men
working times the days worked equals the product of
the new number of men working times the days worked.

1.   Set up an equation (formula) expressing this
     relationship. Let D = the days that 18 men work:

$$(12\ men)(10.5\ days) = (18\ men)(D)$$

2.   Divide both sides by 18 men:

$$\frac{(12\ men)(10.5\ days)}{18\ men} = \frac{(18\ men)D}{18\ men}$$

3. Cancel and evaluate:

$$\frac{(12 \text{ men})(10.5 \text{ days})}{18 \text{ men}} = \frac{(18 \text{ men})D}{18 \text{ men}}$$

$$\frac{126}{18} = D$$

$$7 = D$$

ANSWER

18 men will take 7 days to do the same job that 12 men do in 10.5 days.

CHECK

Substitute 7 days for D in the original equation and evaluate:

$$(12 \text{ men})(10.5 \text{ days}) = (18 \text{ men})(7 \text{ days})$$

$$126 \text{ man days} = 126 \text{ man days}$$

Exercises: Practical Applications of Variation

9-31. If you earn $44 in 8 hours, how much could you earn in 30 hours?

9-32. 8 yd of cotton cloth sells for $2.72. How much will 28 yd cost?

9-33. How much will 25 lb of sugar cost if 10 lb costs $1.89?

9-34. 5 lb of flour costs $0.99. How much will 25 lb cost?

9-35. If your mileage is 15.3 mpg, how many gallons will you need to travel 5586 mi?

9-36. How far can you travel in 8-1/2 hours if you average 55 mph?

9-37. 2-1/2 lb of bananas costs 55¢. How much will 12 lb cost?

9-38. A 3" x 5" photo is to be enlarged to a height of 10". How wide will the photo be?

9-39. The scale on a blueprint is 1/8" = 1'. How long is a room that measures 2-3/16" on the blueprint?

9-40. How long is a hallway that measures 2-1/8" if the scale is 1/4" = 1'?

9-41.   An electric motor delivers 6.5 horsepower and draws 26.5 amperes. How much horsepower would 106 amperes produce?

9-42.   How many bushels of spring wheat will 640 acres yield if 1-1/2 bushels are produced per acre?

9-43.   If 104% equals $176.75, what does 100% equal?

9-44.   If Frigid manufacturing needs 5 workers to produce 42 refrigerators per day, how many workers are needed to produce 1470 refrigerators per day?

9-45.   VIP Learning pays a stock dividend of $3.50 per share every 90 days. How much dividend would be collected on 10 shares in 720 days?

9-46.   A 100cm pulley turning at 1800 RPM drives a 40cm pulley. Find the RPM of the 40cm pulley.

9-47.   What will be the length of shadow cast by a 45' pole if a 14.5' pole casts a shadow of 4.5' (to the nearest tenth of a foot)?

9-48.   A 6.5 oz can of tuna fish costs 89¢. What is the cost per ounce (to the nearest thousandth of a cent)?

9-49.   An imported beverage costs $3.96 per 12-pack. If each can contains 12 oz, what is the cost per ounce?

9-50.   How much will $150,000 of personal liability insurance cost if $1000 costs $0.88?

9-51.   If 6 yd of topsoil costs $36.75 delivered, find the cost of 32 yd.

9-52.   If flying time is 1200 mi in 125 minutes, how many hours will it take to fly 4320 mi?

9-53.   Two snow plows can clear a lot in 45 minutes. How long will 3 snow plows take to clear the same lot?

9-54.   If 67.5 gal of oil flows through a pipe in 18 minutes, how many hours will it take to fill a 3750 gal tank (to the nearest hundredth of an hour)?

9-55.   A pinion gear with 12 teeth meshes with a ring gear of 42 teeth. If the pinion gear turns at 64 RPM, how fast is the ring gear turning?

9-56.   A map is drawn to the scale of 2.25" = 12 mi. Find the actual distance between two towns 6.5" apart on the map.

Chapter 9

Ratio, Proportion, and Variation

••

Solving Applied Welding Problems

Now that you have mastered the fundamental operations
of ratio, proportion, and variation, you are ready to
apply these math skills to typical problems that
welders encounter on the job. After studying the
examples that follow, complete the applied welding
problems.

Example 9–13

PROBLEM

80 lb of hard-facing rod costs $132.00. How much
would 30 lb of the same rod cost?

SOLUTION

1. Set up a proportion to find the cost of 30 lb of
    rod. Let X = the cost of 30 lb of rod:

$$\frac{X}{30} = \frac{\$132.00}{80}$$

2. Multiply both sides by 30 and evaluate for X:

$$\frac{(30)\ X}{30} = \frac{(\$132.00)(30)}{80}$$

$$X = \frac{(\$132.00)(30)}{80} = \$49.50$$

ANSWER

30 lb of rod would cost $49.50.

Example 9–14

PROBLEM

4 men can weld 22 hand trucks in one day. How many
men are needed to weld 110 hand trucks in one day?

SOLUTION

1. Set up a proportion to find the number of men required to weld 110 hand trucks. Let X = the number of men:

$$\frac{X}{110} = \frac{4}{22}$$

2. Multiply both sides by 110 and evaluate for X:

$$(110)\frac{X}{110} = \frac{(4)(110)}{22}$$

$$X = \frac{(4)(110)}{22} = 20$$

ANSWER

20 men are needed to weld 110 hand trucks in one day.

Example 9–15

PROBLEM

A welder earns $88.50 in 6 hours. How many hours will it take her to earn $590.00?

SOLUTION

1. Set up a proportion to find the number of hours needed to earn $590.00. Let X = the hours needed.

$$\frac{X}{\$590.00} = \frac{6}{\$88.50}$$

2. Multiply both sides by $590.00 and evaluate for X:

$$(\$590.00)\frac{X}{\$590.00} = \frac{(6)(\$590.00)}{\$88.50}$$

$$X = \frac{(6)(\$590.00)}{\$88.50} = 40$$

ANSWER

It will take the welder 40 hours to earn $590.00.

Example 9–16

PROBLEM

A welding fixture has a drive gear with 24 teeth that turns the fixture at 85 RPM. The 24-tooth gear meshes with a gear that has 16 teeth. How many revolutions per minute does the gear with 16 teeth turn?

SOLUTION

1. Study the sketch in Figure 9-2.

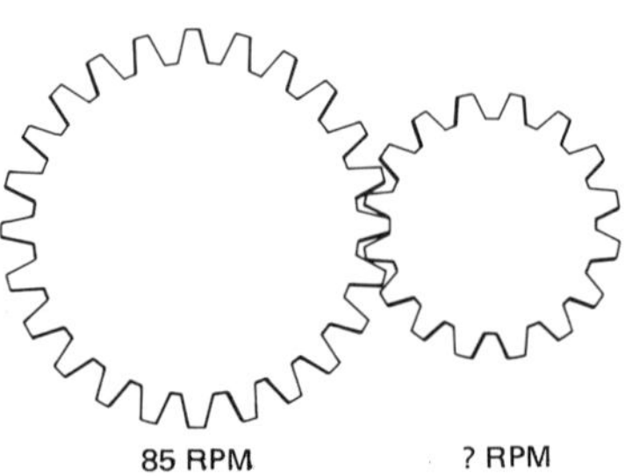

Figure 9-2

<u>Note:</u> The distance the larger gear travels is the
product of the number of teeth times the RPM. This
product equals the number of teeth in the smaller gear
times the RPM.

2.   Set up an equation expressing this relationship.
     Use T for the larger gear and t for the smaller
     gear:

     (T)(RPM) = (t)(RPM)

3.   Substitute the given values:

     (24)(85) = (16)(RPM)

4.   Divide both sides by 16 and evaluate for RPM:

$$\frac{(24)(85)}{16} = \frac{(16)(RPM)}{16}$$

$$\frac{(24)(85)}{16} = RPM$$

$$127.5 = RPM$$

ANSWER

The gear with 16 teeth turns at a rate of 127.5 RPM.

Applied Welding Problems

9-57.   The cost of 1600 lb of steel is $1,344.00.
        What is the cost of 300 lb?

9-58.   1800 lb of angle iron costs $1,116.00. What is
        the cost of 420 lb?

9-59.   If a welder can weld 1488 fillet welds in 24
        hours, how many can he weld in 38 hours?

9-60.   A welder welds 25 hand trucks in 3-1/2 days.
        How many can she weld in 7-1/4 days?

9-61.   A pump used for testing storage tanks can
        deliver 4500 gal per hour. It can fill a
        storage tank in 48 hours. How long would a pump
        that delivered 5500 gal per hour take to fill
        the same tank?

9-62.   A portable welder can operate 132 minutes on
        11 gal of gas. How long can it operate on 48
        gal?

9-63.   A 27" long bar weighs 3 lb 3 oz. How much would
        a 4'9" long bar of similar material weigh?

9-64.   If a machine welds at a rate of 14" per minute,
        how many inches of weld are deposited in an
        8-hour day?

9-65. To fasten a steel beam, 8 rivets with a holding
       force of 60,000 lb each are required. How many
       rivets would be needed for a holding force of
       135,000 lb?

9-66. How many rivets like those in Exercise 9-65
       would be needed for a holding force of 45,000 lb?

9-67. Find the height of a utility tower that casts a
       shadow 75' long when a pole 8' high casts a
       shadow 5' long.

9-68. A welder earns $69.20 in 8 hours. How many
       hours would it take him to earn $890.95?

9-69. If 50 lb of welding rod costs $24.00, how many
       pounds of rod can be bought for $186.00?

9-70. An automatic spot welding machine makes 64,000
       welds in 8 hours. How long will it take to make
       24,000 welds?

9-71. A beam welder welds 18-gage steel at a rate of
       360" in 3 minutes. How many inches will it weld
       in 20 minutes at the same rate?

9-72. If the automatic welder welds at a rate of 85"
       every 30 seconds, how many inches of weld are
       deposited every 60 seconds?

9-73. 200 lb of stainless steel contains 25 lb of
       chromium. How many pounds of chromium are in
       75 lb of stainless steel?

9-74. On a blueprint, 1/4" represents 1'.
       a. How many inches represent 36'?
       b. How many inches represent 28'?
       c. How many inches represent 16'?
       d. How many inches represent 12'?

9-75. If 20 men can complete a section of bridge in
       19 days, how long will 25 men working at the
       same rate take to complete a similar section?

9-76. A belt connects a 20" pulley and a 4" pulley on
       a drill press. The 20" pulley rotates at a rate
       of 210 RPM. What is the speed of the 4" pulley?

9-77. 240 men can complete a barge in 16 days. How
       many days would 165 men working at the same rate
       take to complete the barge?

9-78. A welder uses 25 lb of rod to weld 1000" of
       weld. How many pounds of rod did he use to weld
       656"?

Chapter 9: Ratio, Proportion, and Variation
Applied Welding Problems
Survey Test

Name ____

Course/Sec. ______________ Date ___________________

9-1. A welding shop can produce the following amounts
      of weld with the given amounts of electrodes in
      one week: 1600" of weld with 40 lb of mild
      steel electrodes, 1196" of weld with 46 lb of
      stainless steel electrodes, and 256" with 16 lb
      of iron powder electrodes.

    a. How many inches of weld can be produced with   9-1a. ________________
       3 lb of mild steel electrodes?

    b. How many inches of weld can be produced with   9-1b. ________________
       22 lb of stainless steel electrodes?

    c. How many inches of weld can be produced with   9-1c. ________________
       4 lb of iron powder electrodes?

    d. How many pounds of mild steel electrodes are   9-1d. ________________
       used to produce 640" of weld?

    e. How many pounds of stainless steel electrodes   9-1e. ________________
       are used to produce 832" of weld?

    f. How many pounds of iron powder electrodes   9-1f. ________________
       are used to produce 160" of weld?

9-2. An automatic welding machine can produce 9000" of   9-2. ________________
      weld in 8 hours. How much time would it take to
      weld 7800"?

9-3. A revolving welding fixture makes 875 revolutions   9-3. ________________
      in 4 minutes. How many revolutions will it make
      in 1 hour and 20 minutes?

9-4. A welder can weld 32 jack stands in 8 hours. How   9-4. ________________
      much time would the welder take to weld 450 jack
      stands?

9-5. A turning fixture has a large gear with 74 teeth.   9-5. ________________
      This gear drives a small gear with 40 teeth. If
      the large gear rotates at a rate of 160 RPM, what
      is the rate of the small gear?

9-6.   A welding machine is mounted on a track. The   9-6. _______________
        drive wheel turns 93 times when it moves 440'.
        How many turns will it make in 3960'?

9-7.   If a welding crew welds 840' of metal in 3 days,   9-7. _______________
        how many feet will the crew weld in 25 days at
        the same rate?

9-8.   Two gears that mesh together in a wire feed   9-8. _______________
        welder have a ratio of 2 to 5. The small gear
        turns at a rate of 50 RPM. What is the rate of
        the larger gear?

9-9.   5 welders can build a storage tank framework in   9-9. _______________
        10 days. At the same rate, how long would 8
        welders take to build the same framework?

9-10.   100 lb of high-alloy steel contains 15-1/2 lb   9-10. _______________
        of nickel. At the same ratio of nickel to steel,
        how many pounds of nickel are contained in 37-1/2
        lb of high-alloy steel?

9-11.   On a blueprint, 3/8" equals 1'. How many inches   9-11. _______________
        represent the length of a pipe 15' long?

9-12.   Mild steel electrodes cost $12 for 5 lb. How   9-12. _______________
        many pounds can be purchased for $93?

9-13.   If 15 sheets of steel can be purchased for $420,   9-13. _______________
        how many sheets can be purchased for $1680?

9-14.   If a welder earns $73.60 in 8 hours, how many   9-14. _______________
        hours will it take him to earn $754.50?

9-15.   A submerged arc welding machine can weld 132'   9-15. _______________
        with 11 lb of flux. How many feet can it weld
        with 48 lb of flux?

Chapter 10: Basic Geometry
Preview and Self-Test

• •

Name ____

Course/Sec. ______________   Date ________________________

Chapter 10 covers fundamental definitions, properties,
and skills of basic geometry. Before you turn to the
chapter, complete the Self-Test to determine which
sections in the chapter you need to study carefully.

Self-Test

MEASUREMENT AND PRECISION

Identify the following lettered points on the ruler in
Figure 10-1 to the nearest fractional part of an inch:

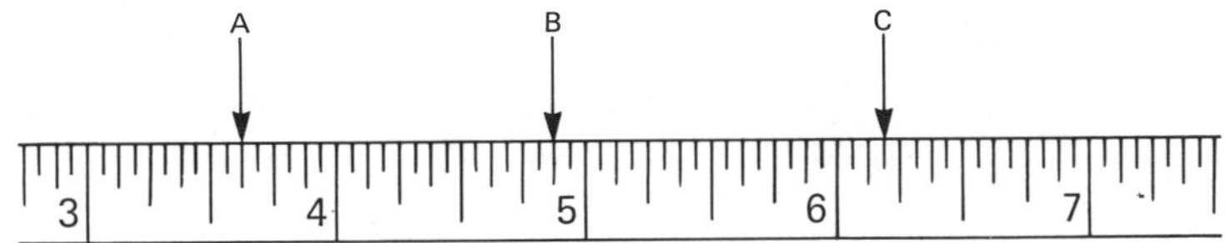

Figure 10-1

10-1.	A = ________________
10-2.	B = ________________
10-3.	C = ________________

Identify the following lettered points on the ruler in
Figure 10-2 to the nearest tenth of a centimeter:

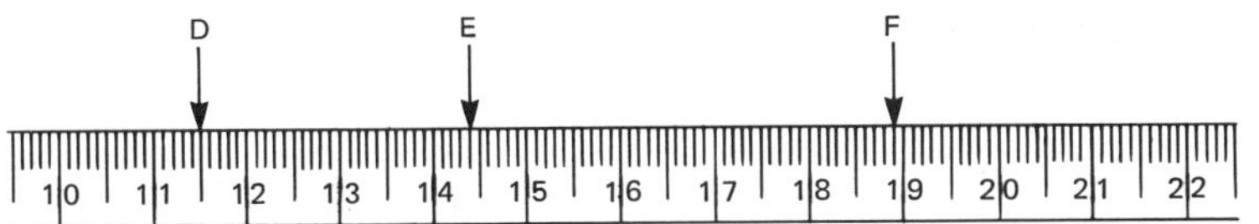

Figure 10-2

10-4.	D = ________________
10-5.	E = ________________
10-6.	F = ________________

Measurement and precision score ________________________

BASIC GEOMETRY CONCEPTS AND FIGURES

10-7. The three types of lines are ________, broken,   10-7. ________________
       and curved.

10-8. Lines that never meet even when extended are   10-8. ________________
       called ________ lines.

261

10-9. A __________ is a plane closed figure bounded by any number of straight lines.

10-9. _______________

Identify the following geometric figures in Figure 10-3.

10-10. G = _______________

10-11. H = _______________

10-12. I = _______________

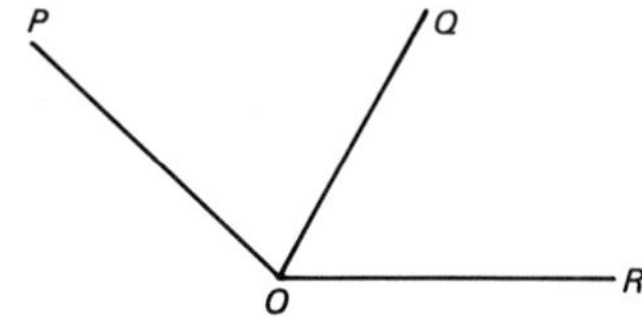

    G   H   I

Figure 10-3

10-13. A polygon with eight sides is called an __________.

10-13. _______________

10-14. A __________ has four sides with opposite sides equal and parallel.

10-14. _______________

10-15. The point where two straight lines intersect is called a __________.

10-15. _______________

                Basic geometry concepts and figures score

____

BASIC PROPERTIES OF ANGLES

10-16. Angles are measured in a __________ direction.

10-16. _______________

10-17. A quarter revolution is equal to a (supplementary or complementary) angle.

10-17. _______________

                Basic properties of angles score

____

MEASUREMENT AND CONSTRUCTION OF ANGLES

Measure the following angles in Figure 10-4 with a protractor.

Figure 10-4

10-18. ∠ ROP

10-18. _______________

10-19. ∠ QOP

10-19. _______________

10-20. Construct a 124° angle with a protractor.

10-20. _______________

                Measurement and construction of angles score

____

After you have checked your answers on the Self-Test, transfer your scores to the Chapter 10 Objectives.

Chapter 10

Basic Geometry

Upon successful completion of this chapter, you will
know the fundamental skills of measurement. You will
understand the basic concepts of geometry, the basic
properties of geometric figures, and the basic properties of angles. You will also be able to measure and
construct angles with a protractor.

Objectives

	Self-Test Scores	If you did poorly on the Self-Test, turn to:
• Measurement and precision	_________	Section 10-3
• Basic geometry concepts and figures	_________	Section 10-4
• Basic properties of angles	_________	Section 10-5
• Measurement and construction of angles	_________	Section 10-6

10–1   Introduction to Geometry

Geometry is one of the oldest branches of mathematics. The word geometry means "earth measure." The ancient Egyptians used geometry to survey their land thousands of years ago, long before the invention of paper, pencils, and calculators.

Two thousand years ago, the Greeks used geometric applications that are still used today. The practical applications of geometry are found in almost every occupation. The applications are obvious in surveying and architecture. Many other occupations also use geometry.

Although geometry can be studied as a mathematical system, we will study geometry mainly for its practical applications.

10–2   Types of Numbers

We count things nearly every day in our work or leisure time. Counting is the process of finding the number of objects. Counting gives us the exact number of objects, whether the number of members in a family or the number of words in this sentence.

In contrast to our counting things, we do not measure things every day. However, many industrial jobs require measurement skills. Measurement is the process of finding the size, height, length, and so on, of something compared to a given measure.

10–3   Measurement and Precision

When we count the number of something, we get an exact number. When we measure the length of something according to a given unit, we get an approximate value. Measurement is approximate because we could always develop a more exact unit of measurement.

Anyone who can count and is willing to use common sense can learn to measure. The unit of measure varies according to the object being measured. The two basic systems of measurement are the English system and the metric system. The basic small unit of the English system is the inch, and the basic small unit of the metric system is the centimeter. The relationship between these systems will be discussed later. The measuring procedure is the same for any system of measurement.

Any person could establish his or her own system of measurement. However, we use standard systems because we want to be able to compare things measured against a standard unit. Thus, a pound of coffee is the same amount whether it is in Brazil, in New York, or in Alexander, North Dakota.

The precision of a measurement depends on the instrument used to measure. The precision of the instrument is determined by the smallest subdivision

of the instrument. By counting the subdivisions of
the ruler, you will know the precision of measurement.
        The longest lines on the 6" ruler in Figure
10-5 indicate 1" divisions. In order of decreasing
length, the other lines indicate 1/2", 1/4", 1/8", and
1/16" divisions.

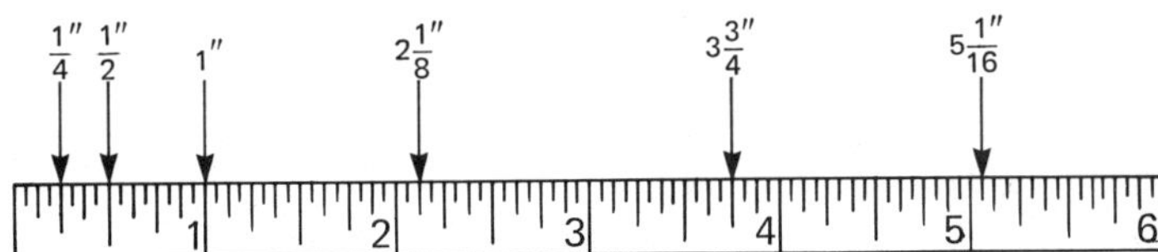

Figure 10-5

        The longest lines on the metric ruler in Figure
10-6 indicate centimeters (abbreviated as cm), and
the shortest lines indicate millimeters (abbreviated
as mm). The medium-length lines indicate 5mm divisions.
The ruler shows that 10mm equal 1cm, and 5mm equal
0.5cm. Converting from one unit of measure to another
is easy in the metric system.

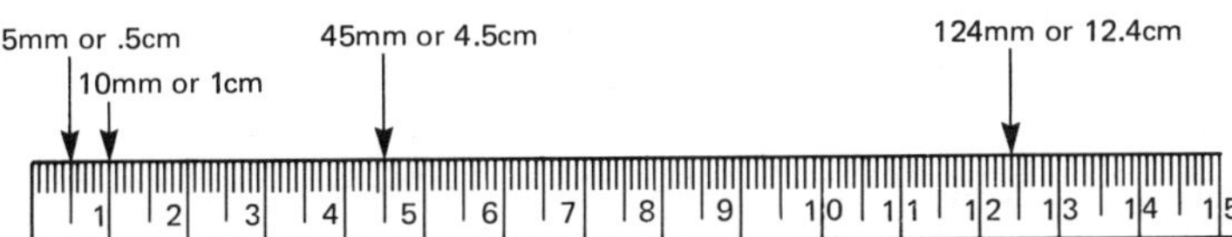

Figure 10-6

        When measuring, decide how precise a measurement
needs to be, then round it to the nearest whole unit of
precision.

Procedure for Measuring

● ●

1. Decide how precise the measurement should be.
2. Measure and then round the measurement to the
    nearest whole unit of precision.

● ●

Example 10-1

PROBLEM

Identify the lettered points on the 6" ruler in Figure
10-7.

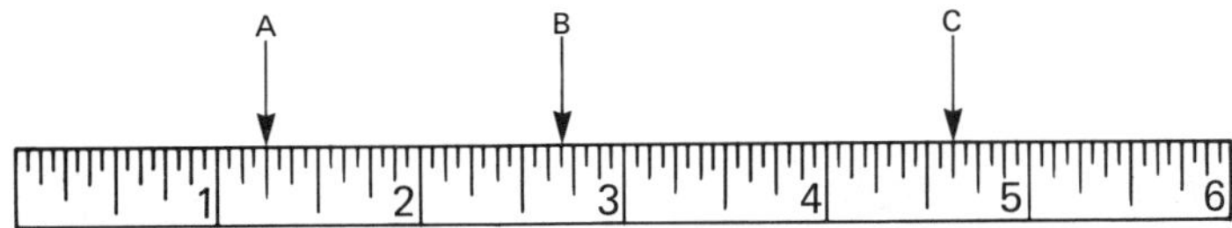

Figure 10-7

SOLUTION

1. Observe that A is at a point that divides the inch
    into 4 equal parts; it is the first such division
    after the 1" mark, or 1-1/4".

2. Observe that B is at a point that divides the inch
    into 16 equal parts; it is the 11th such division
    after the 2" mark, or 2-11/16".

3. Observe that C is at a point that divides the inch
    into 8 equal parts; it is the fifth such division
    after the 4" mark, or 4-5/8".

ANSWER

A = 1-1/4", B = 2-11/16", and C = 4-5/8".

Example 10–2

PROBLEM

Identify the lettered points on the 20cm ruler in
Figure 10-8.

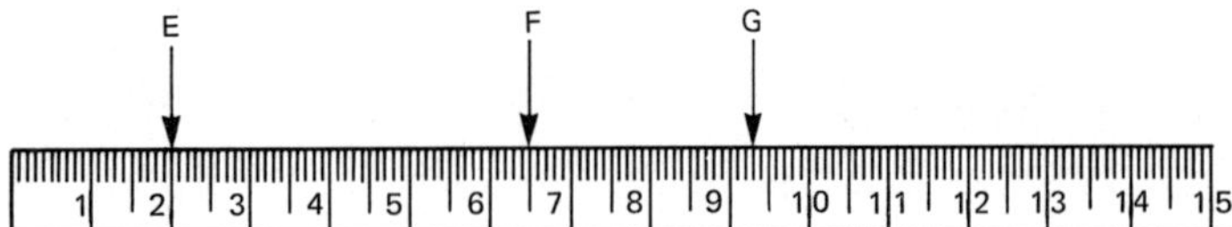

Figure 10-8

SOLUTION

1. Observe that E is at the 2-centimeter (cm) or
    20-millimeter (mm) point.

2. Observe that F is at a point that divides the
    centimeter into 10 equal parts; it is the fifth
    such division after the 6cm mark, or 6.5cm (65mm).

3. Observe that G is at a point that divides the
    centimeter into 10 equal parts; it is the third
    such division after the 9cm mark, or 9.3cm (93mm).

ANSWER

E = 2cm (20mm), F = 6.5cm (65mm), and G = 9.3cm (93mm).

Exercises: Measuring

Identify the following lettered points on the 6" ruler
in Figure 10-9:

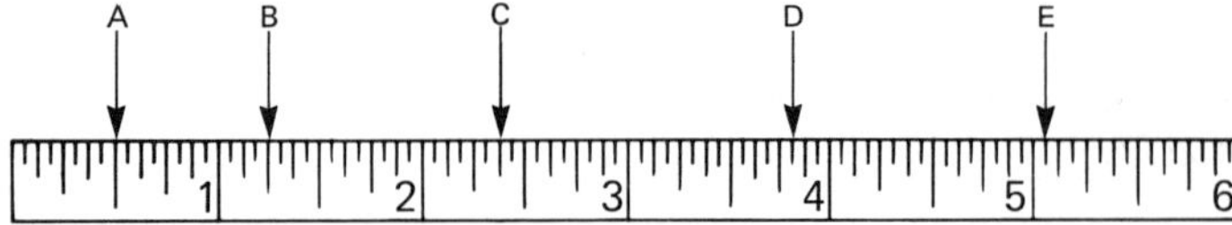

Figure 10-9

10-1. A   10-2. B   10-3. C

10-4. D   10-5. E

Identify the following lettered points on the 35cm ruler in Figure 10-10:

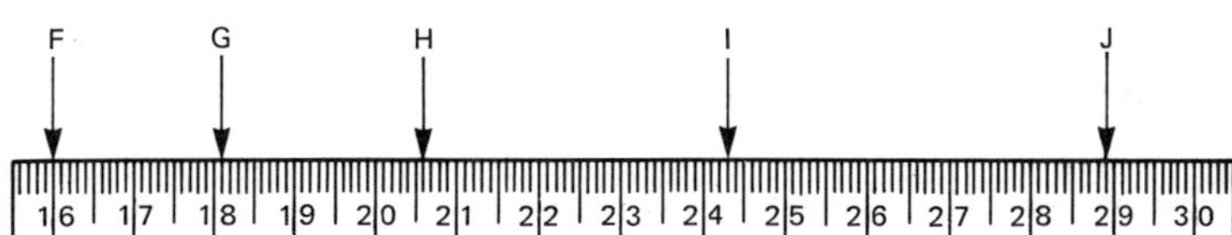

Figure 10-10

10-6. F   10-7. G   10-8. H

10-9. I   10-10. J

10–4 Basic Geometry Concepts and Figures

The three basic concepts of geometry are the point, the line, and the plane. A point is a position or location that has no size or dimension. A capital letter next to a dot generally indicates the position of the point. For example:

• A

A line is a series or set of points that may be extended indefinitely in opposite directions. The extension may result in a straight, broken, or curved line, such as the ones in Figure 10-11. A line has neither width nor depth.

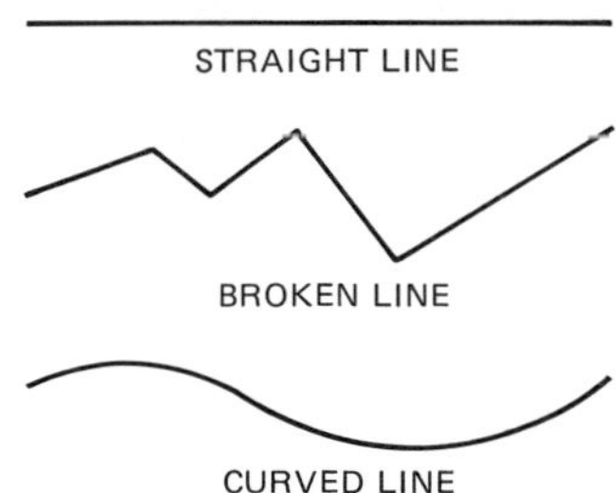

Figure 10-11

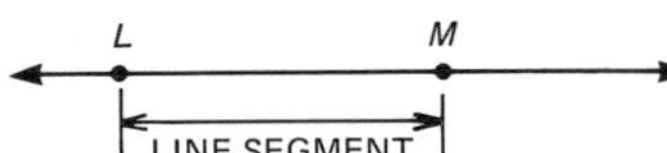

Figure 10-12

A straight line is the shortest distance between two points. When the word line is used, it usually means a straight line. The correct symbol for a line is ◄──►. The arrowheads indicate that the line extends indefinitely in both directions. A line segment has end points and a definite length. Figure 10-12 shows a line segment.

Two lines that never meet even when extended are called parallel lines. The symbol for parallel lines is //. Figure 10-13 shows parallel lines. (Lines that are not parallel and will eventually meet when extended are called oblique lines.)

A plane is a flat surface that has two dimensions: length and width. The wall of a building, the side of

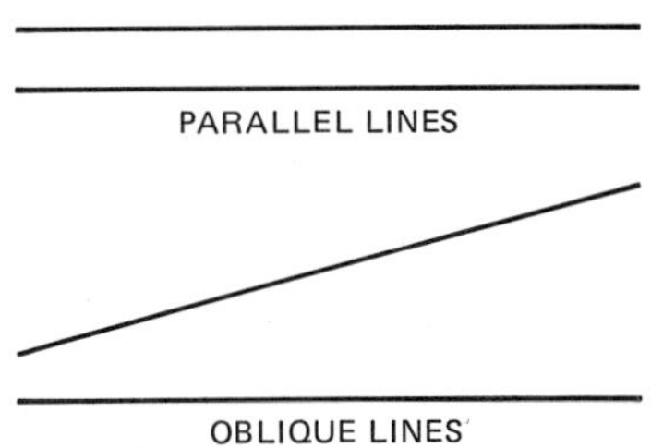

Figure 10-13

a box, and the top of a table are familiar examples of
planes. Figure 10-14 shows a plane.

Plane geometry is concerned with the two dimensions
of length and width. Solid geometry is concerned with
three dimensions: length, width, and height.

The geometric figures we will study in this
chapter are polygons. A polygon is a plane closed
figure bounded by any number of straight lines. (A
plane figure is any figure that can be drawn on a flat
surface.) A regular polygon is a polygon whose sides
are of equal length and whose angles are equal.

The properties of geometric figures can be defined
as follows, in reference to Figure 10-15:

Each line is called a side.
The point where two sides meet is called a vertex.
The distance around a polygon, or the sum of the
  lengths of the sides, is called the perimeter.

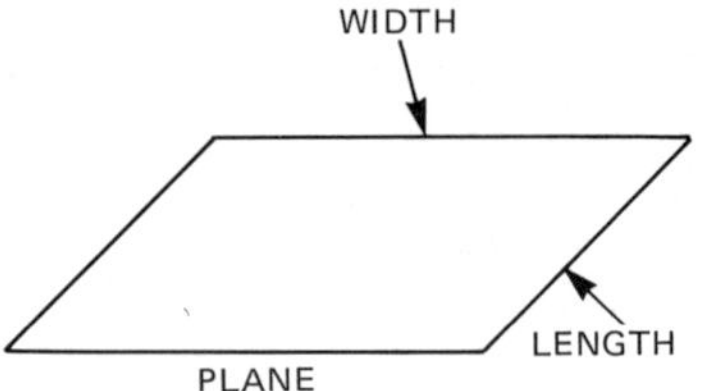

Figure 10-14

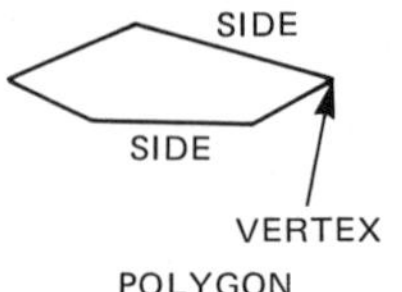

Figure 10-15

A triangle is a polygon with three sides. Three
of the following basic triangles are shown in Figure 10-16:

An equilateral triangle is one in which all sides
  and angles are equal.
An isosceles triangle is one in which two sides
  are equal.
A right triangle is one that contains a right
  (90°) angle (marked by a square, as shown.)

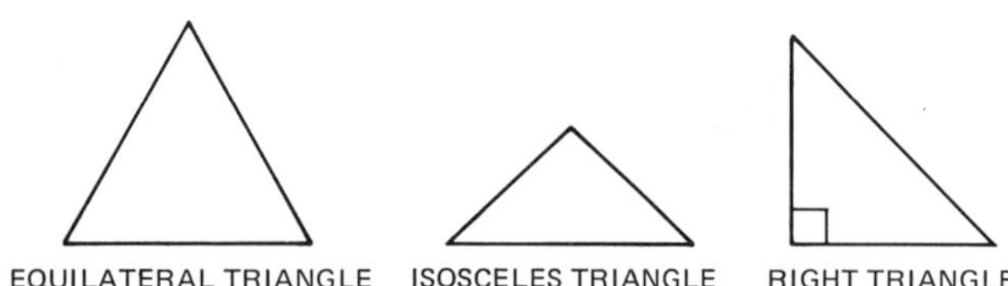

Figure 10-16

A quadrilateral is a polygon with four sides.
Figure 10-17 shows four of the quadrilaterals:

A parallelogram is a four-sided figure whose
  opposite sides are equal and parallel.
A rectangle is a four-sided figure whose opposite
  sides are equal and parallel and whose angles
  are all right angles.
A rhombus is a four-sided figure whose sides are
  all equal and parallel in pairs.
A square is a four-sided figure whose sides are
  all equal and whose angles are all right angles.

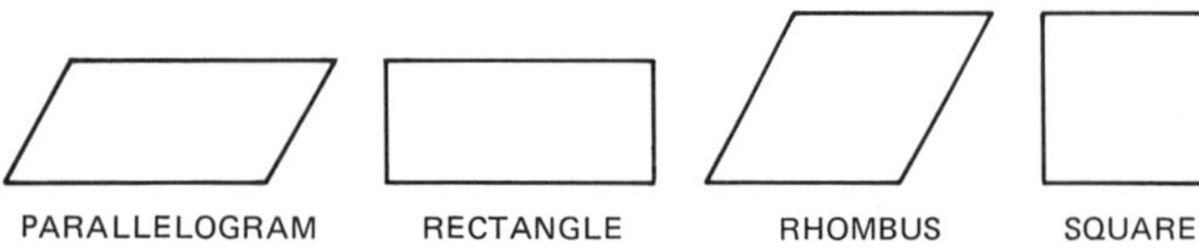

Figure 10-17

The following geometric figures, shown in Figure
10-18, are polygons that are not quadrilaterals:

A <u>pentagon</u> is a five-sided figure.
A <u>hexagon</u> is a six-sided figure.
An <u>octagon</u> is an eight-sided figure (a "stop"
    sign is an octagon).

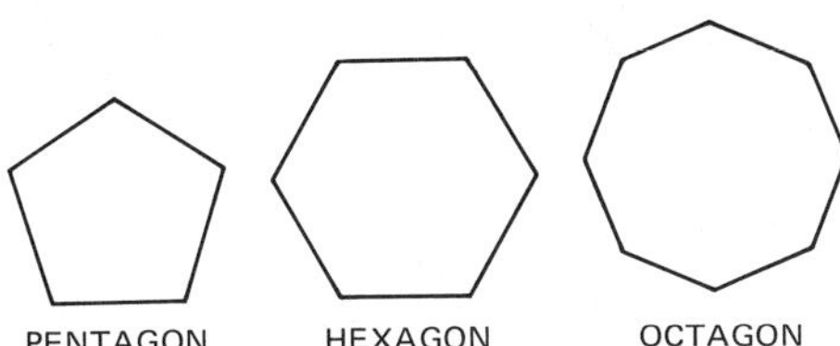

Figure 10-18

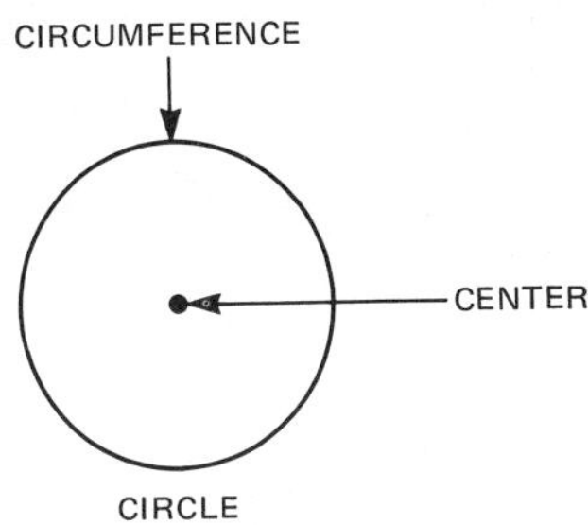

Figure 10-19

A <u>circle</u> is a closed curve in which every point on the <u>curve</u> is equally distant from a fixed point called the center. Figure 10-19 shows a circle with a point at the center. The properties of circles can be defined as follows, in reference to Figure 10-20:

The distance around a circle is called the
    <u>circumference</u>.
A <u>chord</u> is a straight line that connects any two
    points on a circle.
A <u>diameter</u> is a chord that passes through the
    center of a circle.
A <u>radius</u> is half a diameter.

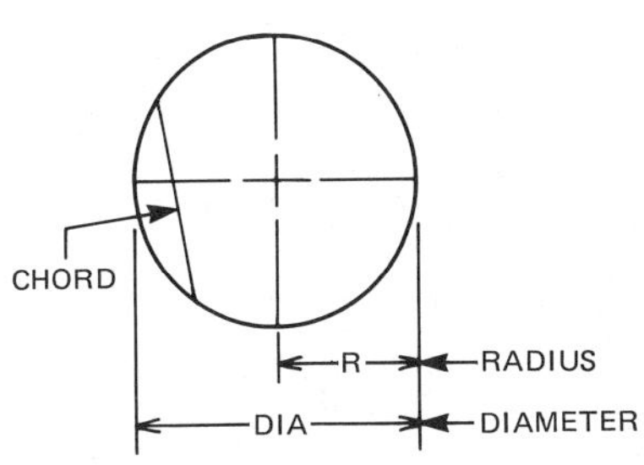

Figure 10-20

Exercises: Defining Basic Geometry Concepts and Figures

Complete the following statements:

10-11. A _________ is a position or location that has
        no size or dimension.

10-12. The three types of lines are straight, _______,
        and curved lines.

10-13. A point is usually defined by a(n) _________
        letter.

10-14. A line _________ has end points and a definite
        length.

10-15. Two lines that never meet are defined as
        _________ _________.

10-16. A <u>plane</u> is a flat surface that has two
        dimensions: _________ and _________.

10-17. _________ geometry is concerned with three
        dimensions: length, width, and height.

10-18. A _________ is a plane closed figure bounded
        by any number of straight lines.

10-19. The point where two sides of a polygon meet
        is called a _________.

10-20. A __________ polygon has sides of equal length and has equal angles.

10-21. A __________ is a polygon with three sides.

Refer to Figure 10-21 and answer the following questions:

10-22. Define the figure shown in part A.

10-23. Define the figure shown in part B.

10-24. Define the figure shown in part C.

10-25. Define the figure shown in part D.

10-26. Define the figure shown in part E.

10-27. Define the figure shown in part F.

10-28. Define the figure shown in part G.

10-29. Define the figure shown in part H.

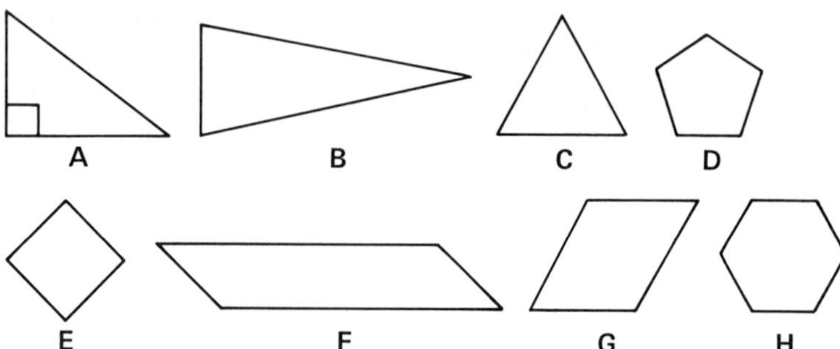

Figure 10-21

Name the following lettered parts of the circle shown in Figure 10-22:

10-30. Name part A.

10-31. Name part B.

10-32. Name part C.

10-33. Name part D.

10-34. Name part E.

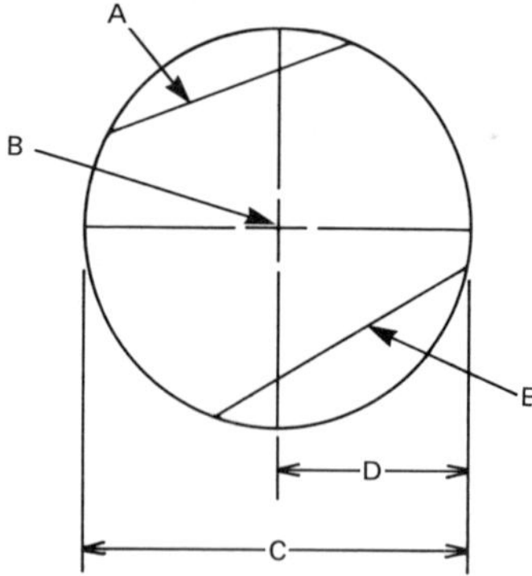

Figure 10-22

10–5 Basic Properties of Angles

Angles are indicated by the symbol $\angle$ and are named by three capital letters. The angles in Figure 10-23 are $\angle COB$, $\angle BOD$, $\angle DOA$, and $\angle COA$. The point of intersection where two straight lines cross is called the <u>vertex</u>. Note that the vertex is always the middle letter.

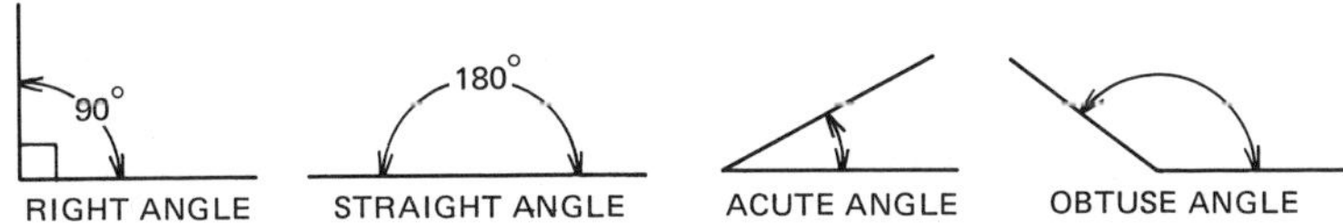

Figure 10-23

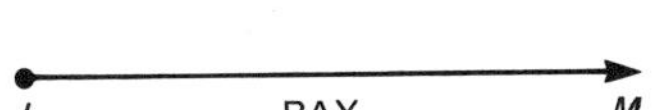

Figure 10-24

A _ray_ is part of a line with one end point. The
symbol for the ray in Figure 10-24 is $\overrightarrow{LM}$. An _angle_
may be formed by rotating (turning) a ray from the end
point, as in Figure 10-25. The arrow indicates the
direction in which the ray has been rotated. The
direction is usually counterclockwise.

A complete rotation, or revolution, from a vertex
is divided into 360 parts. Each part is called a
degree. The small zero above and to the right of the
number in Figure 10-26 is the symbol for degree. Each
degree is divided into 60 equal parts called _minutes_.
Minutes are indicated by a small accent mark above and
to the right of a number: 1° = 60'. Each minute is
divided into 60 _seconds_. Seconds are indicated by two
small accent marks above and to the right of a number:
1' = 60".

Figure 10-27 shows how an angle is formed at
varying degrees of measure:

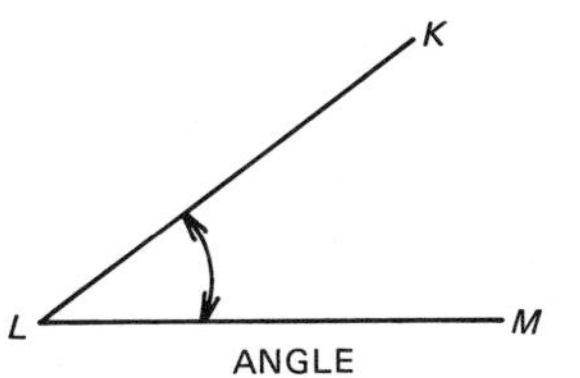

Figure 10-25

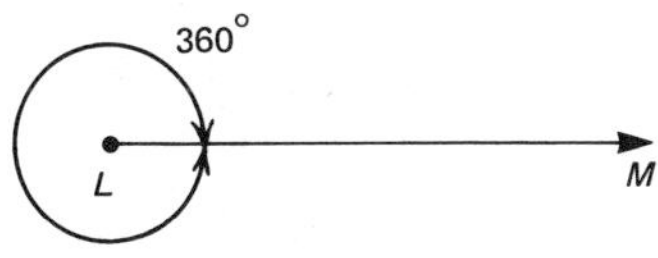

Figure 10-26

> A one-half revolution is 180° and is called a
> _straight angle_.
> A one-fourth revolution is 90° and is called a
> _right angle_; the symbol for right angle is ▫.

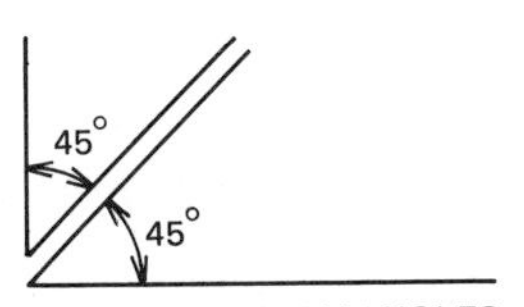

RIGHT ANGLE   STRAIGHT ANGLE   ACUTE ANGLE   OBTUSE ANGLE

Figure 10-27

Lines that form right angles are said to be _perpen-
dicular_ to each other. The symbol for perpendicular
is ⊥.

The names of angles are sometimes related to the
number of degrees in the angle:

> An _acute angle_ contains less than 90°.
> An _obtuse angle_ contains more than 90°.

Other names for angles reflect their relationship to
each other, as shown in Figure 10-28:

> Two angles whose sum equals 90° are called
> _complementary angles_.
> Two angles whose sum equals 180° are called
> _supplementary angles_.

COMPLEMENTARY ANGLES

SUPPLEMENTARY ANGLES

Figure 10-28

**Exercises: Defining Basic Properties
of Angles**

Complete the following statements:

10-35.   The abbreviated symbol for the angle in Figure
10-29 is $\angle PQR$ or ________ .

10-36.   One ray of the angle in Figure 10-29 is $\overrightarrow{QR}$.
What is the other ray?

10-37.   Point Q in Figure 10-29 is called the ________
of the angle.

10-38.   An arrow indicates how an angle is rotated.
Generally, angles are rotated in a ________
direction.

10-39.   A straight angle contains how many degrees?

10-40.   How many minutes does 60" represent?

10-41.   Lines that meet at right angles are said to be
________ to each other.

10-42.   Two angles are ________ when the sum of the two
angles equals 180°.

10-43.   A revolution is divided into 360 equal parts.
These parts are called ________ .

10-44.   An angle of more than 90° is called a(n)
________ angle.

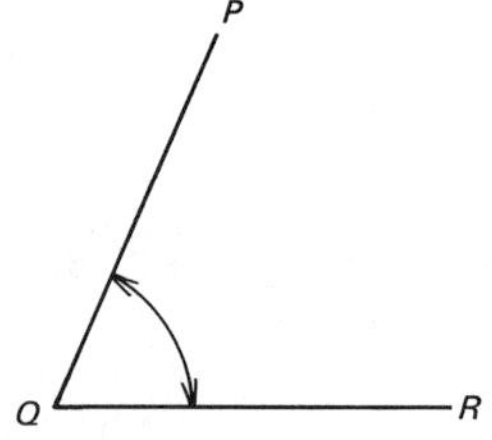

Figure 10-29

10–6   Measurement and Construction of Angles

The size of an angle is measured with a <u>protractor</u>
like the one shown in Figure 10-30. Angles are also
constructed to a desired size with a protractor. Note
that the numbers on the inside run from 0° to 180° in
a counterclockwise direction, and those on the outside
run in a clockwise direction. Each small division
equals 1 degree. The accuracy of measurement or construction depends on the protractor used and on the
person measuring or constructing. Try to be as
accurate as possible according to the requirements of
the problem.

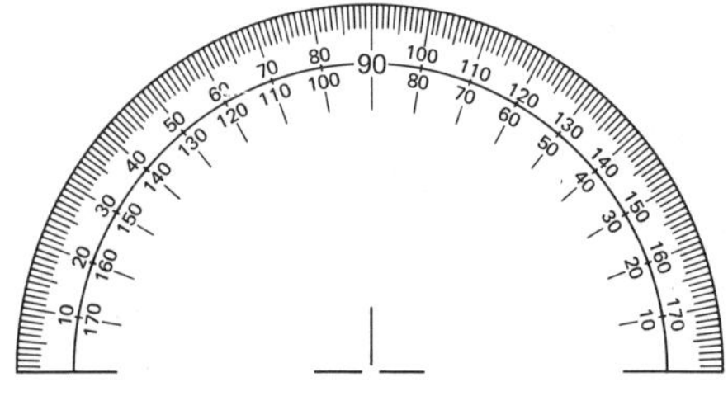

Figure 10-30

Procedure for Measuring Angles

• •

1. Place the base of the protractor along the bottom
   ray of the angle. The center mark at the base of
   the protractor should meet the vertex of the angle.
   If necessary, extend the upper ray of the angle to
   the numbers on the protractor.
2. If the rays of the angle open to the right of the
   vertex, read the size of the angle from the
   numbers that run counterclockwise on the inside
   edge of the protractor.
3. If the rays of the angle open to the left of the
   vertex, read the size of the angle from the
   numbers that run clockwise on the outside edge of
   the protractor.

• •

Example 10–3

PROBLEM

Measure the acute angle *JKL* in Figure 10-31.

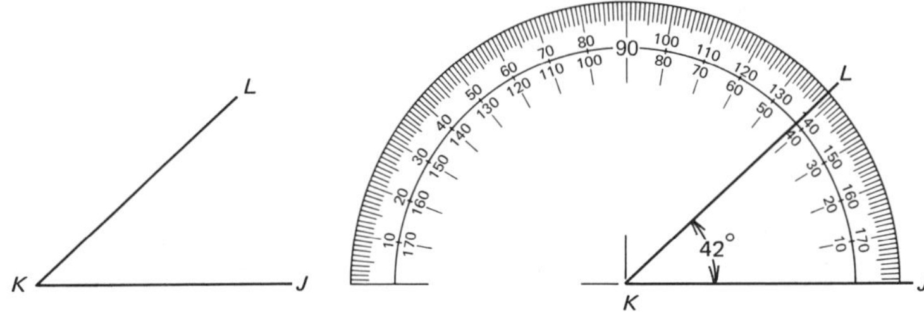

Figure 10–31

SOLUTION

Place the protractor in the position shown in the
figure and read the size of the angle from the numbers
that run counterclockwise on the protractor.

ANSWER

The size of angle *JKL* is 42°.

Example 10–4

PROBLEM

Measure the obtuse angle *PQR* in Figure 10-32.

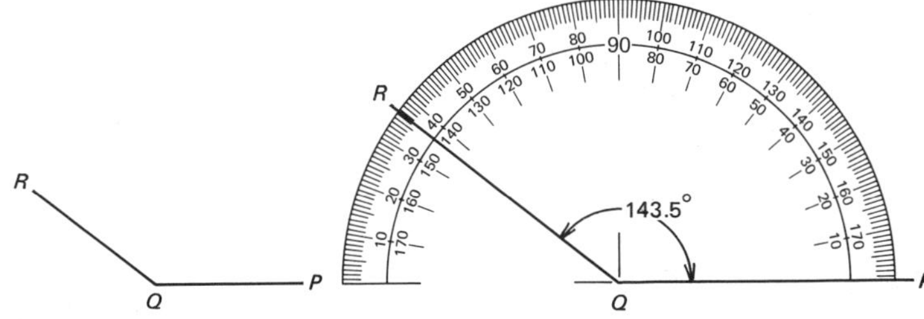

Figure 10–32

SOLUTION

Place the protractor in the position shown in the
figure and read the size of the angle from the numbers
that run counterclockwise on the protractor.

ANSWER

The size of angle *PQR* is 143.5°.

Exercises: Measuring Angles

Refer to Figure 10-33 and answer the following
questions:

10-45.   Measure the angle in part A.

10-46.   Measure the angle in part B.

10-47.   Measure the angle in part C.

10-48.   Measure the angle in part D.

10-49.   Measure the angle in part E.

10-50.   Measure the angle in part F.

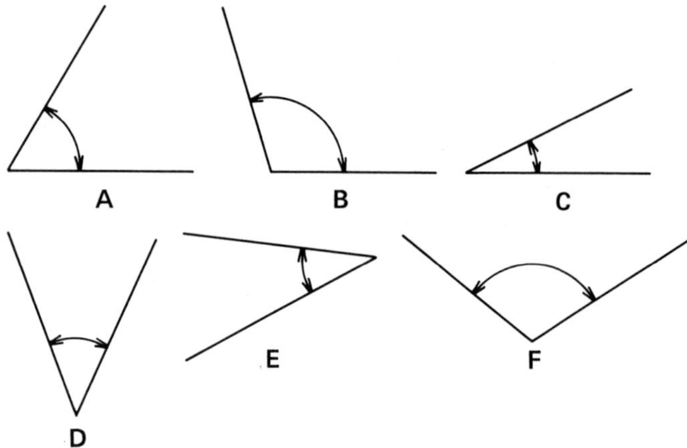

Figure 10-33

Procedure for Constructing Angles

1.   Draw a ray and place the base of the protractor
     along the ray. The center mark at the base of the
     protractor should meet the end point of the ray
     that will be the vertex of the angle.
2.   Find the desired size of the angle on the
     protractor and make a mark at that point just
     beyond the edge of the protractor.
3.   Remove the protractor and draw a ray connecting
     the vertex to the mark you have made.

Example 10–5

PROBLEM

Draw a 67° angle *ABC*.

SOLUTION

1. Draw ray $\overrightarrow{BA}$.

2. Place the protractor on the ray in the position
    shown in Figure 10-34.

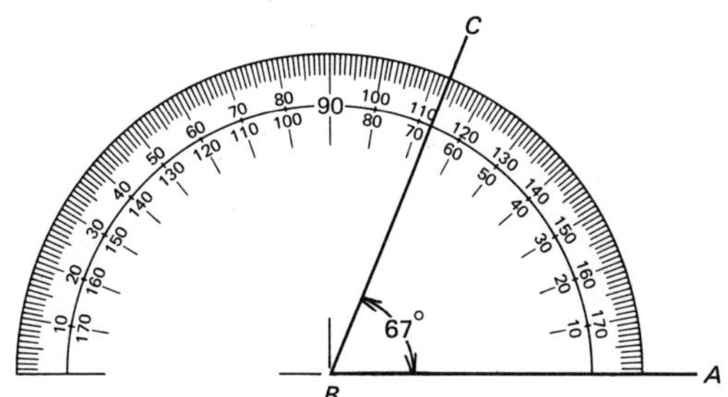

Figure 10-34

3. Find 67° in the counterclockwise numbers and make
    a mark at that point.

4. Remove the protractor and draw $\overrightarrow{BC}$ from the end
    point of $\overrightarrow{BA}$ to the mark you have made. You now
    have ∠*ABC*, which measures 67°.

Example 10–6

PROBLEM

Draw a 139° angle *NRA*.

SOLUTION

1. Draw ray $\overrightarrow{RA}$.

2. Place the protractor on the ray in the position
    shown in Figure 10-35.

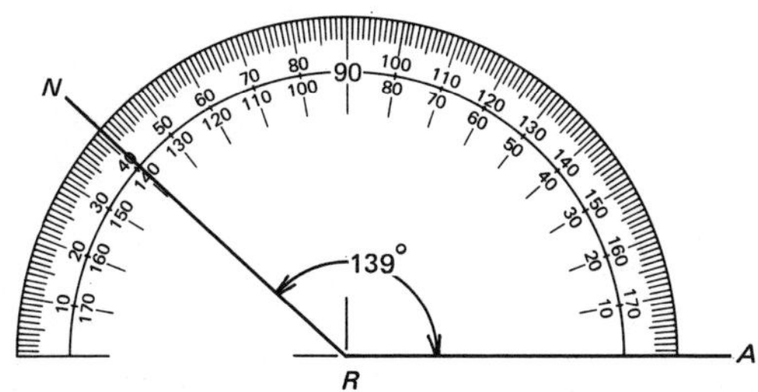

Figure 10-35

3. Find 139° in the counterclockwise numbers and make
    a mark at that point.

4. Remove the protractor and draw $\overrightarrow{RN}$ from the end
    point of $\overrightarrow{RA}$ to the mark you have made. You now
    have $\angle NRA$, which measures 139°.

Exercises: Constructing Angles

Draw the following angles:

10-51.   Draw a 38° angle *RST*.

10-52.   Draw an 84° angle *EFG*.

10-53.   Draw a 125° angle *HIJ*.

10-54.   Draw a 171° angle *KLM*.

Chapter 10

Basic Geometry

Solving Applied Welding Problems

Now that you have mastered the fundamental operations
of basic geometry, you are ready to apply these math
skills to typical problems that welders encounter on
the job. After studying the examples that follow,
complete the applied welding problems.

Example 10–7

PROBLEM

A 2' x 3' opening is to be cut from the 4' x 8' sheet
of steel shown in Figure 10-36. Which dimension is the
vertical dimension?

SOLUTION

The vertical dimension is the straight up-and-down
dimension that corresponds to the width of a plane.

ANSWER

A is the vertical dimension.

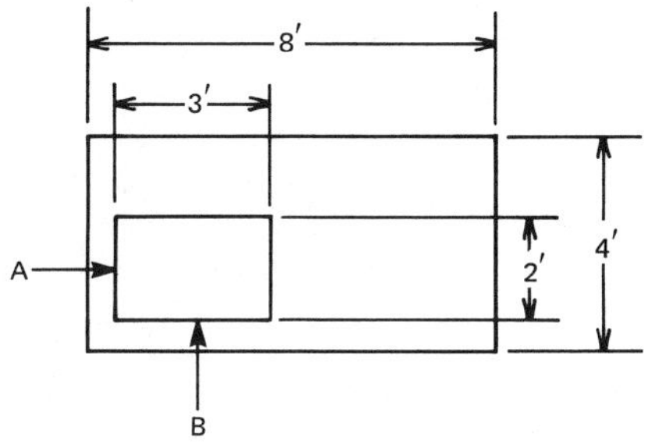

Figure 10-36

Example 10–8

PROBLEM

Which dimension of the steel sheet in Figure 10-36 is
the horizontal dimension?

SOLUTION

The horizontal dimension is the line parallel to the
base line that corresponds to the length of a plane.

ANSWER

B is the horizontal dimension.

Example 10–9

PROBLEM

Metal stops A, B, C, and D are welded to the fixture in Figure 10-37. Which stops are oblique? Which stops are parallel?

SOLUTION

Recall that lines that never meet even when extended are called parallel lines, and lines that are not parallel and will eventually meet when extended are called oblique lines.

ANSWER

Stops C and D are oblique, and stops A and B are parallel.

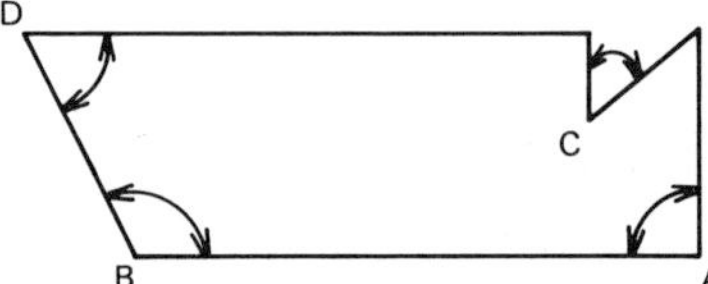

Figure 10-37

Applied Welding Problems

10-55.   The template for a pump box in Figure 10-38 has various angles. Four of them are labeled.
  a. Which one is a right angle?
  b. Which one is an obtuse angle?
  c. Which one is an acute angle?

10-56.   Three different types of triangular gussets, shown in Figure 10-39, are needed to stabilize the pump box.
  a. Which gusset is a scalene triangle (all three sides are unequal)?
  b. Which gusset is an isosceles triangle?
  c. Which gusset is an equilateral triangle?

10-57.   The oil reservoir for a hydraulic log splitter in Figure 10-40 is built to fit into a compact area.
  a. Which side of the reservoir is a rectangle?
  b. Which side of the reservoir is a right triangle?

10-58.   Figure 10-41 shows a snow bucket. Which side (or sides) of the snow bucket is a parallelogram?

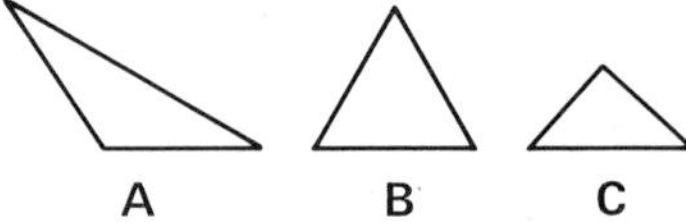

Figure 10-38

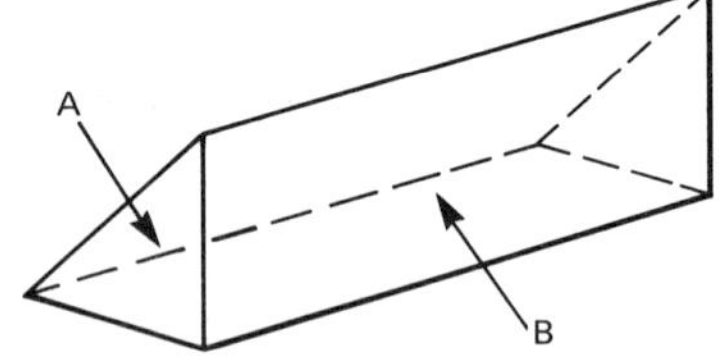

Figure 10-39

Figure 10-40

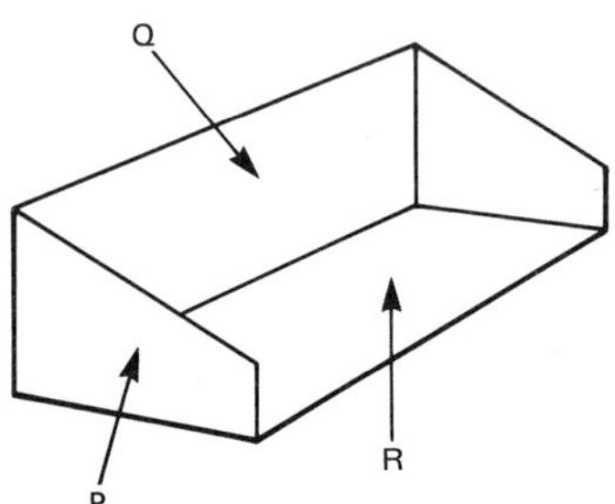

Figure 10-41

10-59. Freeway light poles are being constructed in
        different shapes, like those in Figure 10-42.
        a. Which pole is pentagon shaped?
        b. Which pole is octagon shaped?
        c. Which pole is hexagon shaped?
        d. Which pole is square shaped?

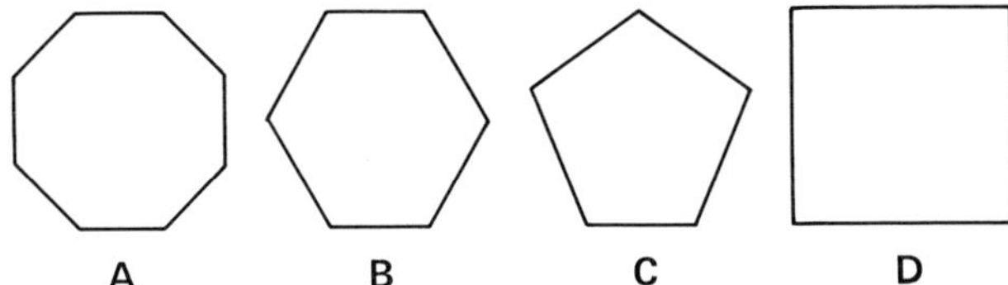

Figure 10-42

10-60. What is the size of the angle formed by the
        electrode and the plate in Figure 10-43?

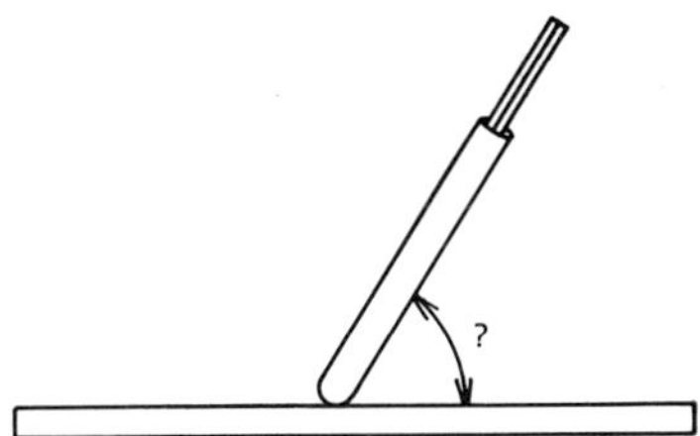

Figure 10-43

Chapter 10: Basic Geometry
Applied Welding Problems
Survey Test

Name ____

Course/Sec. _____________   Date _____________________

Refer to the design for installing steps in Figure
10-44 and complete the following statements:

10-1.   The first floor should be _________ to the
        second floor.

10-1. __________________

10-2.   The angle of the steps at A measures ________
        degrees.

10-2. __________________

10-3.   The angle of the steps at B measures ________
        degrees.

10-3. __________________

10-4.   A is an _________ angle.

10-4. __________________

10-5.   B is an _________ angle.

10-5. __________________

10-6.   Angles A and B are called _________ angles.

10-6. __________________

10-7.   The frame for the steps is a _________ type of
        geometric figure.

10-7. __________________

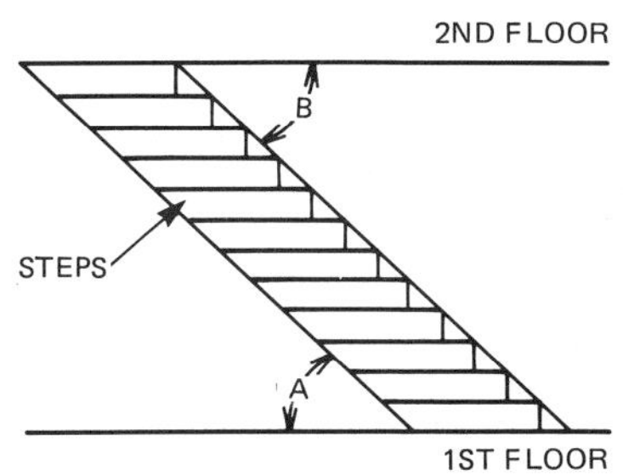

Figure 10-44

Refer to the template in Figure 10-45 and identify the
shape of the lettered parts.

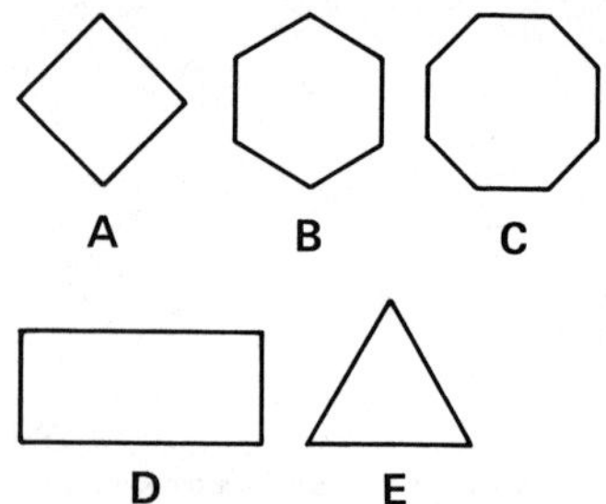

Figure 10-45

10-8.   A = ___________

10-9.   B = ___________

10-10. C = ___________

10-11. D = ___________

10-12. E = ___________

Chapter 11: Perimeter and Area
Preview and Self-Test

Name ____

Course/Sec. _____________ Date _____________________

Chapter 11 covers procedures for finding the perimeters,
circumferences, and areas of basic geometric figures as
well as practical applications of these concepts.
Before you turn to the chapter, complete the Self-Test
to determine which sections in the chapter you need to
study carefully.

Self-Test

FINDING THE PERIMETERS OF POLYGONS

Refer to Figure 11-1 and find the perimeters of
the polygons.

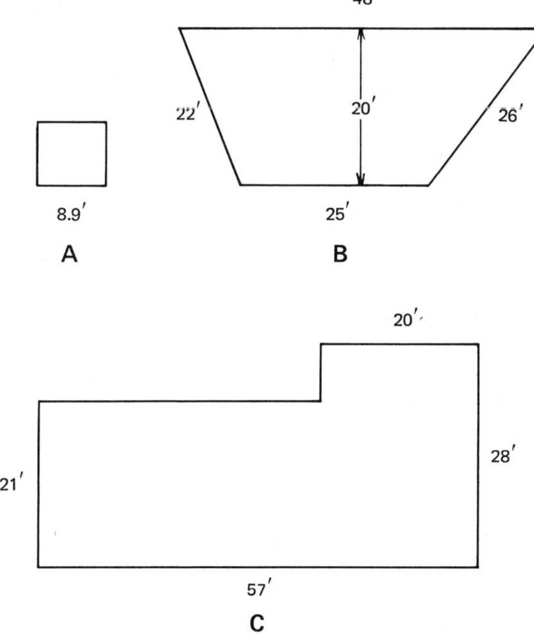

Figure 11-1

11-1. <u>A</u> = _____________________

11-2. <u>B</u> = _____________________

11-3. <u>C</u> = _____________________

Finding the perimeters of polygons score _____________________

283

FINDING THE CIRCUMFERENCES OF CIRCLES

Refer to Figure 11-2 and find the circumferences of
the circles.

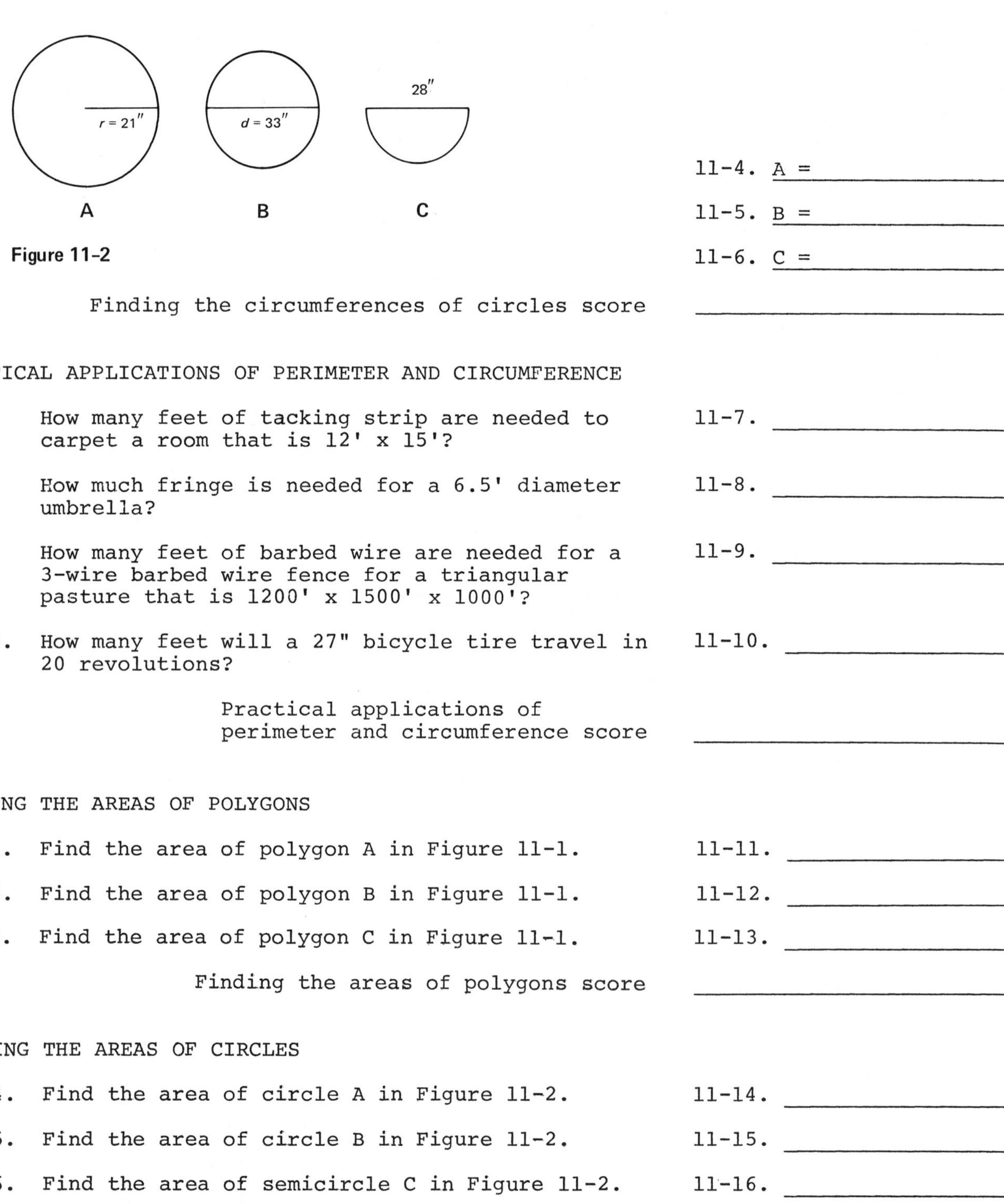

Figure 11-2

11-4. A = _______________

11-5. B = _______________

11-6. C = _______________

            Finding the circumferences of circles score _______________

PRACTICAL APPLICATIONS OF PERIMETER AND CIRCUMFERENCE

11-7.   How many feet of tacking strip are needed to
        carpet a room that is 12' x 15'?

11-7. _______________

11-8.   How much fringe is needed for a 6.5' diameter
        umbrella?

11-8. _______________

11-9.   How many feet of barbed wire are needed for a
        3-wire barbed wire fence for a triangular
        pasture that is 1200' x 1500' x 1000'?

11-9. _______________

11-10. How many feet will a 27" bicycle tire travel in
        20 revolutions?

11-10. _______________

          Practical applications of
        perimeter and circumference score _______________

FINDING THE AREAS OF POLYGONS

11-11. Find the area of polygon A in Figure 11-1.

11-11. _______________

11-12. Find the area of polygon B in Figure 11-1.

11-12. _______________

11-13. Find the area of polygon C in Figure 11-1.

11-13. _______________

         Finding the areas of polygons score _______________

FINDING THE AREAS OF CIRCLES

11-14. Find the area of circle A in Figure 11-2.

11-14. _______________

11-15. Find the area of circle B in Figure 11-2.

11-15. _______________

11-16. Find the area of semicircle C in Figure 11-2.

11-16. _______________

         Finding the areas of circles score _______________

PRACTICAL APPLICATIONS OF AREA

11-17.   Find the cost of carpeting a 27' x 15' living   11-17. ________________
         room if the carpet costs $17.98 per square yard.

11-18.   Find the number of acres to the nearest   11-18. ________________
         hundredth acre in the lot shown in Figure 11-3.
         (43,560 sq ft = 1 acre.)

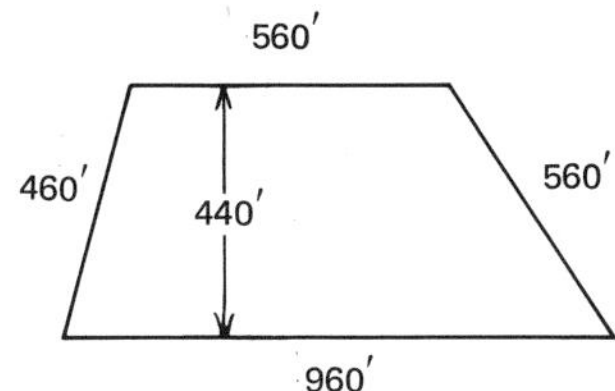

Figure 11-3

11-19.   How many yards of carpet are needed for the   11-19. ________________
         area around the pool in Figure 11-4 (to the
         nearest yard)?

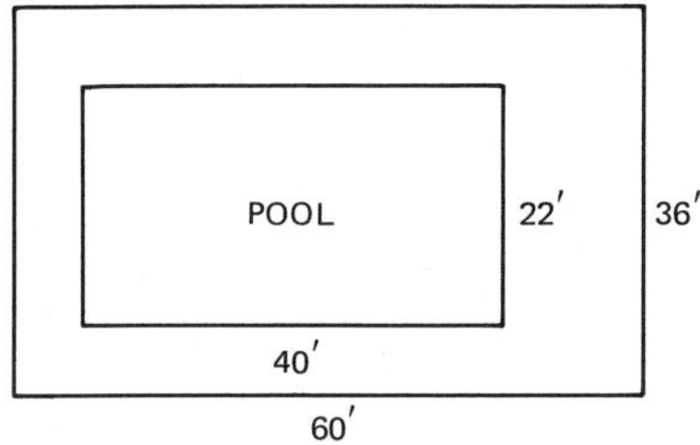

Figure 11-4

11-20.   Find the cost of blacktopping the semicircular   11-20. ________________
         driveway in Figure 11-5 if blacktopping costs
         $5.89 per square yard.

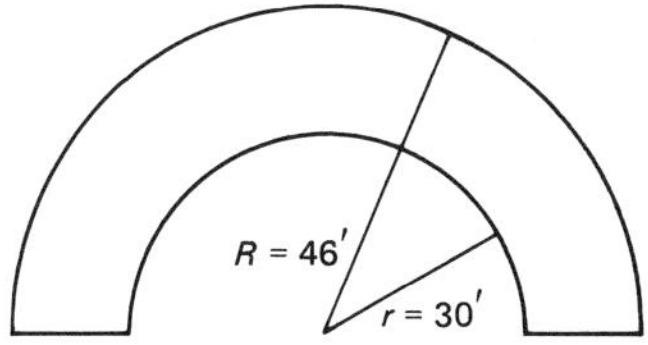

Figure 11-5

                 Practical applications of area score   ________________________

After you have checked your answers on the Self-Test,
transfer your scores to the Chapter 11 Objectives

• •

Perimeter and Area

Upon successful completion of this chapter, you will
be able to find the perimeters of geometric figures, to
find the circumferences of circles, and to make
practical applications of perimeter and circumference.
You will also be able to find the areas of geometric
figures and to make practical applications of area.

Objectives

	Self-Test Scores	If you did poorly on the Self-Test, turn to:
• Finding the perimeters of polygons	__________	Section 11-2
• Finding the circumferences of circles	__________	Section 11-3
• Practical applications of perimeter and circumference	__________	Section 11-4
• Finding the areas of polygons	__________	Section 11-5
• Finding the areas of circles	__________	Section 11-6
• Practical applications of area	__________	Section 11-7

11–1   Introduction to Perimeter and Area

Many occupations require the ability to calculate
perimeter and area. Nearly everyone at some time must
find the distance around or the square units in a plot
of land, a sheet of paneling, a bedspread, or an
irregular-shaped room.
       Learning the basic formulas and how to apply them
will make finding perimeters and areas an easy task.
These skills are not difficult, but first you must
understand the definitions and practice the applications.

11–2   Perimeter

The <u>perimeter</u> of a geometric figure is the distance
around the figure. Perimeter is measured in linear
units, such as inches, feet, yards, centimeters, or
meters. The distance around, or the perimeter of, a
circle is called the <u>circumference</u> of the circle.
Fencing a plot of land, finding the distance traveled
in a revolution of a wheel, sewing a border around a
blanket, and making a frame for a picture are all tasks
that involve the idea of perimeter. The procedure for
finding the perimeter of a polygon is very simple:
determine the lengths of all the sides and then add
them.

Procedures for Finding the Perimeters of Polygons

●●

To find the perimeter of a polygon, add the lengths of
the sides: P = sum of the sides.

●●

Example 11–1

PROBLEM

Find the perimeter of the rectangle in Figure 11-6.

SOLUTION

Apply the formula P = sum of the sides:

   $P = 32' + 17' + 32' + 17' = 98'$

<u>Note:</u> Since opposite sides of a rectangle are equal,
we could have used the formula $P = 2l + 2w$.

ANSWER

The perimeter of the rectangle is 98'.

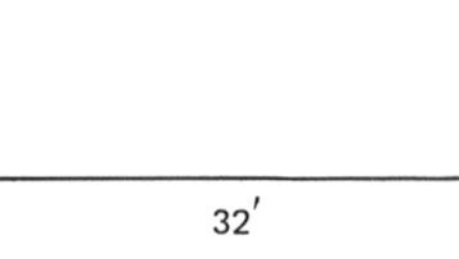

Figure 11-6

Example 11–2

PROBLEM

Find the perimeter of the square in Figure 11-7.

SOLUTION

Apply the formula P = sum of the sides:

$$P = 14m + 14m + 14m + 14m = 56m$$

<u>Note</u>: Since all sides of a square are equal, we could have used the formula $P = 4s$.

ANSWER

The perimeter of the square is 56m.

Figure 11-7

Example 11–3

PROBLEM

Find the perimeter of the polygon in Figure 11-8.

SOLUTION

Apply the formula P = sum of the sides:

$$P = 5cm + 2cm + 3.2cm + 1cm + 2.7cm + 2.2cm$$
$$= 16.1cm$$

ANSWER

The perimeter of the polygon is 16.1cm.

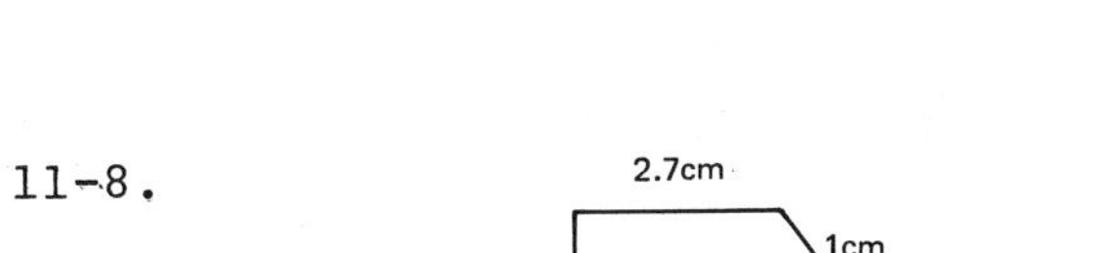

Figure 11-8

Exercises: Finding the Perimeters of Polygons

11-1. Find the perimeter of triangle A in Figure 11-9.

11-2. Find the perimeter of rectangle B in Figure 11-9.

11-3. Find the perimeter of square C in Figure 11-9.

11-4. Find the perimeter of triangle D in Figure 11-9.

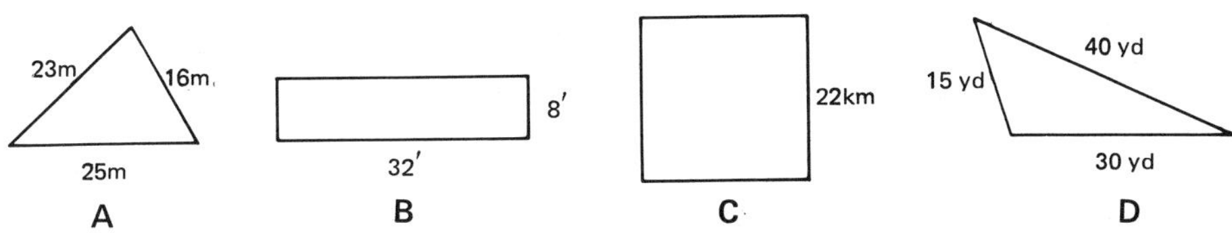

Figure 11-9

11-5.   Find the perimeter of the polygon in Figure 11-10.

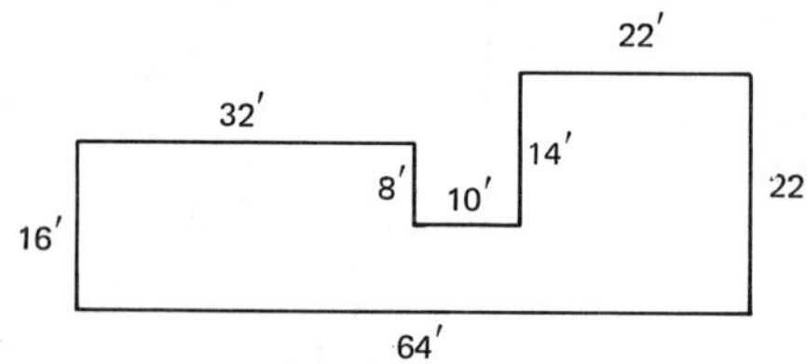

Figure 11-10

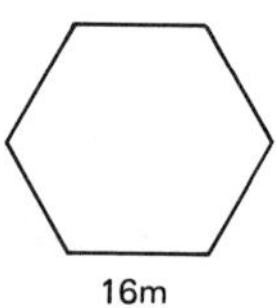

Figure 11-11

11-6.   Find the perimeter of a square whose side is
        14.5m.

11-7.   Find the perimeter of the hexagon in Figure 11-11.

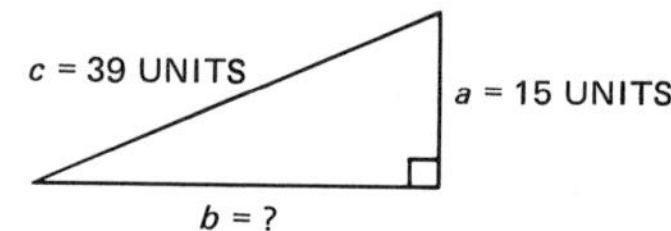

11-8.   Find the perimeter of the right triangle in
        Figure 11-12, given $b = \sqrt{c^2 - a^2}$.

Figure 11-12

11-3   Circumference

The distance around, or the perimeter of, a circle is
called the circumference of the circle. The circumference is found by multiplying the diameter of the
circle by π. π is a Greek letter pronounced "pie."
π is the ratio of the circumference of a circle to its
diameter. The value of π is approximately equal to
3.14159. Generally, 3.14159 is rounded to 3.142. The
formula for finding the circumference of a circle is
$C = d\pi$. If the radius is given, the formula is $C = 2r\pi$,
because 2 times the radius equals the diameter.

Procedure for Finding the Circumferences of Circles

• •

To find the circumference of a circle, multiply the
diameter by π (3.142): $C = d\pi$, or multiply the radius
by 2 and by π: $C = 2r\pi$.

• •

Example 11-4

PROBLEM

Find the circumference of the circle in Figure 11-13.

SOLUTION

1. Recall that the circumference equals the diameter
    times π (3.142).

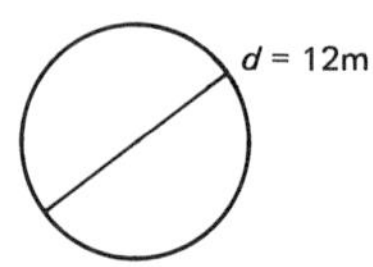

Figure 11-13

2. Apply the formula $C = d\pi$ and substitute the known
    values:

$$C = d\pi$$
$$= (12m)(3.142)$$
$$= 37.704m$$

ANSWER

The circumference of the circle is 37.704m.

Example 11–5

PROBLEM

Find the perimeter of a circle whose radius is 4.6".

SOLUTION

1. Draw a sketch of the circle, as in Figure 11-14.

2. Recall that the perimeter is the circumference, and
    it equals 2 times the radius times π (3.142).

3. Apply the formula $C = 2r\pi$ and substitute the known
    values:

$$C = 2r\pi$$
$$= 2(4.6")(3.142)$$
$$= 28.906"$$

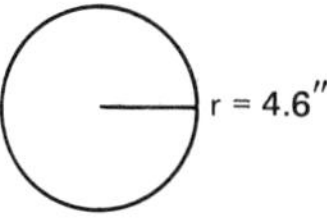

Figure 11-14

ANSWER

The perimeter of the circle is 28.906".

Exercises: Finding the Circumferences of Circles

11-9.   Find the circumference of circle A in Figure
        11-15.

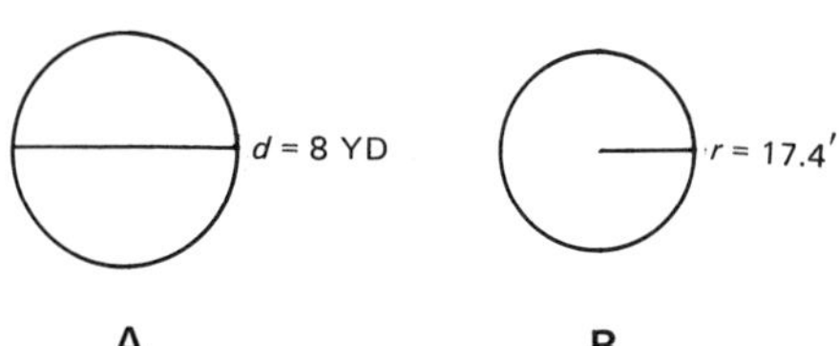

Figure 11-15

11-10. Find the circumference of circle B in Figure
        11-15.

11-11. Find the circumference of a circle whose
        diameter is 8".

11-12. Find the perimeter of a circle whose radius is
        17.6cm.

11–4   Practical Applications for Perimeter and Circumference

Many practical applications of perimeter and circumference are made on the job.

Example 11–6

PROBLEM

How many feet of tack strip are needed for wall-to-wall
carpeting in the room shown in Figure 11-16?

SOLUTION

Note: Tack strip is attached along the perimeter of
a room; therefore, we must find the perimeter of the
given area.

Apply the formula $P = 2l + 2w$ and substitute the known
values:

$$P = 2l + 2w$$
$$= 2(26') + 2(14')$$
$$= 52' + 28'$$
$$= 80'$$

ANSWER

80' of tack strip are needed.

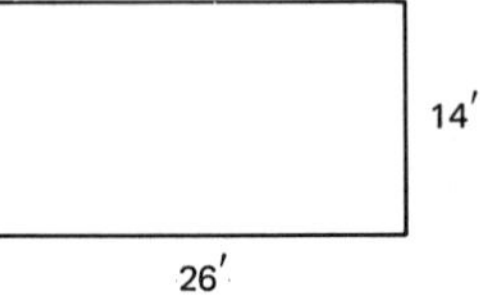

Figure 11-16

Example 11–7

PROBLEM

How many feet of barbed wire are needed to enclose
the plot of land shown in Figure 11-17 if 3 strands of
wire are used? (A strand of wire is a single wire;
a good fence requires at least 3 strands.)

SOLUTION

Note: The wire must go around the perimeter of the
rectangular plot of land.

1. Apply the formula $P = 2l + 2w$ and substitute the
    known values:

$$P = 2l + 2w$$
$$= 2(65 \text{ rods}) + 2(28 \text{ rods})$$
$$= 130 \text{ rods} + 56 \text{ rods} = 186 \text{ rods}$$

2. To find the total amount of wire needed, multiply
    the perimeter by 3:

    total wire = 186 rods x 3 (strands) = 558 rods

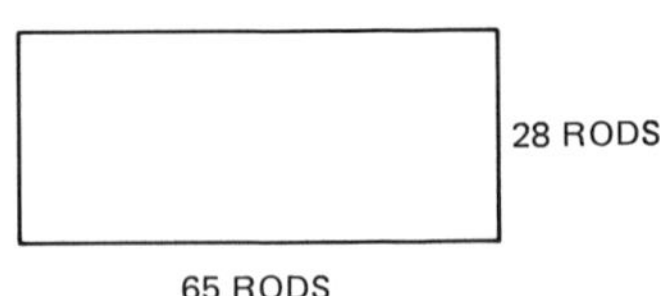

Figure 11-17

3. To convert rods to feet, multiply the total number
    of rods by 16.5' (1 rod = 16.5'):

$$\frac{558 \ \text{rods}}{1} \left(\frac{16.5'}{1 \ \text{rod}}\right) = 9207'$$

ANSWER

9207' of barbed wire are needed.

Exercises: Practical Applications of Perimeter and Circumference

11-13. How much 18" high fencing is needed for the
        flower garden shown in Figure 11-18?

11-14. How many feet will a 26" bicycle wheel travel
        in 1 revolution?

11-15. Find the length of fence needed for the stadium
        in Figure 11-19. (The ends are half-circles.)

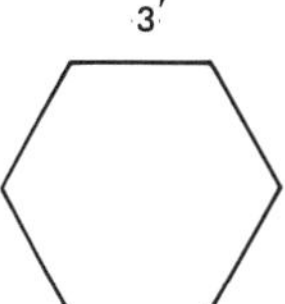

Figure 11-18

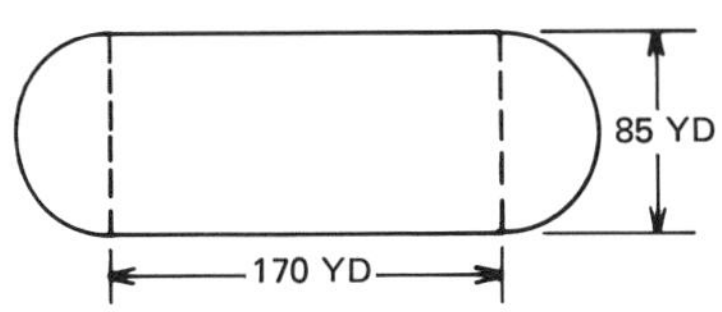

Figure 11-19

11–5   Areas of Polygons

The amount of surface a figure contains, or the amount
of material required to cover the surface, is called
the area of the figure. Area is expressed in square
units. In the metric system, square units are indicated by the exponent 2 (the small number above and to
the right of the unit). The following list compares
abbreviations for some common square units in the
English and metric systems:

Unit	Abbreviation
square feet	sq ft
square yards	sq yd
square centimeters	cm^2
square meters	m^2

To find the area of a polygon, you must know the
values of two dimensions: length and width. The
product (multiplication) of these dimensions yields
the area in square units. Both dimensions must be
expressed in the same unit of measure. For example,
you cannot multiply feet by inches or centimeters by
kilometers. You must first convert the dimensions to
the same unit of measure. The examples illustrate how
to find the areas of polygons. Remember, finding an
area means finding the number of square units (shown
as square blocks or tiles) that will cover a surface.

Procedure for Finding the Areas of Polygons

•••

1. To find the area of a polygon, multiply the length
    by the width: $A = lw$.
2. Express the area in square units.

•••

Example 11–8

PROBLEM

Find the area of the rectangle in Figure 11-20.

SOLUTION

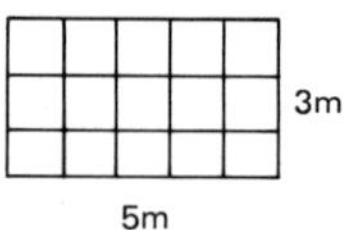

Figure 11-20

Note: To find the number of square meters in the rectangle, we could count them: there are 15 square meters in the rectangle. A more efficient way to find the area is to multiply the length (5m) by the width (3m).

Apply the formula $A = l$ (length) x w (width) and substitute the known values:

$$A = l \text{ x } w$$
$$= (5m)(3m)$$
$$= 15m^2$$

Note: The formula for finding the area of a square is $A = $ side x side, or $A = s^2$.

ANSWER

The area of the rectangle is $15m^2$.

Example 11–9

PROBLEM

Find the area of the parallelogram in Figure 11-21.

SOLUTION

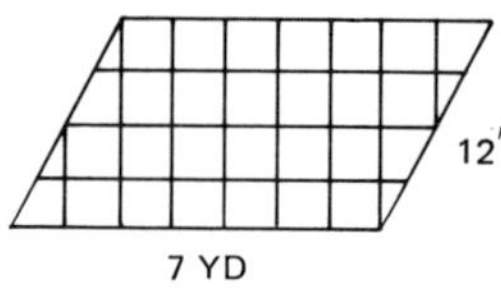

Figure 11-21

Note: To find the number of square yards in the parallelogram, we could move the triangular portion to make a rectangle, as in Figure 11-22.

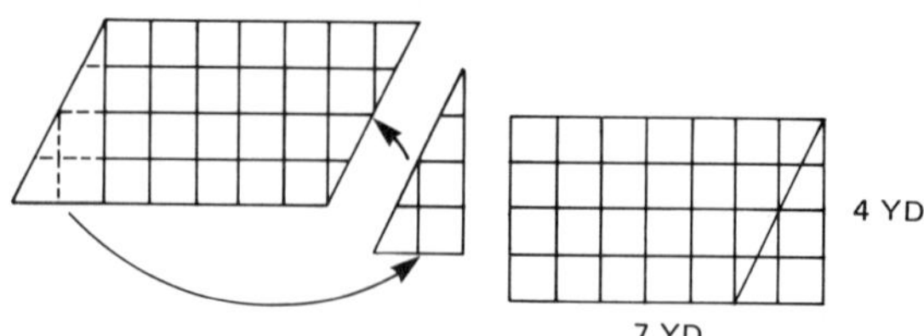

Figure 11-22

Apply the formula for finding the area of a rectangle ($A = l$ x w) and substitute the known values. Remember to convert 12 ft to yards in order to multiply amounts in the same units:

$$A = l \text{ x } w \qquad l = 7 \text{ yd} \qquad w = 12 \text{ ft}$$
$$= \frac{12 \text{ ft}}{1}\left(\frac{1 \text{ yd}}{3 \text{ ft}}\right)$$
$$= 4 \text{ yd}$$
$$= (7 \text{ yd})(4 \text{ yd})$$
$$= 28 \text{ sq yd}$$

ANSWER

The area of the parallelogram is 28 sq yd.

Example 11–10

PROBLEM

Find the area of the triangle in Figure 11-23.

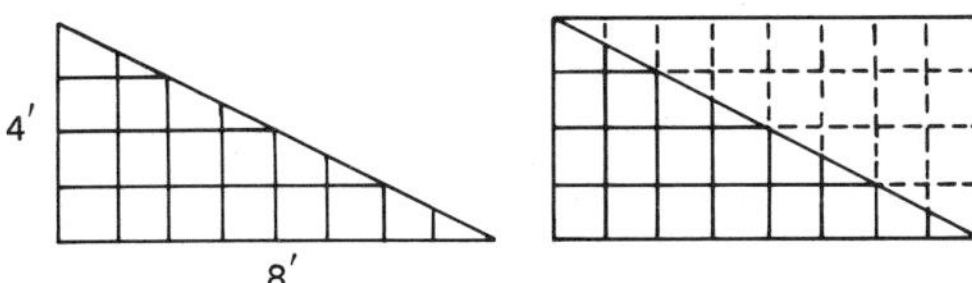

Figure 11-23

SOLUTION

Note: A triangle is actually 1/2 a rectangle, as shown
in the figure, so the area of a triangle is the same as
1/2 the area of a rectangle. The formula for finding
the area of a triangle is A = 1/2 the base times the
height. (Recall that 1/2 = 0.5.)

Apply the formula $A = bh/2$ and substitute the known
values:

$$A = 0.5bh$$
$$= 0.5(8')(4') = 16 \text{ sq ft}$$

ANSWER

The area of the triangle is 16 sq ft.

Example 11–11

PROBLEM

Find the area of the given figure in Figure 11-24.

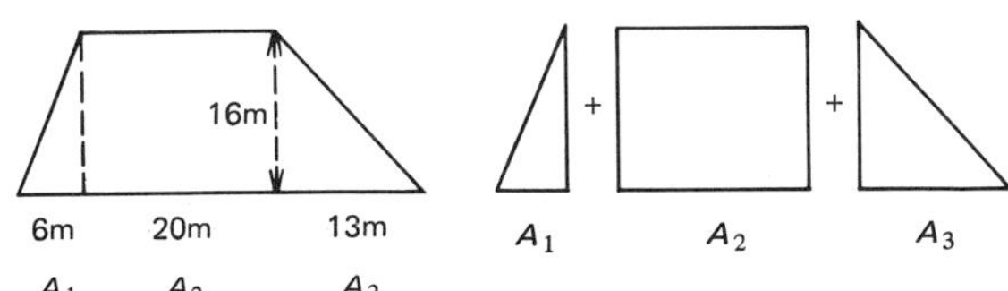

Figure 11-24

SOLUTION

Note: The figure actually consists of a triangle, a
rectangle, and another triangle. To find the area of
the whole figure (in square meters), find the areas of
each part and add them together, as shown in the figure.

1. Apply the formulas for finding the area of a
    triangle and the area of a rectangle and substitute
    the known values:

$$A_1 = 0.5bh \qquad A_2 = lw \qquad A_3 = 0.5bh$$
$$\quad = 0.5(6m)(16m) \qquad = (20m)(16m) \qquad = 0.5(13m)(16m)$$
$$\quad = 48m^2 \qquad\qquad = 320m^2 \qquad\qquad = 104m^2$$

2. To find the total area, add the areas of each part:

$$\text{total area} = A_1 + A_2 + A_3$$
$$= 48m^2 + 320m^2 + 104m^2$$
$$= 472m^2$$

ANSWER

The area of the geometric figure is $472m^2$.

Exercises: Finding the Areas of Polygons

11-16. Find the area of the square in Figure 11-25.

11-17. Find the area of rectangle A in Figure 11-26.

11-18. Find the area of rectangle B in Figure 11-26.

11-19. Find the area of rectangle C in Figure 11-26.

11-20. Find the area of rectangle D in Figure 11-26.

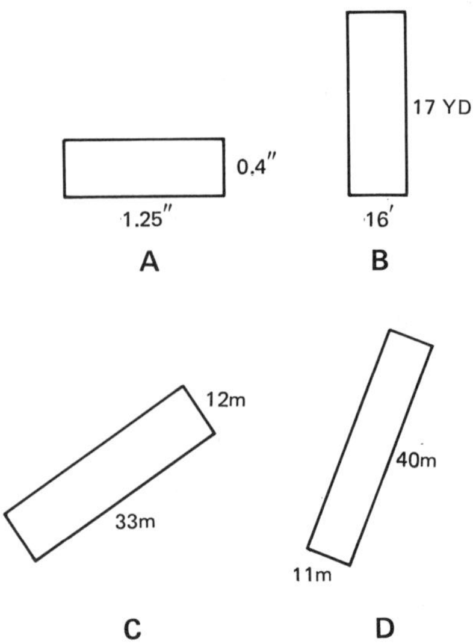

Figure 11-26

Figure 11-25

11-21. Find the area of triangle A in Figure 11-27.

11-22. Find the area of triangle B in Figure 11-27.

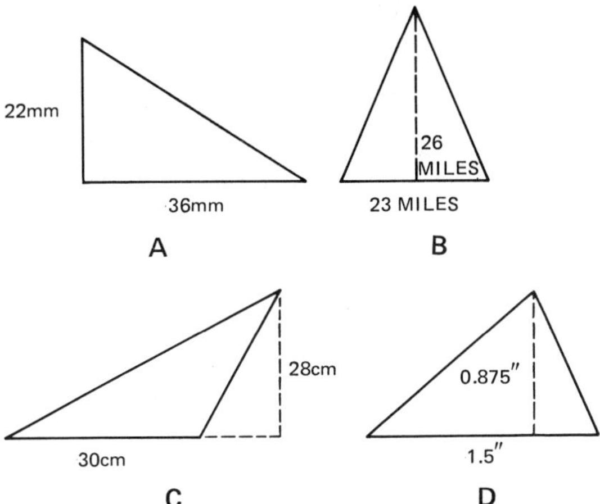

Figure 11-27

11-23. Find the area of triangle C in Figure 11-27.

11-24. Find the area of triangle D in Figure 11-27.

11-25. What is the area of a triangle when the base is 288" and the height is 4 yd?

11-26. Find the area of polygon A in Figure 11-28.

11-27. Find the area of polygon B in Figure 11-28.

11-28. Find the area of polygon C in Figure 11-28.

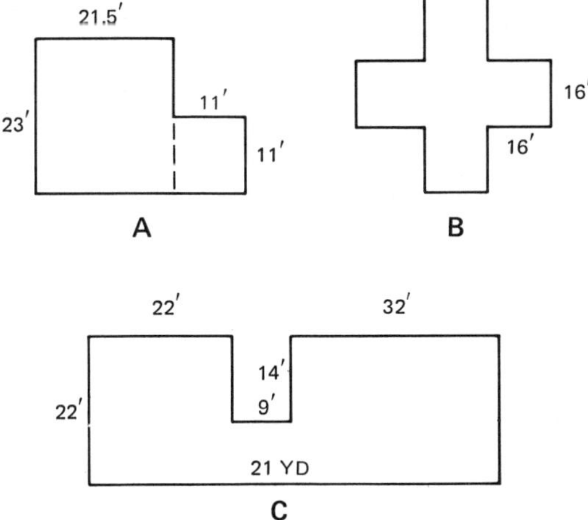

Figure 11-28

11-6 Areas of Circles

The area of a circle is found by multiplying π by the radius by the radius, or π by the radius squared, which is written as $A = \pi r^2$. If the diameter is given, the area equals 0.7854 times the diameter times the diameter, or $A = 0.7854d^2$.   This formula is actually a form of $A = \pi r^2$ because 2 times the radius equals the diameter:

$$2r = d$$
$$r = \frac{d}{2}$$

Substituting $d/2$ for r in the formula, we can solve for A as follows:

$$A = \pi r^2$$
$$= \pi \left(\frac{d}{2}\right)^2$$
$$= \frac{\pi d^2}{4}$$
$$= \frac{3.1416}{4}d^2$$
$$= 0.7854d^2$$

Note how conveniently 7854 is located in the upper left-hand corner of a calculator keyboard.

Procedure for Finding the Areas of Circles

••

1. To find the area of a circle, multiply π by the radius squared: $A = \pi r^2$, or multiply 0.7854 by the diameter squared: $A = 0.7854d^2$.
2. Express the area in square units.

••

Example 11–12

PROBLEM

Find the area of the circle in Figure 11-29.

SOLUTION

Apply the formula $A = 0.7854d^2$ and substitute the known values:

$$A = 0.7854d^2$$
$$= 0.7854(12m)(12m) = 113.10m^2$$

ANSWER

The area of the circle is $113.10m^2$.

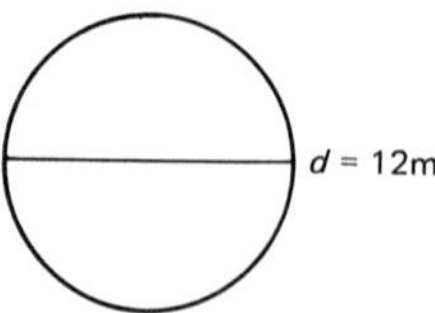

Figure 11–29

Example 11–13

PROBLEM

Find the area of a circle with a radius of 12.5'.

SOLUTION

Apply the formula $A = \pi r^2$ and substitute the known
values:

$$A = \pi r^2$$
$$= (3.142)(12.5')(12.5') = 490.9375 \text{ sq ft}$$

ANSWER

The area of the circle is 490.9375 sq ft.

Exercises: Finding the Areas of Circles

11-29. Find the area of circle A in Figure 11-30.

11-30. Find the area of circle B in Figure 11-30.

11-31. What is the area of circle C in Figure 11-30?

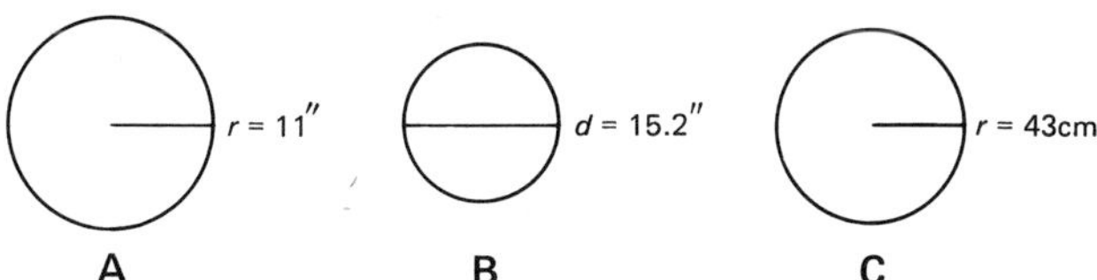

Figure 11-30

11-32. What is the area of a circle with a radius of
0.375"?

11-33. Find the area of a circle with a 75.6' diameter.

11-34. Find the radius of a circle whose area equals
the circumference.

11-7   Practical Applications of Area

Finding the area of a circle is sometimes necessary to
solve certain problems. The examples illustrate such
problems.

Example 11-14

PROBLEM

How many square yards of sod are necessary to cover the
football field in Figure 11-31? The ends of the field
are semicircles (half-circles).

SOLUTION

Find the area of the rectangle (60 yd x 120 yd) and add

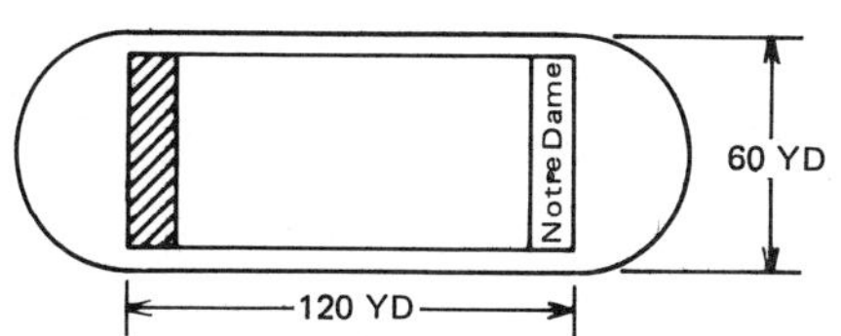

Figure 11-31

the areas of the 2 semicircles or the area of 1 circle
with a diameter of 60 yd:

$$A = lw + 0.7854d^2$$
$$= (120 \text{ yd})(60 \text{ yd}) + 0.7854(60 \text{ yd})^2$$
$$= 7200 \text{ sq yd} + 2827.44 \text{ sq yd} = 10,027.44 \text{ sq yd}$$

ANSWER

10,027.44 sq yd of sod are needed.

Example 11–15

PROBLEM

How many square feet of blacktop are needed to cover
the semicircular driveway in Figure 11–32?

SOLUTION 1

Find the area of the larger semicircle (radius = 66')
and subtract the area of the smaller semicircle
(radius = 50'):

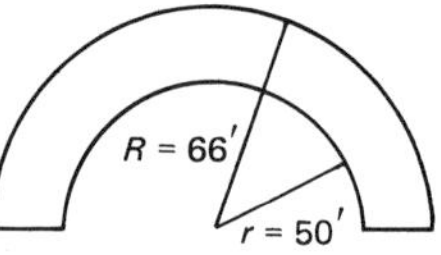

Figure 11–32

$$A = \frac{\pi R^2}{2} - \frac{\pi r^2}{2}$$

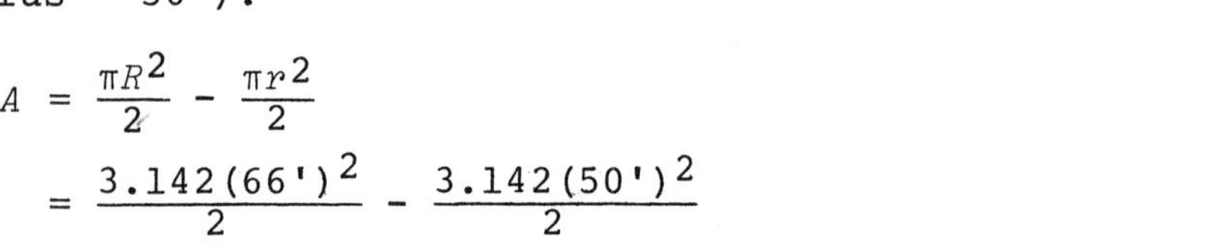

$$= \frac{3.142(66')^2}{2} - \frac{3.142(50')^2}{2}$$

$$= 6843.276 \text{ sq ft} - 3927.500 \text{ sq ft} = 2915.78 \text{ sq ft}$$

SOLUTION 2

The formula used above could be rewritten and solved
as follows:

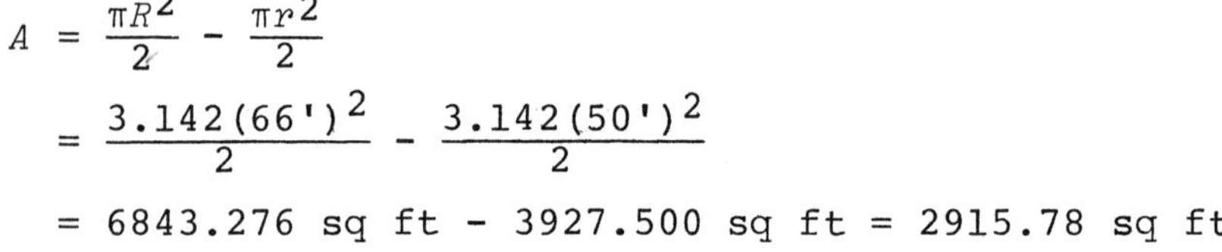

$$A = \frac{\pi}{2}(R^2 - r^2)$$

$$= 1.571(66 \text{ ft}^2 - 50 \text{ ft}^2)$$
$$= 1.571(4536 \text{ ft}^2 - 2500 \text{ ft}^2)$$
$$= 1.571(1856 \text{ sq ft}) = 2915.78 \text{ sq ft}$$

ANSWER

2915.78 sq ft of blacktop are needed.

Exercises: Practical Applications of Area

11-35.   What is the surface area of a swimming pool
         with a 24' diameter?

11-36.   What is the surface area of a 4.125" diameter
         piston in a V-8 engine?

11-37.   How many square yards of sod are needed for the
         odd-shaped lot in Figure 11-33?   (The bottom
         portion is a semicircle.)

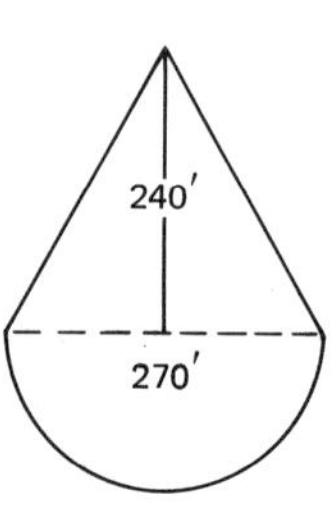

Figure 11-33

Chapter 11

Perimeter and Area

●●

Solving Applied Welding Problems

Now that you have mastered the fundamental operations
of finding perimeter and area, you are ready to apply
these math skills to typical problems that welders
encounter on the job. After studying the examples that
follow, complete the applied welding problems.

Example 11–16

PROBLEM

The deck of a barge under construction is 40' wide and
80' long. How many square feet of steel are required
to cover the deck?

SOLUTION

Multiply the length by the width to get the square
footage:

$$A = lw = 80' \times 40' = 3200 \text{ sq ft}$$

ANSWER

3200 sq ft of steel are needed.

Example 11–17

PROBLEM

What is the area of the triangular piece of metal in
Figure 11-34?

SOLUTION

Multiply 1/2 by the base by the height:

$$A = 1/2 \, bh = 1/2 \times 12" \times 10" = 60 \text{ sq in}$$

ANSWER

The area is 60 sq in.

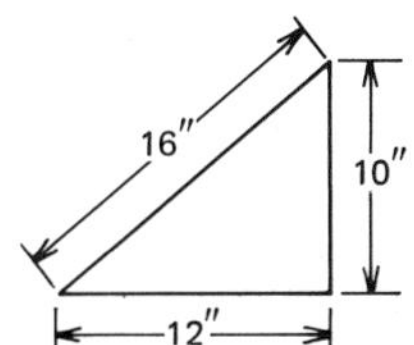

Figure 11-34

Example 11–18

PROBLEM
How many inches of weld are required to weld the pipe
to the base plate in Figure 11-35?

SOLUTION

Find the circumference of the pipe:

$$C = d\pi = 8" \times 3.142 = 25.136"$$

ANSWER

25.136" of weld are required.

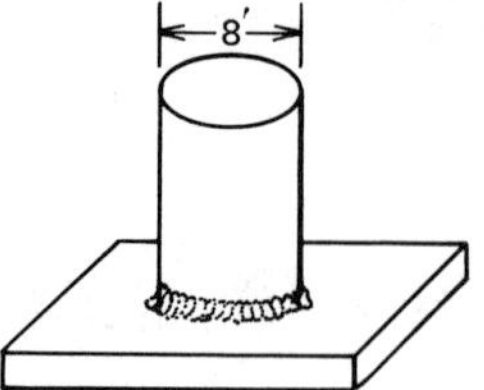

Figure 11-35

Example 11–19

PROBLEM

How many square inches of metal are required to cap the
end of the 12" diameter pipe in Figure 11-36?

SOLUTION

Find the area of the circle by multiplying 0.7854 by
the diameter squared:

$$A = 0.7854d^2 = 0.7854(12")(12") = 0.7854(144 \text{ sq in})$$
$$= 113.0976 \text{ sq in}$$

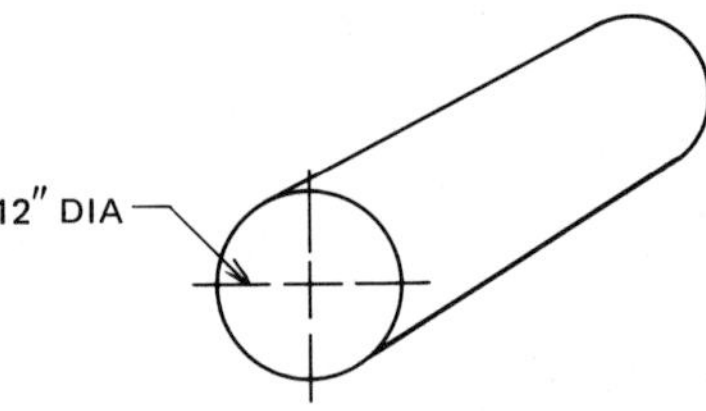

Figure 11-36

ANSWER

113.0976 sq in of metal are needed.

Example 11–20

PROBLEM

How many square inches of metal are needed to fabricate
the stainless steel tube in Figure 11-37?

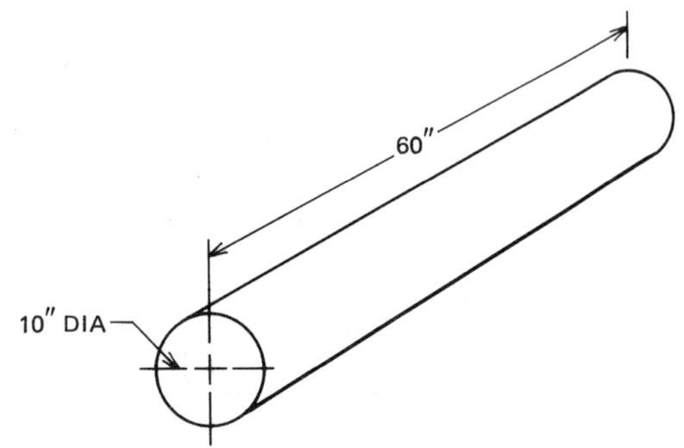

Figure 11-37

SOLUTION

To find the area, determine the circumference and
multiply this dimension by the length:

$$A = d\pi(60") = 10" \times 3.142 \times 60" = 1885.2 \text{ sq in}$$

ANSWER

1885.2 sq in of metal are needed.

Applied Welding Problems

11-38. A transport trailer is shown in Figure 11-38.
    a. How many square feet of mild steel are
        required to cover the floor of the trailer?
        (Area = length x width.)
    b. How many 4' x 8' sheets of steel are needed
        for the floor?
    c. How many square feet of steel are needed for
        one side of the trailer?

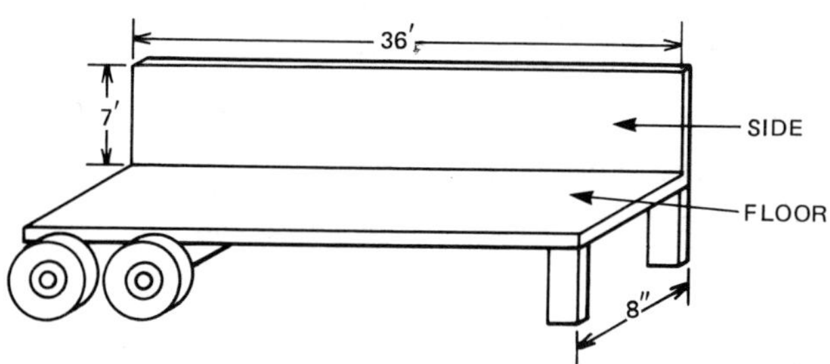

Figure 11-38

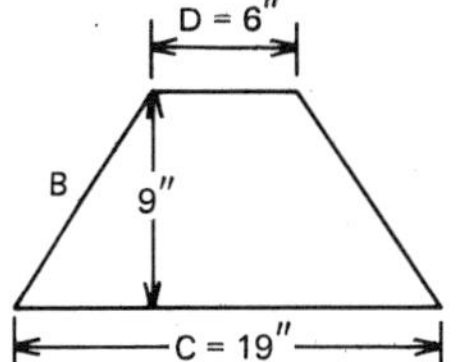

Figure 11-39

11-39. The end view of a casting, shown in Figure 11-39,
    revealed an imperfection that was removed. How
    many square inches of metal were removed?
    (Area = B(C + D)/2.)

11-40. A fireproof canvas curtain is needed to shield
    workers from arc flash. The curtain must be
    185' long and 8' high. At $0.95 per square yard,
    how much will the curtain cost?

11-41. Specifications require 350 support gussets like
    the ones in Figure 11-40 for a concrete form.
    How many square inches of metal are needed for
    350 gussets? (Area = 1/2 x base x height.)

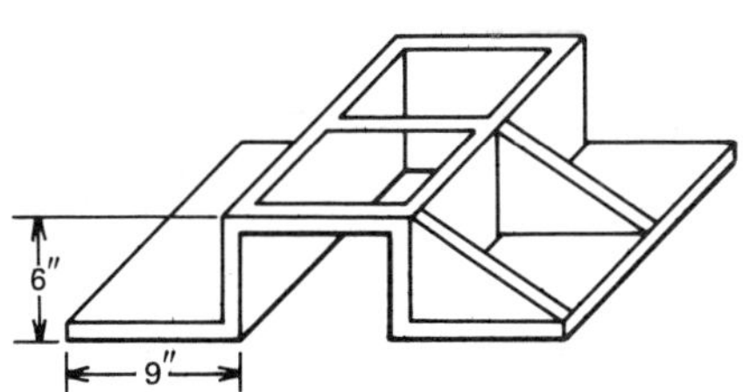

Figure 11-40

11-42. Welding electrode containers like the one in
    Figure 11-41 are to be fabricated. They have two
    sides A, one end C, and a bottom B.
    a. How many square inches of metal are needed
        for bottom B?
    b. How many square inches of metal are needed
        for the two sides A?
    c. How many square inches of material are
        needed for end C?

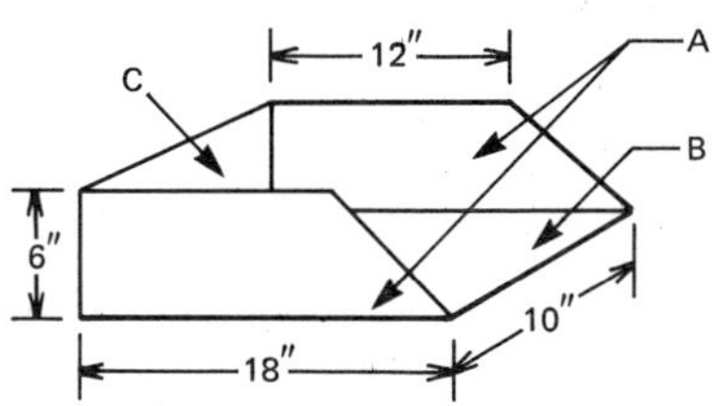

Figure 11-41

11-43. How many square feet of deck plate are required
        to cover the 24" wide catwalk in Figure 11-42?

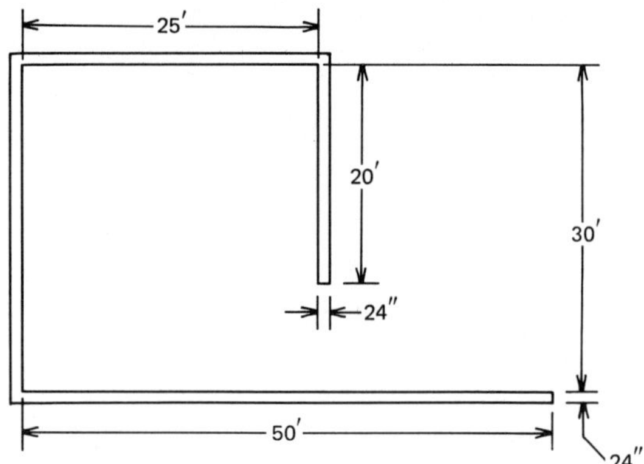

Figure 11-42

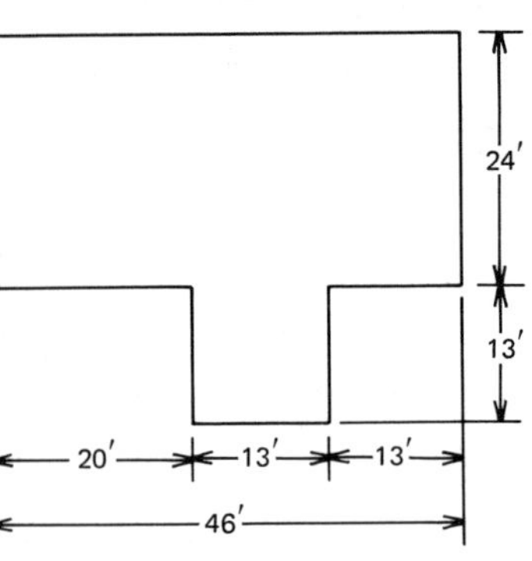

Figure 11-43

11-44. The rent for a welding shop whose dimensions
        are shown in Figure 11-43 is $0.58 per square
        foot per month. How much rent is paid in 1 year?

11-45. How many square feet of steel are needed to
        build a storage tank with a bottom but no top,
        as shown in Figure 11-44?

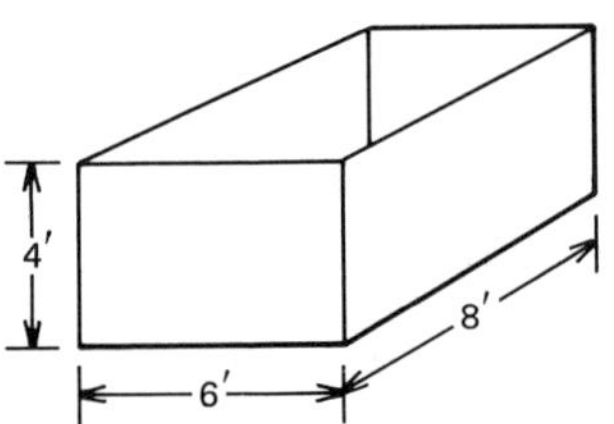

11-46. How many square feet of 1/4" metal are required
        to build the support column in Figure 11-45?
        (Area = length x width.)

Figure 11-44

11-47. Refer to Figure 11-46.
        a. How many square feet of metal are needed to
            cover walkway A (not including the outside
            frame)?
        b. How many square feet of metal are needed to
            cover platform B (not including the outside
            frame)?
        c. How many feet of 1" square tubing are needed
            for the outside frame of the platform and
            the walkway?

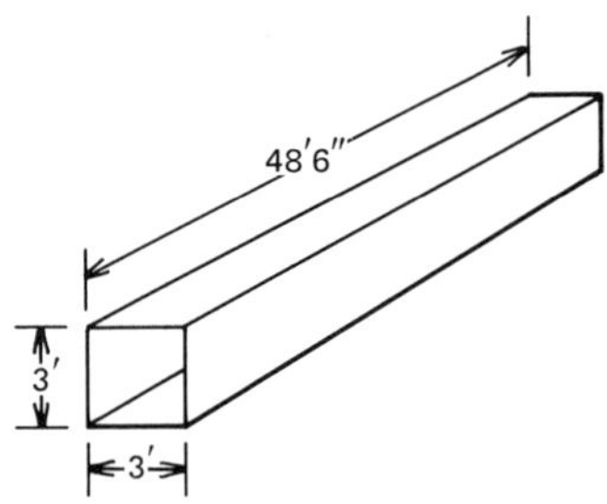

Figure 11-45

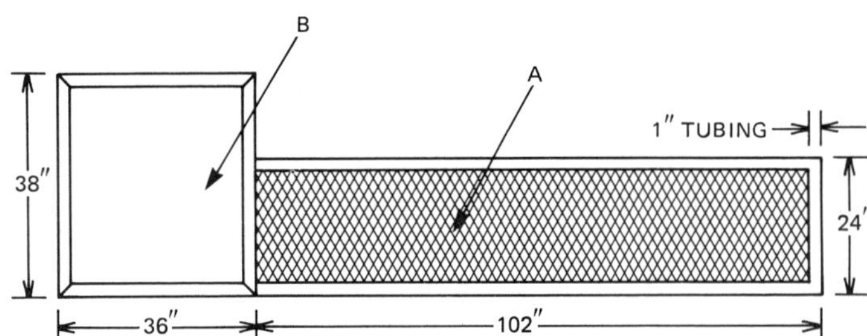

Figure 11-46

11-48. How many square inches of metal are needed for
        the axle bracket on the utility trailer shown
        in Figure 11-47?

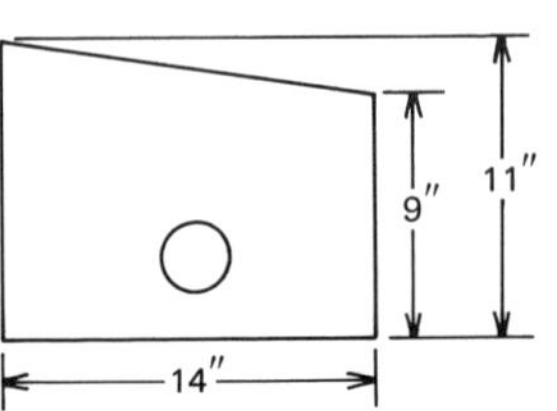

Figure 11-47

11-49. What is the circumference of the center hole in the spacer shown in Figure 11-48?

11-50. What length of material is needed to fabricate the pipe hanger in Figure 11-49?

11-51. The acid dip tank in Figure 11-50 needs two reinforcing bands welded on all sides. How many feet of banding are needed?

11-52. Figure 11-51 shows an electric utility pole cover.
  a. How many square inches of material are required for part A?
  b. How many square inches of material are needed for part B?
  c. How many square inches of material are required for 20 pole covers?

11-53. 3 different-sized ramps are needed for loading and unloading trucks. How long should side C be on ramp No. 1 in Figure 11-52? ($C = \sqrt{A^2 + B^2}$.)

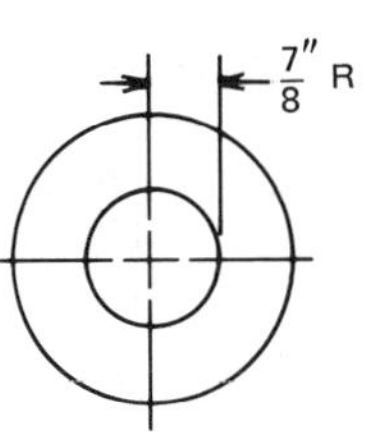

Figure 11-48

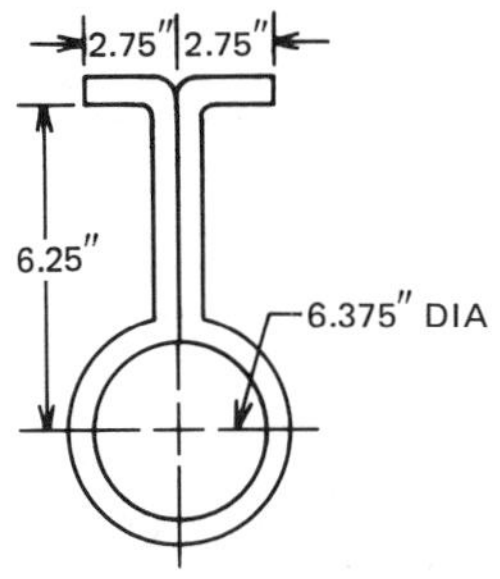

Figure 11-49

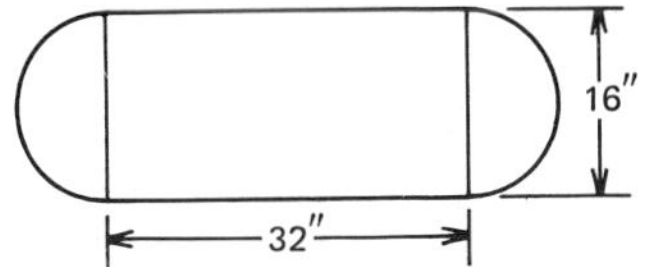

Figure 11-50

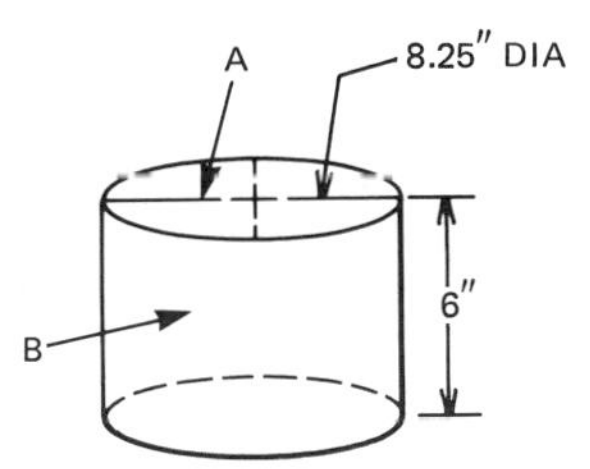

Figure 11-51

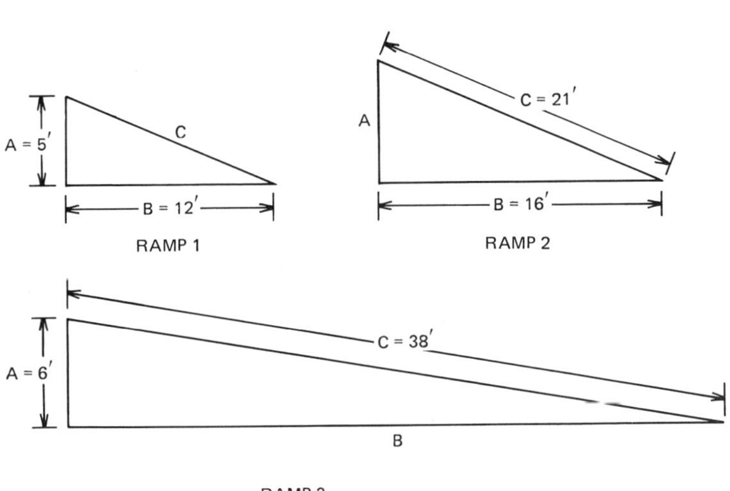

Figure 11-52

11-54. How long is side A of ramp No. 2 in Figure 11-52? ($A = \sqrt{C^2 - B^2}$.)

11-55. How long is side B of ramp No. 3 in Figure 11-52? ($B = \sqrt{C^2 - A^2}$.)

11-56. How many gussets like the one in Figure 11-53 can be cut from a 4' x 8' sheet? (Area = 1/2 x base x height.)

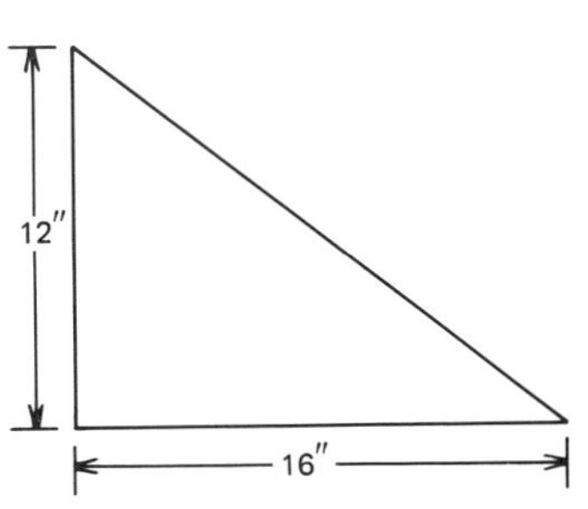

Figure 11-53

11-57.   A guard is needed around the 2 idler pulleys in
         Figure 11-54. The guard requires 2" of
         clearance. What length of material is needed
         for the guard? (Circumference = diameter x π.)

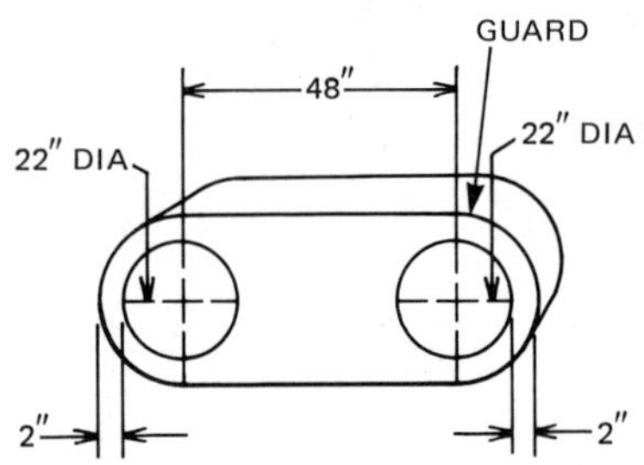

Figure 11-54

11-58.   How many sheets of 4' x 8' steel are needed to
         build the tube in Figure 11-55? (Circumference
         = diameter x π.)

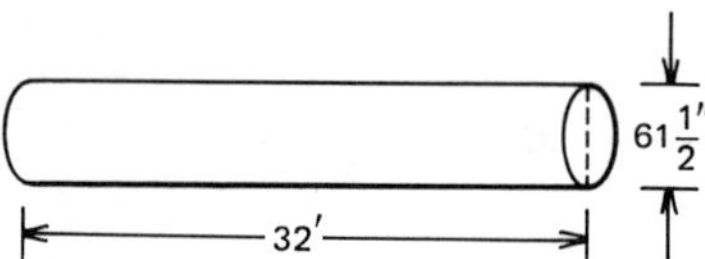

Figure 11-55

11-59.   How many inches of metal are needed to make the
         hanger bracket in Figure 11-56?

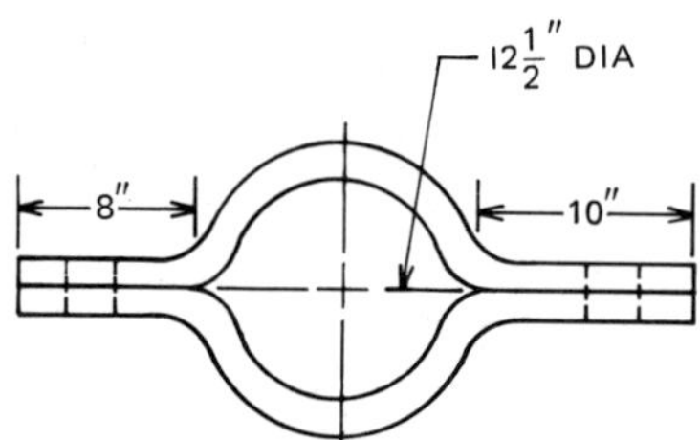

Figure 11-56

Applied Welding Problems
Survey Test

Name ____

Course/Sec. _________________   Date ___________________________

11-1. How many square feet make up 6 sheets of steel   11-1. ________________
       that measure 4' x 8'?

11-2. Refer to Figure 11-57.

    a. How many square inches of metal are needed   11-2a. ________________
        for the sides of the tray?

    b. How many square inches of metal are needed   11-2b. ________________
        for the bottom of the tray?

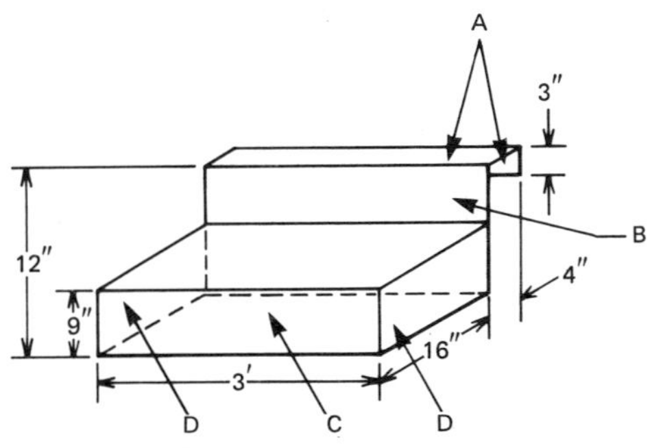

Figure 11–57

11-3. Figure 11-58 shows a window box.

    a. How many square inches of metal are needed   11-3a. ________________
        for part A?

    b. How many square inches of metal are needed   11-3b. ________________
        for part B?

Figure 11-58

c. How many square inches of metal are needed for side C?

11-3c. ______________

d. How many square inches of metal are needed for side D?

11-3d. ______________

e. How many square inches of metal are needed for the bottom?

11-3e. ______________

f. What is the total number of square inches of metal required?

11-3f. ______________

g. How many square feet of metal are needed? (1 sq ft = 144 sq in.)

11-3g. ______________

11-4. How many inches of weld are needed around the tubes to weld them to the header pipe in Figure 11-59?

11-4. ______________

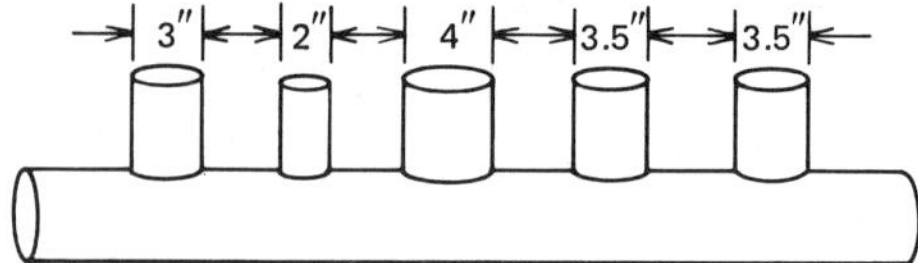

Figure 11-59

11-5. How many inches of metal are required to make the S hook in Figure 11-60?

11-5. ______________

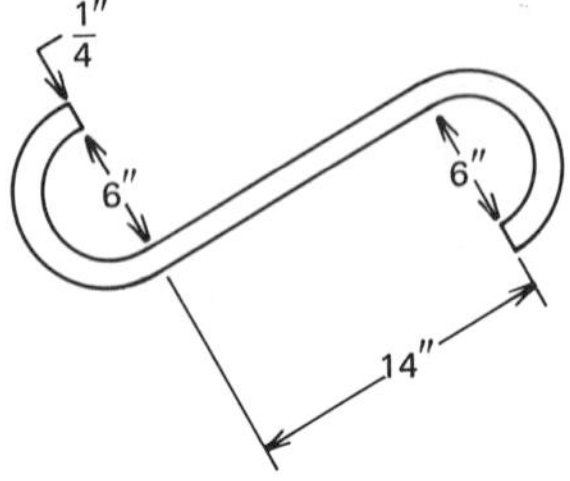

Figure 11-60

11-6. What is the area of the support gussets A in Figure 11-61?

11-6. ______________

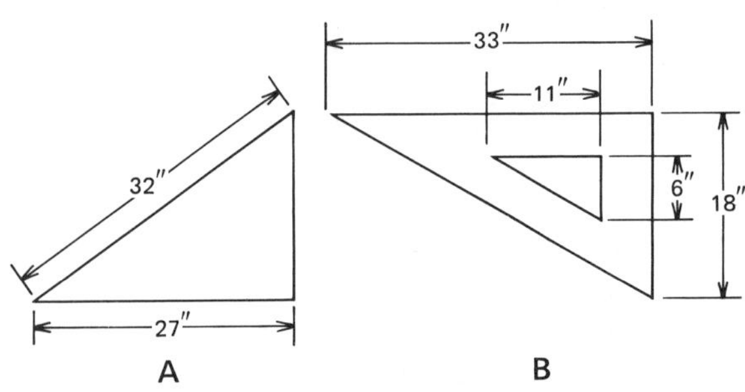

Figure 11-61

11-7. What is the area of the support gusset B?

11-7. ______________

11-8. A storage tank is shown in Figure 11-62.

    a. How many square inches of metal are wasted due to opening A?

11-8a. ___________

    b. How many square inches of metal are wasted due to opening B?

11-8b. ___________

    c. How many square inches of metal are wasted due to opening C?

11-8c. ___________

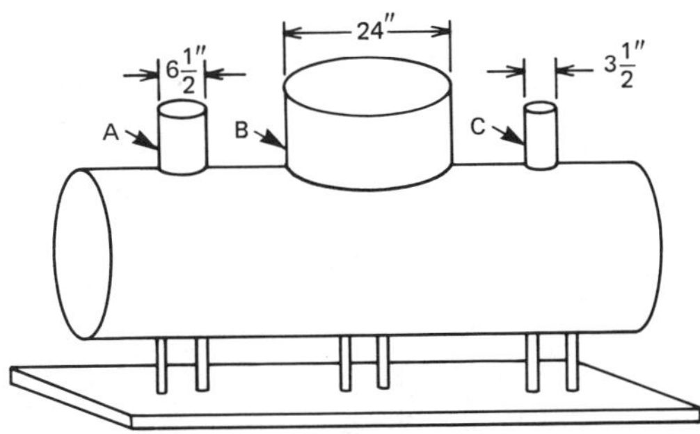

Figure 11-62

11-9. The storage tank in Figure 11-63 is anchored to a mounting frame. How long are the holding bands?

11-9. ___________

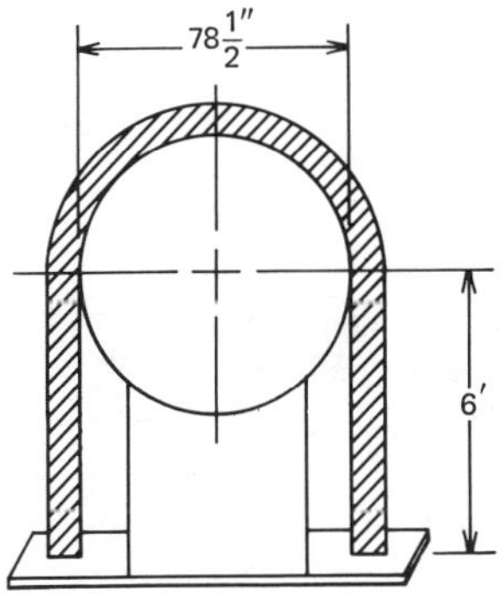

Figure 11-63

11-10. An ornamental iron fence is to be built around the yard shown in Figure 11-64. How many feet of fence will be needed?

11-10. ___________

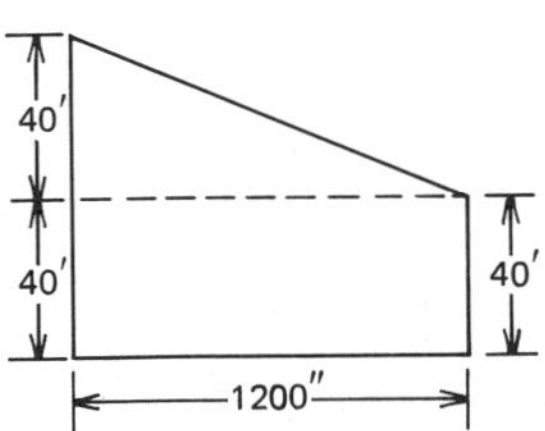

Figure 11-64

11-11. Three circular holes were cut in the sheet of
metal shown in Figure 11-65. How many square
inches of metal were removed?

11-11. _______________

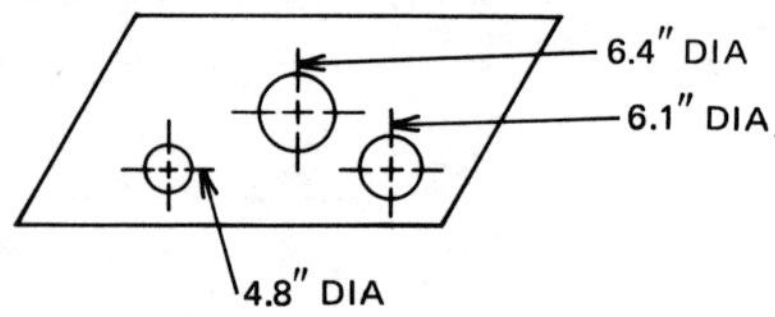

Figure 11-65

Chapter 12: Volume
Preview and Self-Test

Name ____

Course/Sec. _______________   Date _______________________

Chapter 12 covers definitions and properties of volume,
procedures for finding the volumes of geometric solids,
and practical applications of volume. Before you turn
to the chapter, complete the Self-Test to determine
which sections in the chapter you need to study carefully.

Self-Test

DEFINITIONS AND PROPERTIES OF VOLUME

12-1. A _________ is a solid geometric figure with   12-1. _______________
       parallel edges and a uniform cross section.

12-2. When all the lateral faces of a solid are   12-2. _______________
       perpendicular to the bases, they are called

       _________ _________.

12-3. A _________ face is any plane surface of a solid   12-3. _______________
       figure.

                Definitions and properties of volume score   _______________

VOLUMES OF RECTANGULAR SOLIDS

12-4. Find the volume of solid A in Figure 12-1.   12-4. _______________

12-5. Find the volume of solid B in Figure 12-1.   12-5. _______________

Figure 12-1

12-6.   How many cubic inches of cheese are there in 16   12-6. _______________
        cubes of cheese if each is a 3/4" cube?

                      Volumes of rectangular solids score   _______________

VOLUMES OF CYLINDERS

12-7.   Find the volume of cylinder A in Figure 12-2.   12-7. _______________

12-8.   Find the volume of cylinder B in Figure 12-2.   12-8. _______________

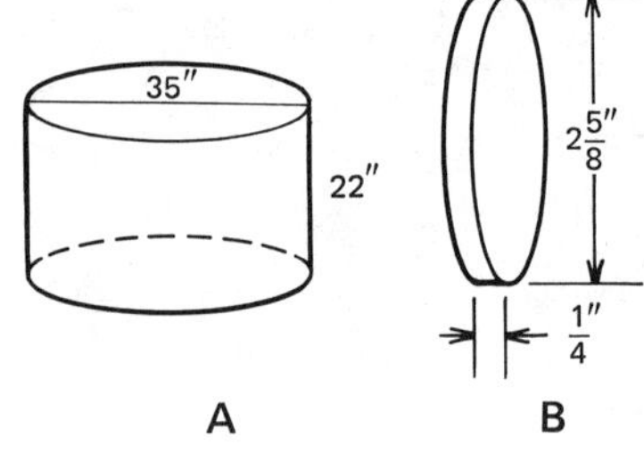

Figure 12-2

12-9.   Find the volume of a 4" pipe that is 10' long.   12-9. _______________

                          Volumes of cylinders score   _______________

VOLUMES OF PYRAMIDS AND CONES

12-10.   Find the volume of cone A in Figure 12-3.   12-10. _______________

12-11.   Find the volume of cone B in Figure 12-3.   12-11. _______________

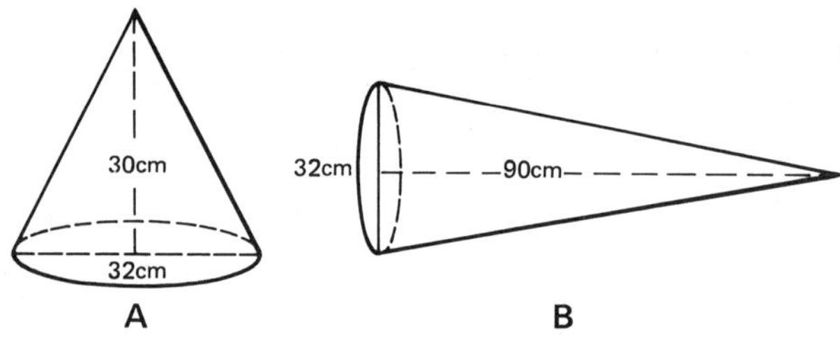

Figure 12-3

12-12.   Find the volume of pyramid A in Figure 12-4.   12-12. _______________

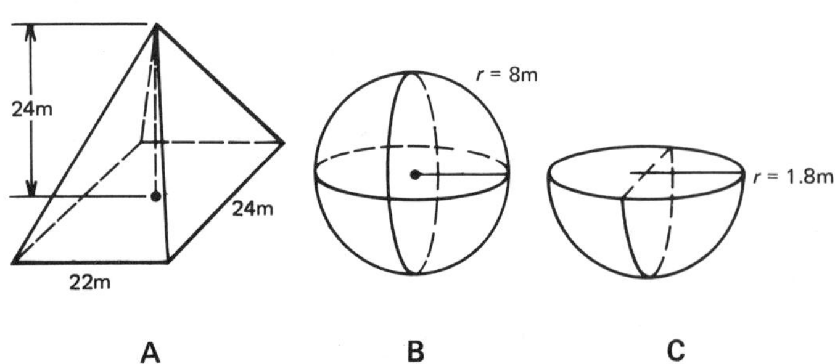

Figure 12-4

                  Volumes of pyramids and cones score   _______________

VOLUMES OF SPHERES AND HEMISPHERES

12-13.   Find the volume of sphere B in Figure 12-4.   12-13. _______________

12-14.   Find the volume of hemisphere C in Figure 12-4.   12-14. _______________

12-15.   Find the volume of a 4.5' sphere.   12-15. _______________

                Volumes of spheres and hemispheres score   _______________

PRACTICAL APPLICATIONS OF VOLUME

12-16.   A classroom is 35' wide, 40' long, and 10.5'   12-16. _______________
          high. If the air in the room is exchanged once
          every 90 seconds, how many cubic feet are moved
          each minute?

12-17.   A 3 lb coffee can is 6-1/8" in diameter and 7"   12-17. _______________
          high. How much does 1 cu in of coffee weigh?

12-18.   The distance between shutoff valves of an 8"   12-18. _______________
          pipe in an oil refinery is 48'. How many
          gallons (to the nearest hundredth) of crude oil
          are in the section? (1 gal = 231 cu in.)

12-19.   How much water will a 2-1/2" fish bobber   12-19. _______________
          displace when it is submerged?

12-20.   The nose cone of a rocket is 12.5' long and   12-20. _______________
          4.5' in diameter. How many cubic feet does it
          contain?

                Practical applications of volume score   _______________

After you have checked your answers on the Self-Test,
transfer your scores to the Chapter 12 Objectives.

Chapter 12

Volume

Upon successful completion of this chapter, you will
know the definitions and properties of volume. You will
be able to find the volumes of basic geometric shapes
and to make practical applications of volume.

Objectives

	Self-Test Scores	If you did poorly on the Self-Test, turn to:
• Definitions and properties of volume	__________	Section 12-2
• Finding the volumes of rectangular solids	__________	Section 12-3
• Finding the volumes of cylinders	__________	Section 12-4
• Finding the volumes of pyramids and cones	__________	Section 12-5
• Finding the volumes of spheres and hemispheres	__________	Section 12-6
• Practical applications of volume	__________	Section 12-7

12–1   Introduction to Volume

The ability to calculate volume is necessary for many
occupations. This ability is also important in our
everyday lives. The energy shortage, for example, has
caused nearly everyone to become more concerned with
volume. We want to have enough "room," or volume, in a
car and still get good mileage. Smart shoppers are
concerned with volume in an indirect way because they
want to get the most value per ounce, pound, kilogram,
quart, liter, and so on.

In this chapter, we learn to calculate the volumes
of regular solids and of some odd or irregular solids as
well. The more we learn about volume, the more applications we will find.

12–2   Basic Definitions and Properties of Volume

A solid has three dimensions: length, width, and
height. Every solid has a surface area and a volume.
The formulas used to find volumes are similar to the
formulas used to find perimeters and areas.

The volume of a solid can be thought of as the
amount the object "holds." Volume is defined as the
enclosed space within an object.

A prism (solid) is a solid geometric figure with
parallel edges and uniform cross sections. The names
of prisms are usually derived from the shapes of their
bases. Some examples and properties of prisms are
shown in Figure 12-5.

A face or lateral face is any plane surface of a
solid figure. The bases are the top and the bottom
lateral faces. The lateral edge of a prism or solid is
the line where two sides of a solid meet or come
together.

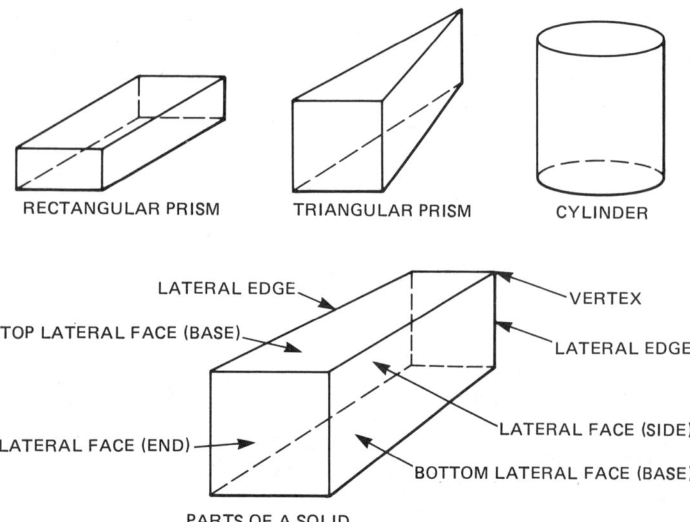

Figure 12-5

The <u>vertex</u> is the point where three sides or lateral faces meet. Prisms whose lateral faces are all perpendicular ($\perp$) to the bases are called <u>right</u> <u>prisms</u>. If the lateral faces are not perpendicular, the prisms are called <u>oblique</u> <u>prisms</u>.

12–3   Volumes of Rectangular Solids

The volume of a rectangular solid is found by finding the area of the base and multiplying it by the height. Volume is always measured in cubic (three) units because three dimensions (length, width, and height) are used to find volume.

The length, width, and height of a solid must be expressed in the same unit of measure before we can find the volume. For example, the dimensions may be measured in feet, yards, and meters, but we must convert all the dimensions to a common measure before calculating the volume.

Volume is expressed in cubic units. In the metric system, cubic units are indicated by the exponent 3. Examples of abbreviations for some common units in both the English and the metric systems are shown in the following list:

Unit	Abbreviation
cubic inches	cu in
cubic yards	cu yd
cubic centimeters	cm^3
cubic meters	m^3

Procedure for Finding the Volumes of Rectangular Solids

••

1. To find the volume of a rectangular solid, multiply the area of the base by the height: $V = lwh$.
2. Express the volume in cubic units.

••

Example 12–1

PROBLEM

Find the volume of the rectangular solid in Figure 12-6.

SOLUTION

<u>Note</u>: All the dimensions of the solid are given in feet, so we do not need to make any conversions. We could count the number of cubic feet, but it is much more efficient to use the formula for finding volume.

1. Recall that to find the volume of a rectangular solid, we must find the area of the base and

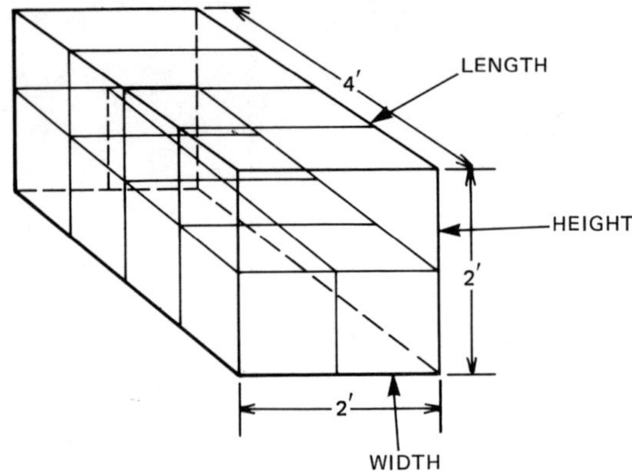

Figure 12-6

multiply it by the height. The formula is $V = Bh$, where B = the area of the base, and $B = l \times w$. Thus, the formula could be written $V = lwh$.

2. Apply the formula and substitute the known values:

$V = lwh$
$\quad = (4')(2')(2')$
$\quad = 16 \text{ cu ft}$

ANSWER

The volume is 16 cu ft.

Example 12–2

PROBLEM

Find the volume of a 10 yd x 8 yd x 3 yd rectangular solid.

SOLUTION

1. Draw a sketch of the solid, as shown in Figure 12-7. First draw the front face and then project (draw) the other lateral edges so they are parallel to each other. Draw dotted or dashed lines for the edges that are not seen.

2. Apply the formula $V = lwh$ and substitute the known values:

$V = lwh$
$\quad = (10 \text{ yd})(8 \text{ yd})(3 \text{ yd})$
$\quad = 240 \text{ cu yd}$

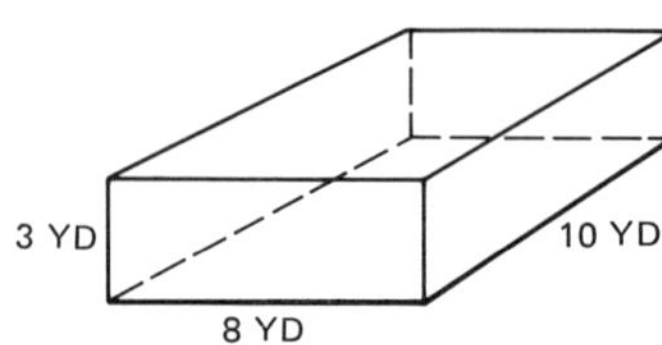

Figure 12-7

ANSWER

The volume is 240 cu yd.

Exercises: Finding the Volumes of Rectangular Solids

12-1.   Find the volumes of the rectangular solids in
        Figure 12-8. Drawing sketches like those in the
        figure will help you understand the process of
        finding volume.

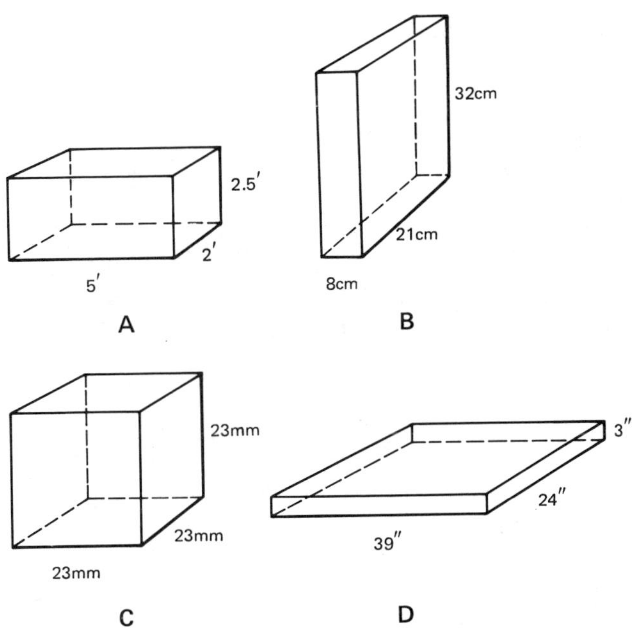

Figure 12-8

12-2.   Find the volume of a 15" cube. (All dimensions
        of a cube are the same.) Draw a sketch as in
        Exercise 12-1.

12-3.   Find the volume of a space 30 yd x 45 yd x 7.5
        yd.

12–4   Volumes of Cylinders

The volume of a cylinder is found by finding the area
of the base and multiplying it by the height.

Procedure for Finding the Volumes of Cylinders

1.   To find the volume of a cylinder, multiply the area
     of the base by the height: $V = 0.7854d^2h$, or
     $V = \pi r^2 h$.
2.   Express the volume in cubic units.

Example 12–3

PROBLEM

Find the volume of the cylinder in Figure 12-9.

SOLUTION

<u>Note</u>: The diameter and the height are given in meters, so we do not need to make any conversions.

1. To find the volume of a cylinder, we must find the area of the base (which is the same as finding the area of a circle) and multiply it by the height. The formula is $V = Bh$, where B = the area of the base, and $B = 0.7854d^2$. Thus, the formula can be written $V = 0.7854d^2h$.

2. Apply the formula and substitute the known values:

$$V = 0.7854d^2h$$
$$= 0.7854(28m)(28m)(26m)$$
$$= 16009.5936m^3$$

ANSWER

The volume is $16009.5936m^3$.

<u>Note</u>: We could round the volume to $16009.59m^3$ because accuracy cannot be created by calculation.

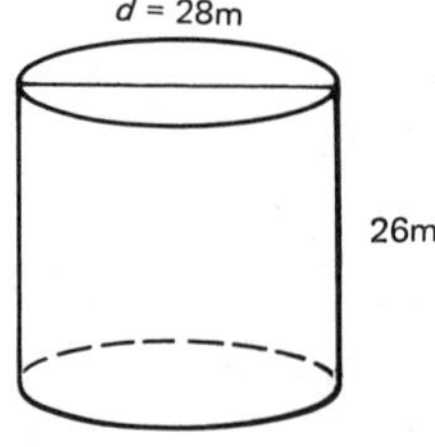

Figure 12-9

Example 12–4

PROBLEM

Find the volume of the cylinder in Figure 12-10.

SOLUTION

1. Convert the 4" radius to feet (recall that 12" = 1' and that 12"/12" = 1'/12"):

$$1' = \frac{1'}{12"} \quad \text{so} \quad \left(\frac{4"}{1}\right)\left(\frac{1'}{12"}\right) = \frac{1'}{3}$$

2. To find the volume, find the area of the base and multiply it by the height. The formula is $V = \pi r^2h$:

$$V = \pi r^2h$$
$$= \pi\left(\frac{1'}{3}\right)\left(\frac{1'}{3}\right)(30')$$
$$= 3.1416\left(\frac{10}{3} \text{ cu ft}\right)$$
$$= 10.472 \text{ cu ft}$$

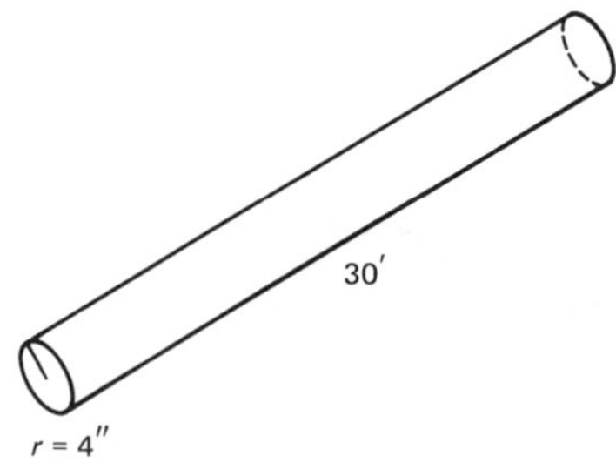

Figure 12-10

ANSWER

The volume is 10.47 cu ft (rounded to the nearest hundredth because accuracy cannot be created by calculation).

Exercises: Finding the Volumes of Cylinders

12-4.   Find the volumes of the cylinders in Figure
        12-11.

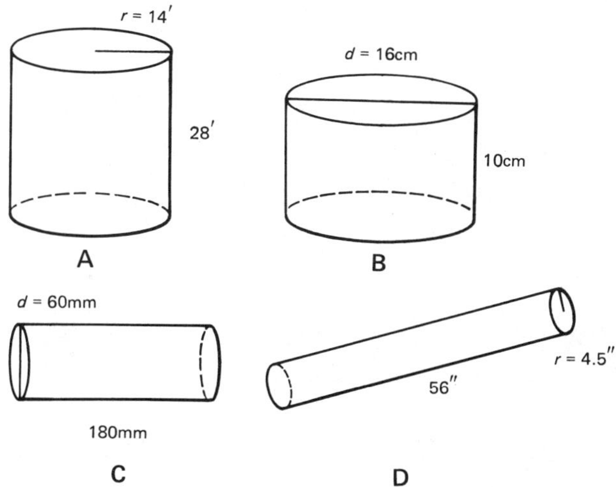

Figure 12-11

12-5.   Find the volume of a cylinder with a radius of
        12" and a length of 80'.

12-6.   If the diameter of a cylinder is 48" and the
        length is 7', how many cubic feet will the
        cylinder hold?

12–5   Volumes of Pyramids and Cones

Other common types of solids are pyramids and cones.
Examples are shown in Figure 12-12.

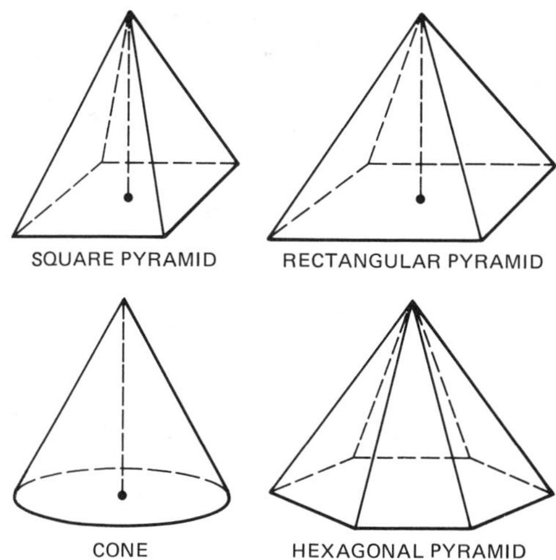

Figure 12-12

Whereas the base of a cone is a circle, the base of a pyramid can be any polygon. A pyramid is identified by the shape of its base. Figure 12-12 shows a rectangular pyramid and a hexagonal pyramid.

The volume of a pyramid or cone is one-third of the area of the base multiplied by the height:

$$V = \frac{1}{3}Bh$$

where B is the area of the base. The validity of this formula can be shown by pouring the contents of a cone-shaped container into a cylinder with the same diameter and height. The contents of three cones are needed to fill the cylinder, as shown in Figure 12-13. The same is true for pyramids.

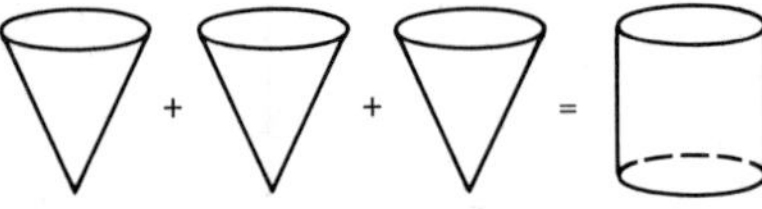

Figure 12-13

Procedure for Finding the Volumes of Pyramids and Cones

1. To find the volume of a pyramid or a cone, multiply 1/3 the area of the base times the height:

$$V = \frac{1}{3}Bh \quad \text{or} \quad V = \frac{Bh}{3}$$

2. Express the volume in cubic units.

Example 12–5

PROBLEM

Find the volume of the pyramid in Figure 12-14.

SOLUTION

1. Recall that to find the volume of a pyramid, we must multiply 1/3 the area of the base by the height. The formula is $V = Bh/3$, where $B = lw$. Thus, the formula can be written $V = lwh/3$.

2. Apply the formula and substitute the known values:

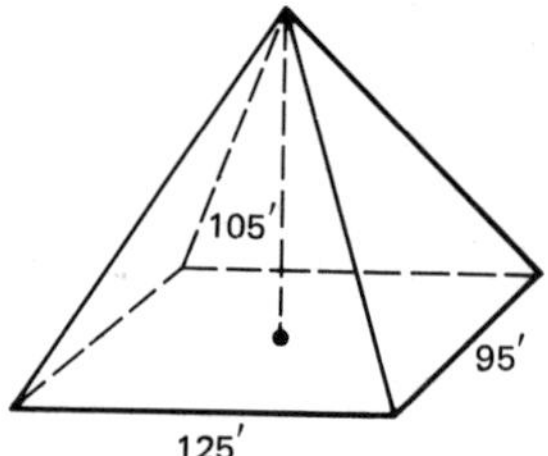

Figure 12-14

$$V = \frac{lwh}{3}$$
$$= \frac{95' \times 125' \times \overset{35}{\cancel{105'}}}{\cancel{3}}$$
$$= 415{,}625 \text{ cu ft}$$

ANSWER

The volume is 415,625 cu ft.

Example 12–6

PROBLEM

Find the volume of the cone in Figure 12-15.

SOLUTION

Apply the formula $V = Bh/3$, where $B = \pi r^2$:

$$V = \frac{\pi r^2 h}{3}$$

$$= \frac{3.1416\,(14m)\,(14m)\,(44m)}{3}$$

$$= 9031.0528m^3$$

ANSWER

The volume is $9031.05m^3$ (rounded to the nearest hundredth).

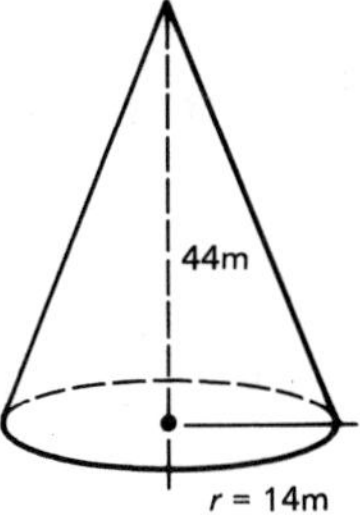

Figure 12-15

Exercises: Finding the Volumes of Pyramids and Cones

12-7.   Find the volumes of the pyramids and cones in Figure 12-16.

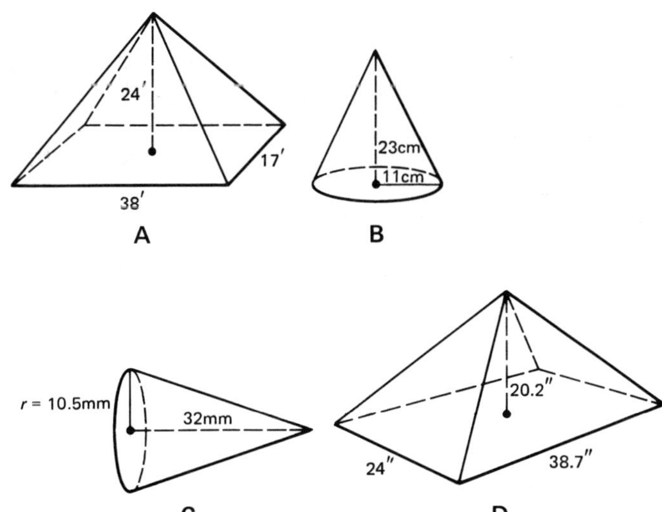

Figure 12-16

12-8.   Find the volume of a square pyramid with a 24' square base and a height of 16'.

12-9.   What is the volume of a 12.75cm-diameter cone with a height of 34.6cm?

12–6   Volumes of Spheres and Hemispheres

A _sphere_ is a solid bounded by a curved surface, where every point is equally distant from the center point within the solid. A ball is an example of a sphere.

The surface area of a sphere, as shown in Figure 12-17, is equal to four times the area of the great circle (the largest circle) of the sphere. The formula is:

$$S = 4\pi r^2 \quad \text{or} \quad S = \pi d^2$$

The volume of a sphere is equal to one-third the surface area multiplied by the radius. The formulas are:

$$V = \frac{4\pi r^3}{3}, \quad V = \frac{\pi d^3}{6}, \quad \text{or} \quad V = 0.5236 d^3$$

A _hemisphere_ is half a sphere. Half a grapefruit or an orange is an example of a hemisphere. The formulas for finding the surface area and volume of a hemisphere are the same as those for a sphere, but multiplied by one-half:

$$S = 2\pi r^2 \quad \text{or} \quad S = \frac{\pi d^2}{2}$$

$$V = \frac{2\pi r^3}{3}, \quad V = \frac{\pi d^3}{12}, \quad V = 0.2618 d^3$$

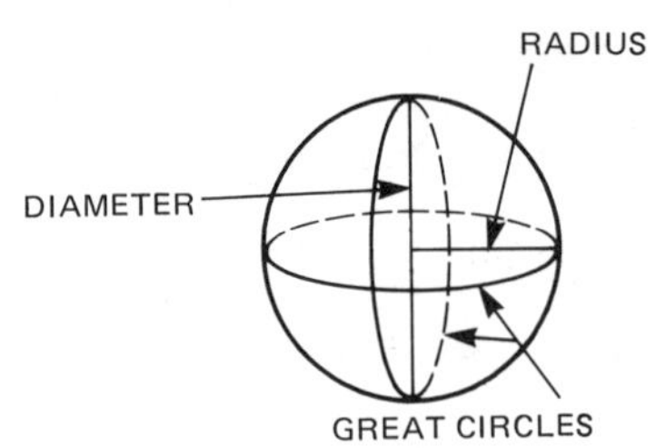

Figure 12-17

Procedure for Finding the Surface Areas and Volumes of Spheres and Hemispheres

1. To find the volume of a sphere, multiply 1/3 the surface area times the radius. The formula for finding the surface area is:

    $$S = 4\pi r^2 \quad \text{or} \quad S = \pi d^2$$

    The formula for finding the volume is:

    $$V = \frac{4\pi r^3}{3}, \quad V = \frac{\pi d^3}{6}, \quad \text{or} \quad V = 0.5236 d^3$$

2. To find the surface area and volume of a hemisphere, multiply the formulas above by 1/2:

    $$S = 2\pi r^2 \quad \text{or} \quad S = \frac{\pi d^2}{2}$$

    $$V = \frac{2\pi r^3}{3}, \quad V = \frac{\pi d^3}{12}, \quad \text{or} \quad V = 0.2618 d^3$$

3. Express the surface area in square units and the volume in cubic units.

Example 12–7

PROBLEM

Find the surface area and volume of sphere A in Figure 12–18.

SOLUTION

1. Because the diameter is given, use the following formulas:

$$S = \pi d^2 \quad \text{and} \quad V = \frac{\pi d^3}{6}$$

2. Substitute the known values and evaluate:

$$S = (3.1416)(2.5")^2 = 19.635 \text{ sq in}$$

$$V = \frac{(3.1416)(2.5")^3}{6} = 8.18125 \text{ cu in}$$

ANSWER

The surface area is 19.64 sq in, and the volume is 8.18 cu in (rounded to the nearest hundredth).

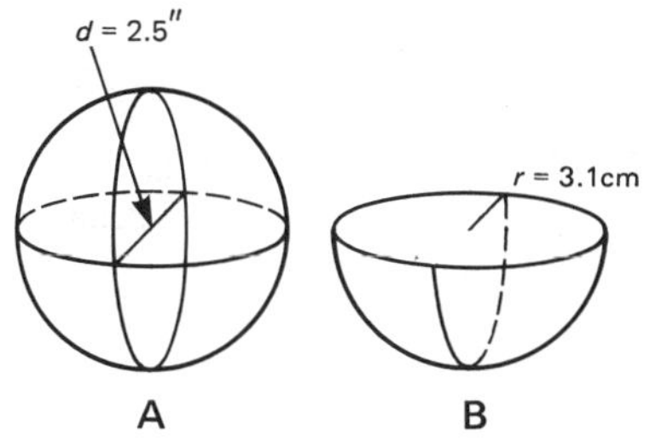

Figure 12-18

Example 12–8

PROBLEM

Find the surface area and volume of hemisphere B in Figure 12–18.

SOLUTION

1. Because the radius is given, use the following formulas:

$$S = 2\pi r^2 \quad \text{and} \quad V = \frac{2\pi r^3}{3}$$

2. Substitute the known values and evaluate:

$$S = 2(3.1416)(3.1\text{cm})^2 = 60.381552\text{cm}^2$$

$$V = \frac{2(3.1416)(3.1\text{cm})^3}{3} = 62.39427040\text{cm}^3$$

ANSWER

The surface area is 60.38cm^2, and the volume is 62.39cm^3 (rounded to the nearest hundredth).

Exercises: Finding the Surface Areas and Volumes of Spheres and Hemispheres

12-10. Refer to Figure 12-19.
   a. Find surface area of sphere A.
   b. Find volume of sphere A.

12-11. a. Find surface area of sphere B.
        b. Find volume of sphere B.

12-12. a. Find surface area of sphere C.
        b. Find volume of sphere C.

12-13. a. Find surface area of hemisphere D.
        b. Find volume of hemisphere D.

12-14. a. Find surface area of hemisphere E.
        b. Find volume of hemisphere E.

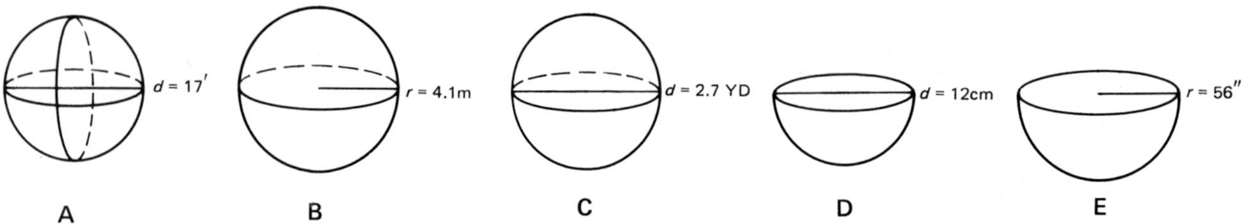

Figure 12-19

12-15. Find the surface area of a 20" (diameter)
        sphere.

12-16. How many cubic yards will a 53 yd (diameter)
        hemisphere hold?

12–7   Practical Applications of Volume

Numerous applications of surface area and volume are
made in business and industry. The solids are not
always of the regular shapes we have studied. However,
they often are a combination of two or more of the
regular types of solids. Therefore, when finding the
area or volume of an irregular solid, we can add or
subtract various regular solids.
     The examples show some of the applications of
volume. To solve the problems, you need to know the
facts used to calculate area and volume. We cannot give
examples of every kind of application. But if you know
how to find the surface areas and volumes of regular
solids, and if you use common sense, you will be able to
solve numerous surface area and volume problems.

Example 12–9

PROBLEM

a. How many bushels of grain will a 7.5' x 32' truck
    trailer hold if it is filled to a depth of 4.5'?
    (1 cu ft = 0.8 bushel.)

b. Find the weight of the load if the grain weighs
    60 lb per bushel.

SOLUTION

1. Draw a sketch of the trailer, as in Figure 12-20.
    Remember to draw the front face first and to make
    all projected lines parallel. Label the dimensions.

2. Apply the formula for finding the volume of a
    rectangular solid and substitute the known values:

$$V = Bh \quad \text{or} \quad V = lwh$$
$$= 32' \times 7.5' \times 4.5'$$
$$= 1080 \text{ cu ft}$$

3. Since 1 cu ft = 0.8 bu, multiply 1080 cu ft by
    0.8 bu/1 cu ft:

$$\left(\frac{1080 \text{ cu ft}}{1}\right)\left(\frac{0.8 \text{ bu}}{1 \text{ cu ft}}\right) = 864 \text{ bu}$$

4. Each bushel weighs 60 lb, so multiply 864 bu by
    60 lb/1 bu:

$$\left(\frac{864 \text{ bu}}{1}\right)\left(\frac{60 \text{ lb}}{1 \text{ bu}}\right) = 51,840 \text{ lb}$$

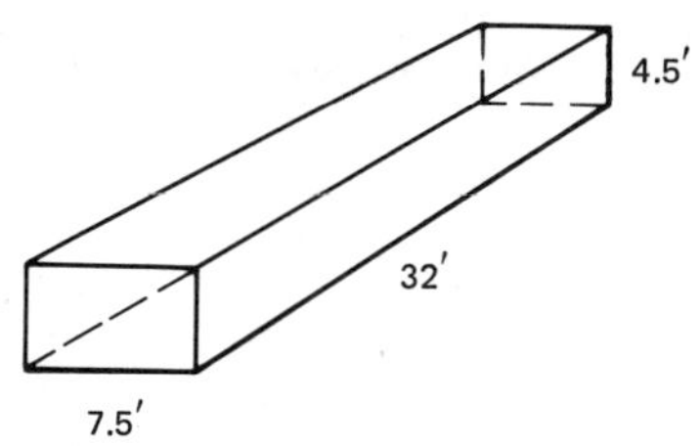
Figure 12-20

ANSWER

a. The trailer will hold 864 bushels of grain.
b. The load will weigh 51,840 lb.

Example 12–10

PROBLEM

How many gallons are needed to fill a 30' x 55' pool if
it is 6' deep on one end and 3' deep on the other?
(1 cu ft = 7.5 gal.)

SOLUTION

1. Draw a sketch of the pool, as shown in Figure 12-21.
    Think of the front face as the base.

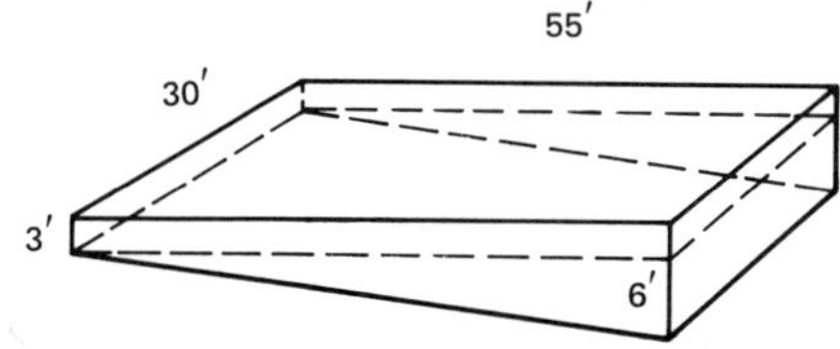

Figure 12-21

2. Note that the figure consists of a rectangular
    solid on the top (marked by the dotted lines) and a
    triangular solid on the bottom. To find the volume,
    add the volume of the rectangular solid to the
    volume of the triangular solid. The formula is:

$$V = Bh, \quad \text{where } B = lw, \ + \frac{1}{2}bl$$

(We consider the width of the pool the height in this formula.)

3. Apply the formula and substitute the known values:

$$V = (lw + \tfrac{1}{2}bl)h$$
$$= (55' \times 3' + 0.5 \times 6' \times 55')30'$$
$$= (165 \text{ sq ft} + 165 \text{ sq ft})(30')$$
$$= (330 \text{ sq ft})(30')$$
$$= 9900 \text{ cu ft}$$

4. Since each cubic foot holds 7.5 gal, multiply 9900 cu ft by 7.5 gal/cu ft:

$$\left(\frac{9900 \text{ cu ft}}{1}\right)\left(\frac{7.5 \text{ gal}}{1 \text{ cu ft}}\right) = 74,250 \text{ gal}$$

ANSWER

74,250 gal are needed to fill the pool.

Example 12–11

PROBLEM

The Alaskan pipeline is 48" in diameter. How many gallons are there in a 1/2 mi section? (1 cu ft = 7.5 gal.)

SOLUTION

1. Draw a sketch, as shown in Figure 12-22, and label the dimensions. (The jagged lines indicate that the drawing is continuous but has not been drawn to scale.)

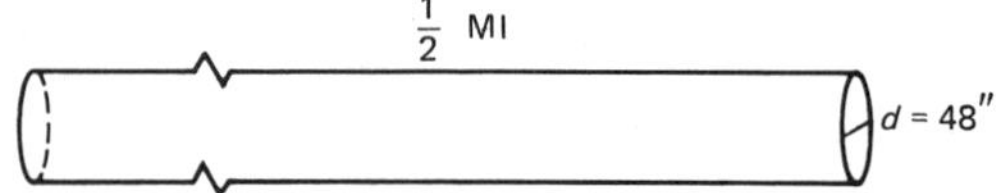

Figure 12-22

2. Note that the pipeline is a cylinder with a diameter of 48", or 4':

$$\left(\frac{48''}{1}\right)\left(\frac{1'}{12''}\right) = 4'$$

and a length of 1/2 mi, or 2640':

$$\left(\frac{0.5 \text{ mi}}{1}\right)\left(\frac{5280'}{1 \text{ mi}}\right) = 2640'$$

3. Apply the formula $V = Bh$, where $B = 0.7854d^2$ and h = the length, and substitute the known values:

$$V = 0.7854d^2l$$
$$= (0.7854)(4')^2(2640')$$
$$= 33175.296 \text{ cu ft}$$

4. Since each cubic foot contains 7.5 gal, multiply 33175.296 cu ft by 7.5 gal/cu ft:

$$\left(\frac{33175.296 \text{ cu ft}}{1}\right) \left(\frac{7.5 \text{ gal}}{1 \text{ cu ft}}\right) = 258{,}814.72 \text{ gal}$$

ANSWER

There are 258,814.72 gal in a 1/2 mi section.

Example 12–12

PROBLEM

Creek Valley Elementary School classrooms are hexagonal shaped. Each of the 6 sides is 20' long. The ceilings are 12' high. Find the number of cubic feet of air that must circulate every minute if the air is exchanged each minute.

SOLUTION

1. Draw a sketch, as shown in Figure 12-23A, and label the dimensions.

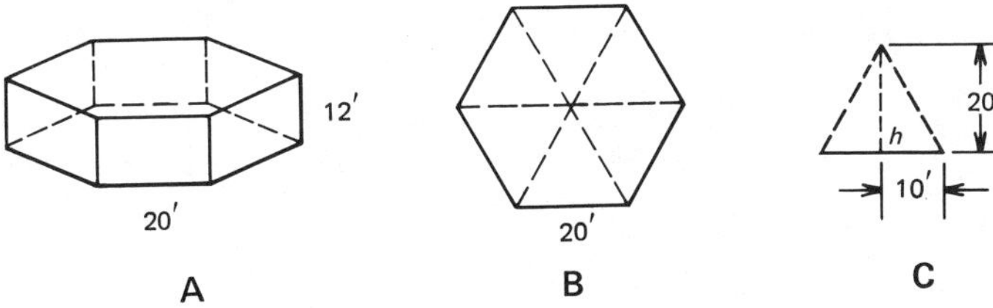

Figure 12-23

2. Note that the base (floor) is a hexagon made up of 6 equilateral triangles, as shown in part B. To determine the area of the base, find the area of one of the triangles and multiply it by 6. Before you can find the area, however, you must find the height of the triangles by using the Pythagorean Theorem. Sketch the triangle, as shown in part C of Figure 12-23, and draw the height perpendicular to the center of the base.

3. Then apply the Pythagorean Theorem to find the height:

$$h^2 = (20)^2 - (10)^2 = 400 - 100$$
$$\sqrt{h^2} = \sqrt{300}$$
$$h = 17.321'$$

4. Find the area of one of the triangles (A_1) by using the formula $A = bh/2$:

$$A_1 = \frac{1}{2}bh$$
$$= \frac{1}{2}(20')(17.321')$$
$$= 173.205 \text{ sq ft}$$

5   Find the area of the base of the hexagon by multi-
    plying the area of one triangle by 6:

$$A = 6A_1$$
$$= (6)(173.205 \text{ sq ft})$$
$$= 1039.23 \text{ sq ft}$$

6. Finally, determine the volume of the hexagonal solid
    (the classroom) by applying the formula $V = Bh$,
    where $B = 6A_1$, and substituting the known values:

$$V = Bh, \text{ where } B = 6A_1$$
$$= 6A_1 h$$
$$= (1039.23 \text{ sq ft})(12')$$
$$= 12470.766 \text{ cu ft}$$

ANSWER

12,471 cu ft of air (rounded to the nearest unit) must
circulate each minute.

Example 12–13

PROBLEM

How much weight will two pontoons of a boat, as shown
in Figure 12-24, support (displace) if the pontoons are
22' long and submerged to the dotted line?   (1 cu ft of
water supports 62.5 lb.)

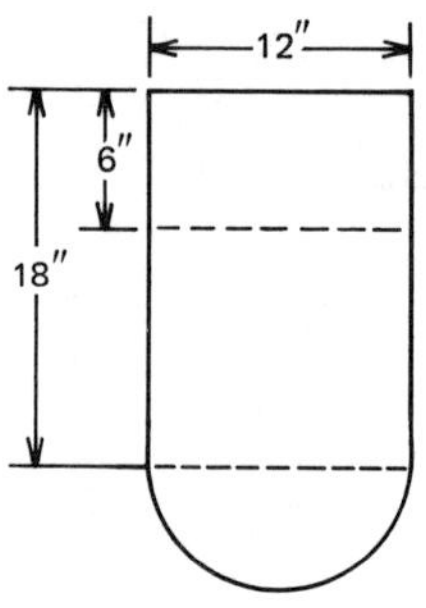

Figure 12-24

SOLUTION

1. Note that the pontoon base is a rectangle with a
    semicircular end.   The submerged portion is a 12" x
    12" square with a semicircular bottom portion whose
    diameter is 12".   To find the volume of the two
    submerged portions of the pontoons, the formula is:

$$V = 2Bl, \text{ where } B = S^2 + \frac{0.7854d^2}{2} \text{ and } l = \text{the length}$$

2. Apply the formula and substitute the known values
    (convert all dimensions to feet):

$$V = 2\left(S^2 + \frac{0.7854d^2}{2}\right)l$$
$$= (2)\left[(1' \times 1') + \frac{0.7854(1')^2}{2}\right](22')$$
$$= (2)(1 \text{ sq ft} + 0.3927 \text{ sq ft})(22')$$
$$= 61.28 \text{ cu ft}$$

3. Since each cubic foot will support 62.5 lb, multiply
    61.28 cu ft by 62.5 lb/1 cu ft:

$$\left(\frac{61.28 \text{ cu ft}}{1}\right)\left(\frac{62.5 \text{ lb}}{1 \text{ cu ft}}\right) = 3830 \text{ lb}$$

ANSWER

Two pontoons will support 3830 lb.

Example 12–14

PROBLEM

Figure 12-25 shows a water tower. Find its capacity in gallons and the number of square feet of surface to be painted. Water is stored only in the sphere, and the circular portion where the sphere is joined to the cylinder is not painted. (1 cu ft = 7.5 gal.)

SOLUTION

1. Recall that the formula for finding the volume of a sphere is:

$$V = \frac{\pi d^3}{6}$$

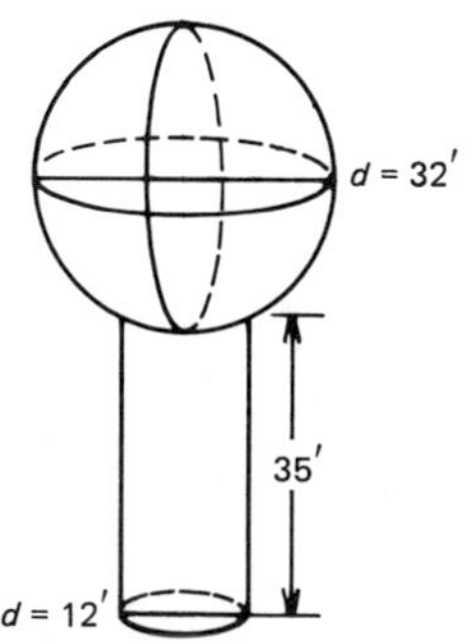

Figure 12-25

2. Apply the formula and substitute the known values:

$$V = \frac{\pi (32')^3}{6}$$

$$= 17157.3248 \text{ cu ft}$$

3. To find the capacity of the sphere, multiply the volume by 7.5 gal/1 cu ft:

$$\text{capacity} = \left(\frac{17157.32 \text{ cu ft}}{1}\right)\left(\frac{7.5 \text{ gal}}{1 \text{ cu ft}}\right)$$

$$= 128679.9360 \text{ gal}$$

4. To find the surface area that must be painted, add the surface area of the sphere ($S = \pi d^2$) to the area of the cylinder ($S = \pi dh$) and subtract from their sum the area of the circle where the cylinder joins the sphere ($0.7854d^2$). Let d_1 represent the diameter of the sphere, and d_2 the diameter of the circle where the cylinder joins the sphere:

$$A = \pi d_1^2 + \pi d_2 h - 0.7854 d^2$$

$$= \pi (32')^2 + \pi (12')(35') - 0.7854(12')^2$$
$$= 3216.9984 \text{ sq ft} + 1319.4720 \text{ sq ft} - 113.0976 \text{ sq ft}$$
$$= 4423.3728 \text{ sq ft}$$

ANSWER

The capacity of the water tower is 128,680 gal, and the surface area to be painted is 4423 sq ft (rounded to the nearest unit).

Exercises: Practical Applications of Volume

Solve the following:

12-17.   How many 3/4 cup servings are contained in the pan shown in Figure 12-26?   (1 cup = 14.4 cu in.)

12-18.   How many cubic yards of concrete are needed for the driveway shown in Figure 12-27?

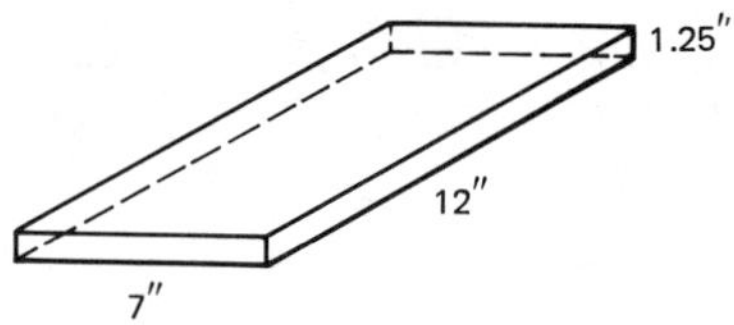

Figure 12-26

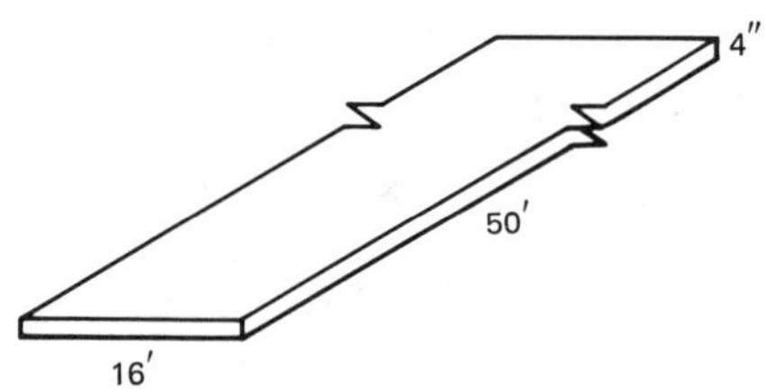

Figure 12-27

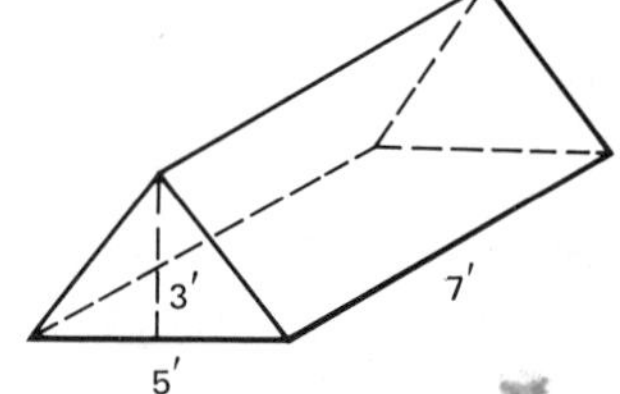

Figure 12-28

12-19.   Find the volume of the tent shown in Figure 12-28.

12-20.   Find the capacity of the cylindrical water tank shown in Figure 12-29.

12-21.   How many yards must be excavated for an 80' x 55' home if the excavation is 12' deep and must be extended 2 yd beyond the walls on all sides (not the bottom)?

12-22.   How many cubic feet are there in a car trunk with the dimensions given in Figure 12-30?

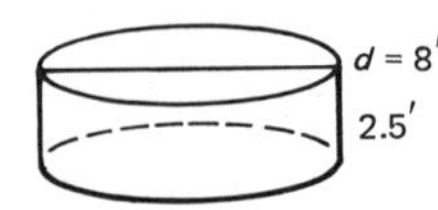

Figure 12-29

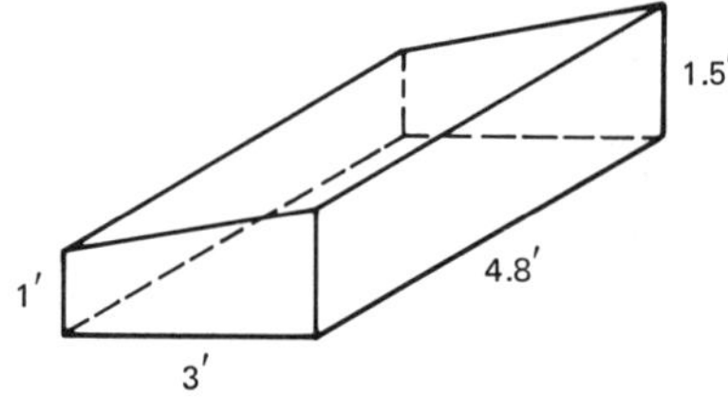

Figure 12-30

12-23.   A classroom is 36' long x 24' wide x 8' high. How many cubic feet are there for each student if there are 34 students in the classroom?

12-24.   How many 1" cubes of cheese can be cut from the wedge shown in Figure 12-31?

12-25.   How many gallons of water will an 18" hemispherical-shaped bathroom sink hold?

12-26.   How much water will a 2.5" lead ball displace?

12-27.   How many gallons of diesel fuel can be stored in the auxiliary tank shown in Figure 12-32?

Figure 12-31

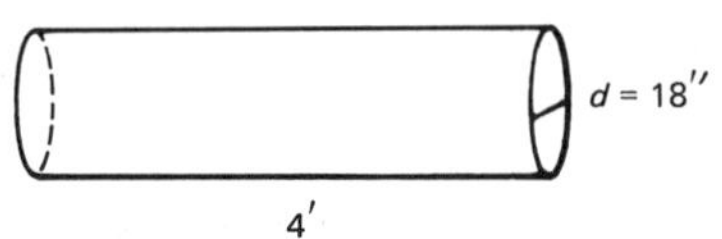

Figure 12-32

12-28.   How many gallons of water are contained in a
         50' length of 5/8" hose?

12-29.   What is the land surface of the earth, in square
         miles, if the diameter of the earth is 8000 mi?
         (Land occupies 1/4 of the earth's surface.)

12-30.   How many gallons of gasoline can be stored in
         a cylindrical tank 42" in diameter and 8' long?
         (1 cu ft = 7.5 gal.)

12-31.   A landscape company stores its pulverized dirt
         in the storage unit shown in Figure 12-33. How
         many cubic yards will the unit hold?

12-32.   Wheat is piled in a cone-shaped pile because of
         lack of storage. How many bushels are in a pile
         6' high with a diameter of 32'? (1 cu ft = 0.8
         bushel.)

12-33.   How many gallons of water can be stored in the
         stock water tank shown in Figure 12-34? (1 gal
         = 231 cu in)

12-34.   How many cubic yards of coal can be stored in
         the loading dock shown in Figure 12-35?

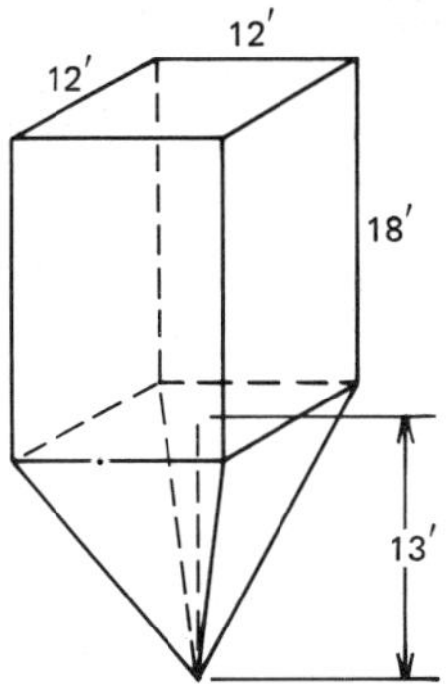

Figure 12-33

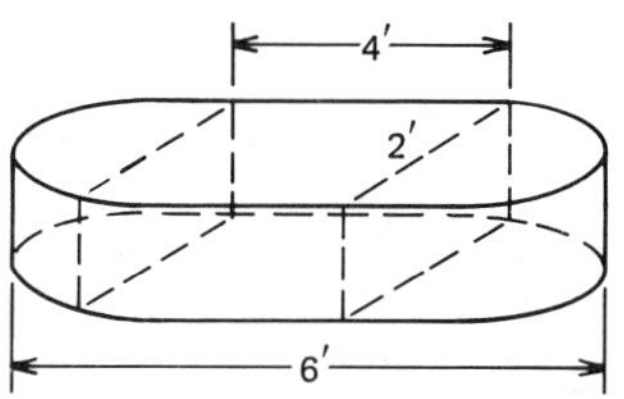

Figure 12-34

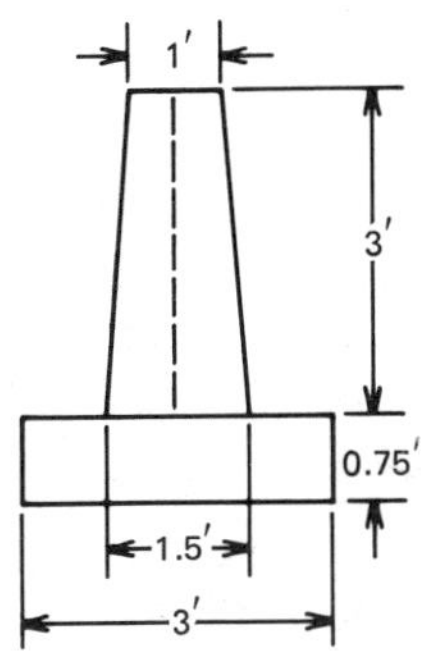

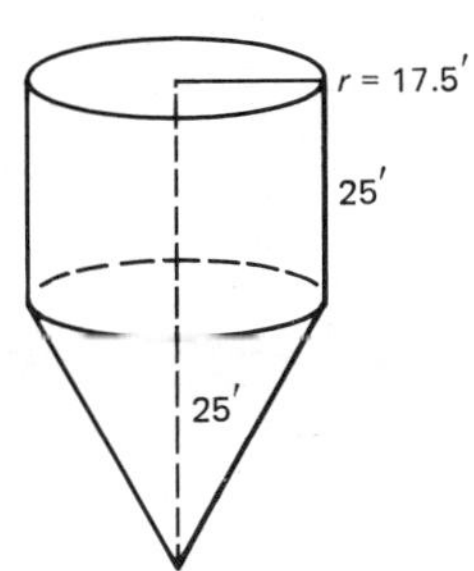

Figure 12-35

Figure 12-36

12-35.   How many cubic yards of concrete are needed for
         a mile of the highway divider shown in Figure
         12-36?

12-36.   How many cups of soft ice cream would you
         receive in the cone shown in Figure 12-37 if
         the cone also is filled with ice cream?
         (1 cup = 14.4 cu in)

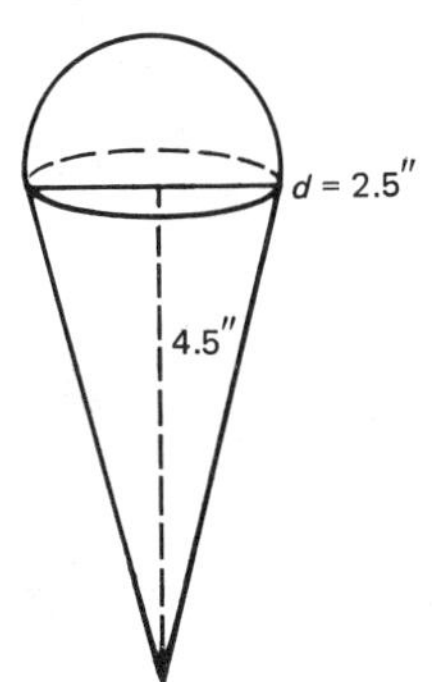

Figure 12-37

Chapter 12

Volume

• •

Solving Applied Welding Problems

Now that you have mastered the fundamental operations
on volume, you are ready to apply these math skills to
typical problems that welders encounter on the job.
After studying the examples that follow, complete the
applied welding problems.

Example 12–15

PROBLEM

What is the capacity in cubic inches of the flux
container shown in Figure 12-38?

SOLUTION

1. Recall that the volume of a rectangular solid equals
    width multiplied by length multiplied by height.

2. Apply the formula and substitute the known values:

    $V = w$ x l x h
       = 14" x 36" x 18"
       = 9072 cu in

ANSWER

The capacity is 9072 cu in.

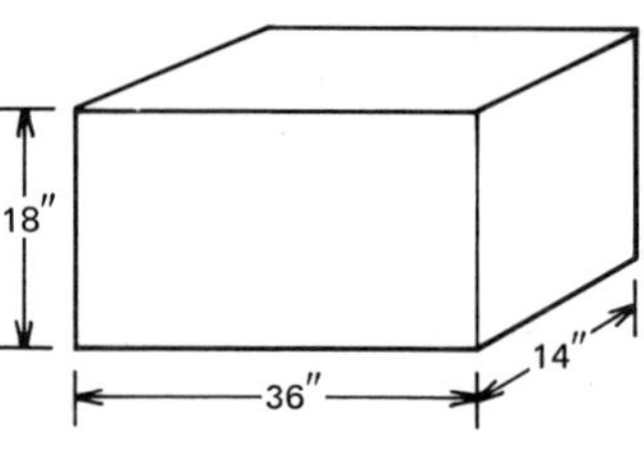

Figure 12-38

Example 12–16

PROBLEM

What is the volume of the cylindrical water storage
tank shown in Figure 12-39?

SOLUTION

1. Recall that the volume of a cylinder equals the
    area of the base circle multiplied by the height.

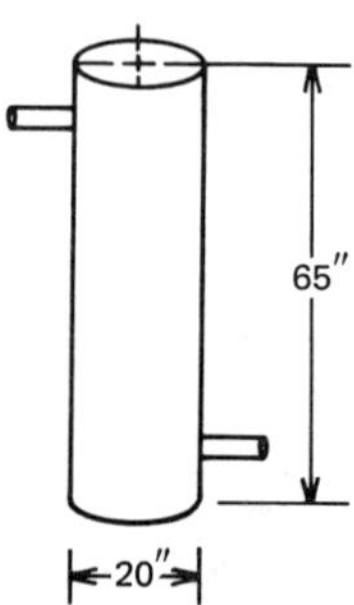

Figure 12-39

2.   Apply the formula and substitute the known values:

$$V = 0.7854d^2h$$
$$= 0.7854(20)^2(65)$$
$$= 20,420.4 \text{ cu in}$$

ANSWER

The volume is 20,420.4 cu in.

Example 12–17

PROBLEM

How many gallons will the tank in Example 12-16 hold?

SOLUTION

1.   Note that 1 gal contains 231 cu in.

2.   Divide 231 into 20,420.4 to determine the number of gallons the tank will hold:

$$\text{capacity in gallons} = \frac{20,420.4}{231}$$
$$= 88.4$$

ANSWER

The tank will hold 88.4 gal.

Applied Welding Problems

12-37.   How many cubic inches are in a casting that measures 12" x 12" x 12"?

12-38.   How many cubic inches are contained in a 4' x 8' x 1/2" sheet of steel?

12-39.   Figure 12-40 shows a 12" length of square tubing.
    a.   How many cubic inches of metal does the tubing contain?
    b.   How many cubic inches does the opening represent?

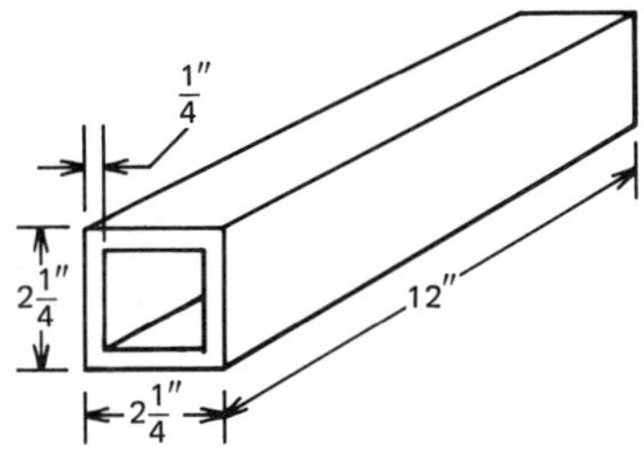

Figure 12–40

12-40. Figure 12-41 shows a metal plate with two
openings.
  a. How many cubic inches of metal were removed
      from the plate when opening A was cut into
      it?
  b. How many cubic inches of metal were removed
      when opening B was cut into it?

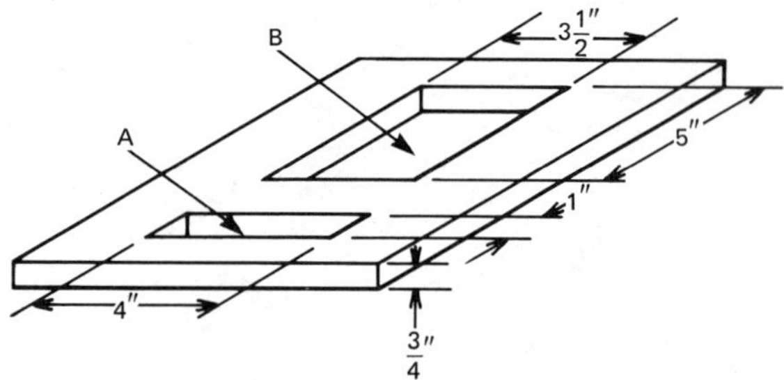

Figure 12-41

12-41. What is the volume of a pipe that is 12" long
and 6" in diameter?

12-42. One cubic inch of cast iron weighs 2 oz. How
many pounds does the casting shown in Figure
12-42 weigh? (1 lb = 16 oz.)

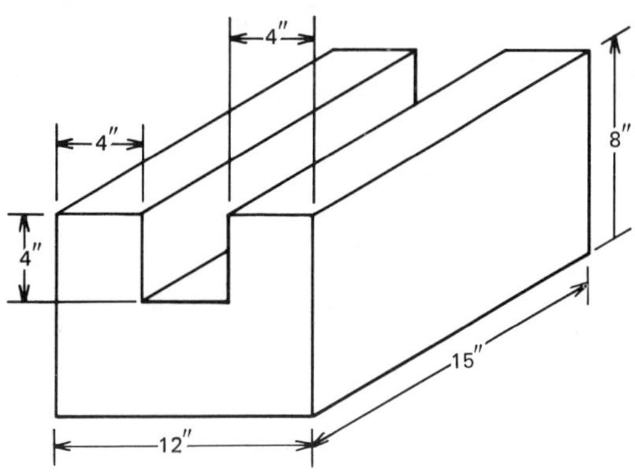

Figure 12-42

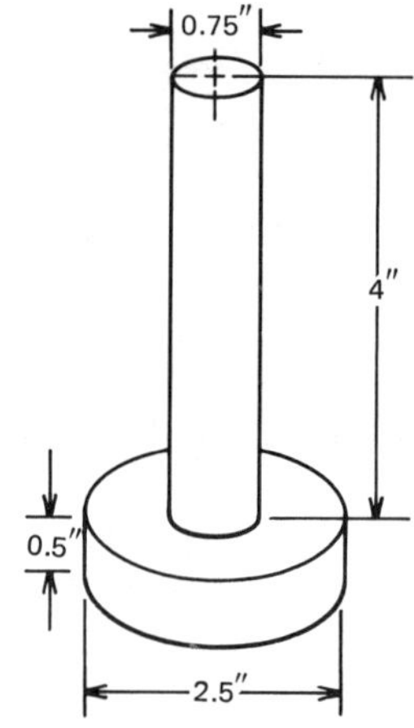

Figure 12-43

12-43. The metal storage area in a building is 24' x
16' x 8'. How many cubic feet of metal storage
does the building contain?

12-44. How many cubic feet of water will a 49" x 38"
x 13" rectangular cooling tank hold?

12-45. What is the volume of a cylindrical length of
steel that is 8" long and 2" in diameter?

12-46. What is the volume of metal in the cast iron
idler wheel and shaft shown in Figure 12-43?

12-47. Figure 12-44 shows a cylindrical fuel tank.
Find its capacity (in gallons) given the

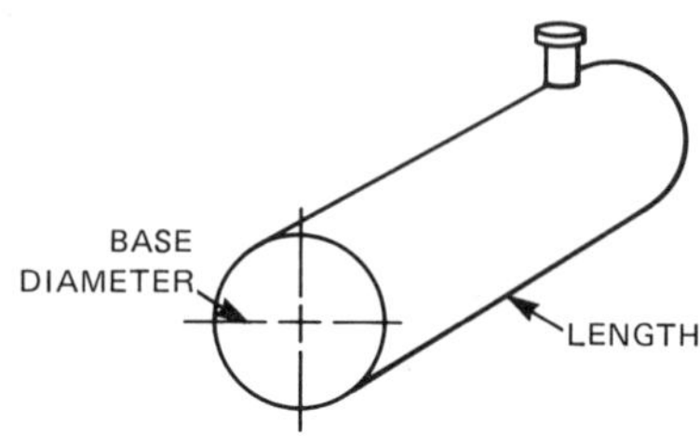

Figure 12-44

following dimensions (1 cu ft = 7.5 gal;
1 gal = 231 cu in):
a.   Base diameter = 24" and length = 30".
b.   Base diameter = 14" and length = 60".
c.   Base diameter = 30" and length = 15".

12-48.   Figure 12-45 shows a rectangular fuel tank.
Find its capacity (in gallons) given the
following dimensions:
a.   Length = 32", width = 12", and height = 15".
b.   Length = 18", width = 9", and height = 26".
c.   Length = 76", width = 24", and height = 18".

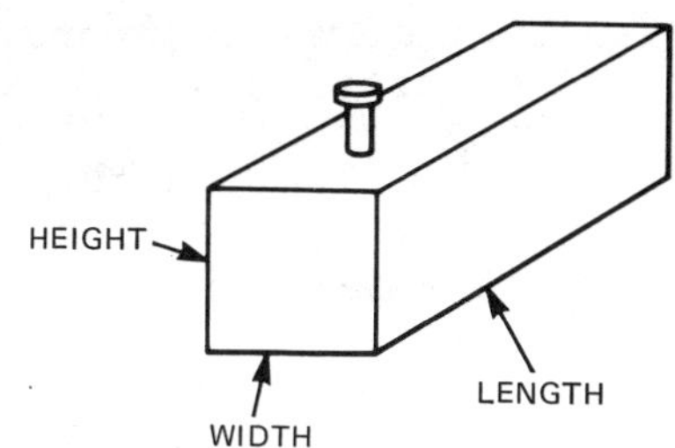

Figure 12-45

12-49.   Figure 12-46 shows a special tank with two
parts, A and B, and a filler pipe, C.
a.   What is the capacity of part A?
b.   What is the capacity of part B?
c.   What is the total capacity of the tank
including filler pipe C?

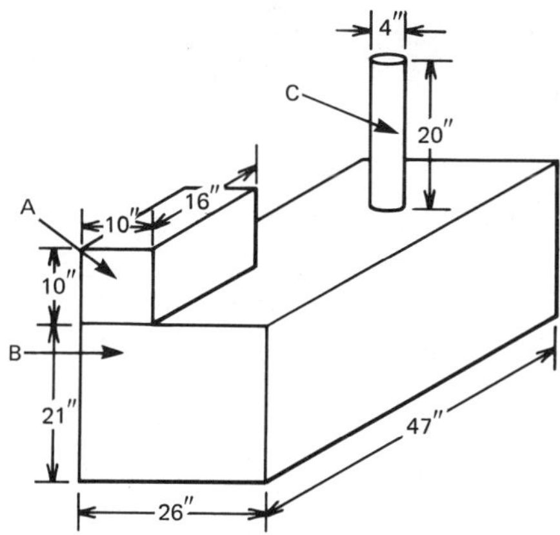

Figure 12-46

12-50.   Figure 12-47 shows a special tank with two
parts, D and E, and two filler pipes, F and G.
a.   What is the capacity of part D?
b.   What is the capacity of part E?
c.   What is the total capacity of the tank including
the two filler pipes?

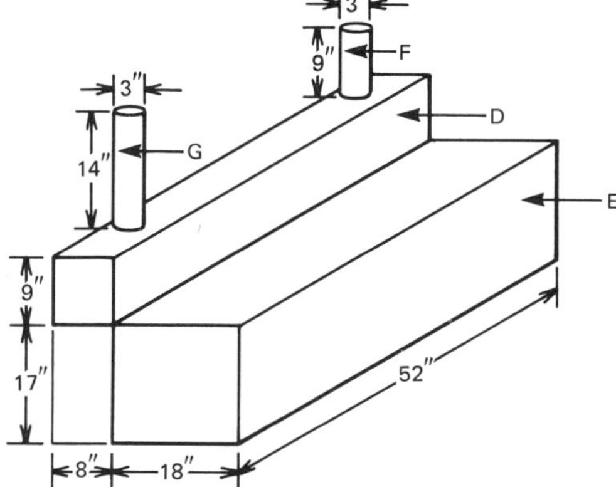

Figure 12-47

12-51.   Figure 12-48 shows a special tank with two
         parts, H and I, and a filler neck, J.
         a.   What is the capacity of part H?
         b.   What is the capacity of part I?
         c.   What is the total capacity of the tank
              including the filler neck J?

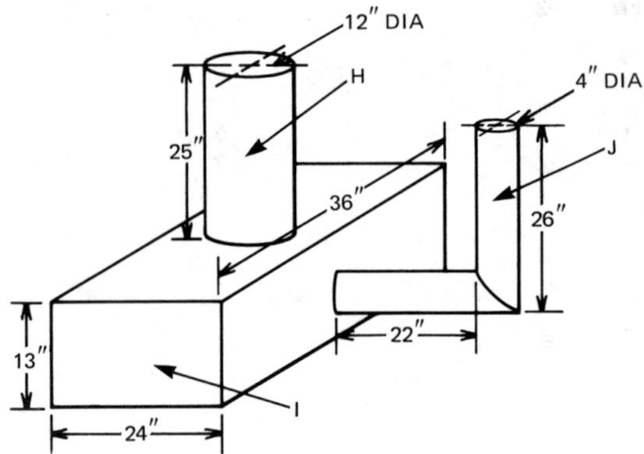

Figure 12-48

12-52.   Figure 12-49 shows a fermentation unit with
         three parts: A, B, and C.
         a.   What is the capacity of part A?
         b.   What is the capacity of part B?
         c.   What is the total capacity of the unit
              including part C?

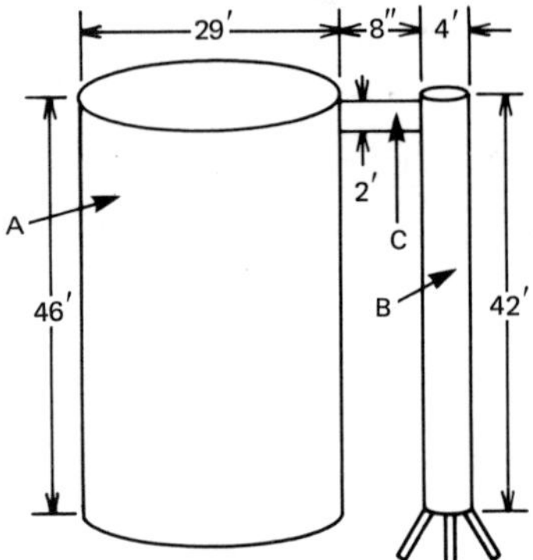

Figure 12-49

Chapter 12: Volume
Applied Welding Problems Survey Test

Name ____

Course/Sec. _______________   Date _____________________

12-1.   How many cubic inches of space are contained in
        a metal box that is 26" long, 14" wide, and 12"
        high?

12-1. _____________________

12-2.   The container shown in Figure 12-50 is used in a
        galvanizing process. In compartment A will be
        a mild acid. In compartment B will be a water
        rinse. In compartment C will be an acid
        neutralizer.

   a.   What is the capacity (in cubic inches) of
        compartment A?

12-2a. _____________________

   b.   What is the capacity of compartment B?

12-2b. _____________________

   c.   What is the capacity of compartment C?

12-2c. _____________________

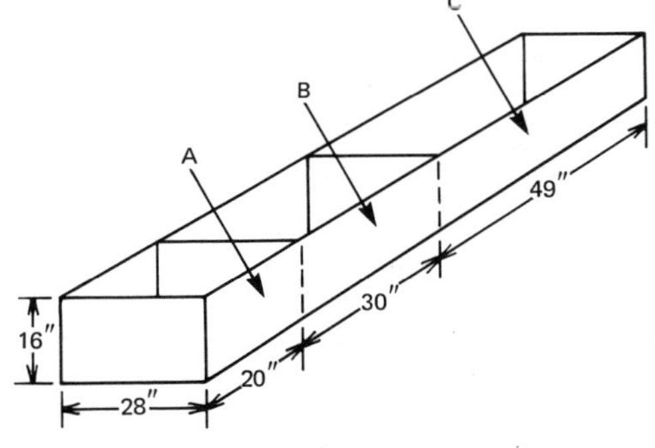

Figure 12-50

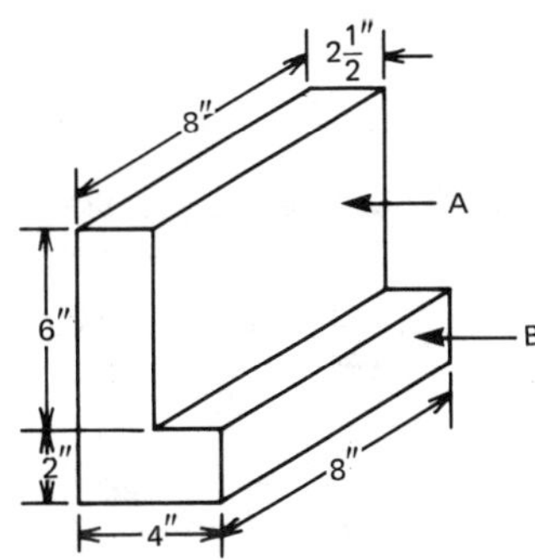

Figure 12-51

12-3.   The container shown in Figure 12-51 is a gas tank
        for a lawn mower.

   a.   What is the capacity (in cubic inches)
        of part A?

12-3a. _____________________

   b.   What is the capacity of part B?

12-3b. _____________________

12-4.   What is the volume of a natural gas pipe that is 50" long and 6" in diameter?

12-4. ______________

12-5.   What is the volume in cubic inches of a water main pipe that is 100' long and 6" in diameter?

12-5. ______________

12-6.   How many gallons will a 100' length of 6" crude oil pipe hold? (1 gal = 231 cu in.)

12-6. ______________

12-7.   The storage tank shown in Figure 12-52 has three compartments to contain three different chemicals for a plating shop.

    a.   What is the capacity, in gallons, of compartment A (to the nearest gallon)?

12-7a. ______________

    b.   What is the capacity of compartment B?

12-7b. ______________

    c.   What is the capacity of compartment C?

12-7c. ______________

    d.   What is the volume, in cubic inches, of the whole tank?

12-7d. ______________

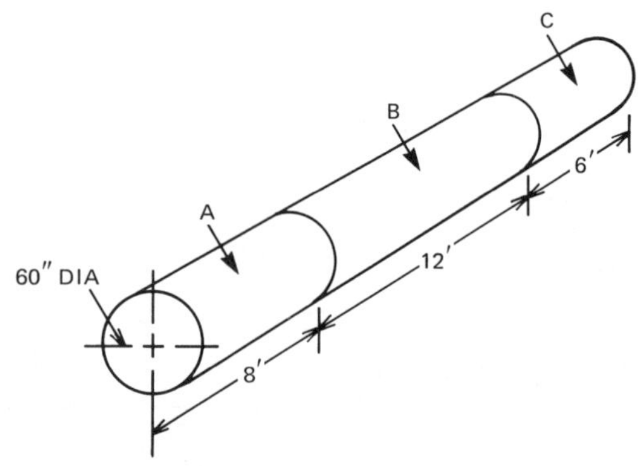

Figure 12-52

12-8.   Figure 12-53 shows a casting made of two parts: A and B.

    a.   How many cubic inches of cast iron make up part A?

12-8a. ______________

    b.   How many cubic inches of cast iron make up part B?

12-8b. ______________

    c.   The casting weighs 2 oz per cubic inch. What is the total weight of the casting?

12-8c. ______________

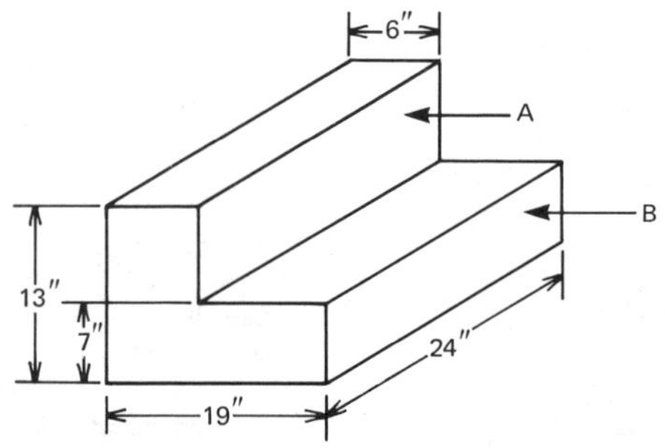

Figure 12-53

12-9. How many gallons does the cooling tank shown in   12-9. ___________________
       Figure 12-54 contain?   (1 cu ft = 7.5 gal.)

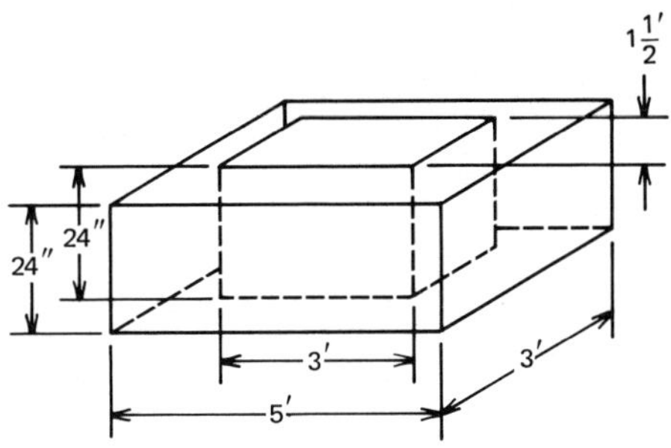

Figure 12-54

Chapter 13: The Metric System
Preview and Self-Test

Name ____

Course/Sec. ______________ Date ____________________

Chapter 13 covers the basic definitions and properties of the metric system, including length, area, volume, weight, and temperature. It also covers metric and English system conversions. Before you turn to the chapter, complete the Self-Test to determine which sections in the chapter you need to study carefully.

Self-Test

BASIC UNITS OF THE METRIC SYSTEM

13-1. Which prefix in the metric system indicates 1000?   13-1. ________________

13-2. Which Latin prefix indicates 0.01?   13-2. ________________

                    Basic units of the metric system score   ________________

CONVERTING FROM ONE METRIC UNIT TO ANOTHER

13-3.   567mm = ________ cm   13-3. ________________

13-4.   184cm = ________ m   13-4. ________________

13-5.   4.3kg = ________ mg   13-5. ________________

13-6.   Convert 50500m^2 to hectares.   13-6. ________________

13-7.   How many hectares are in a plot of land 0.5km x 650m?   13-7. ________________

13-8.   28.4dℓ = ________ ℓ 13-8. ________________

13-9.   0.87kℓ = ________ dℓ 13-9. ________________

13-10.   How many m^3 can be excavated from a pit that measures 55m x 33.4m x 14.8dm?   13-10. ________________

13-11.   4860mg = ________ g   13-11. ________________

13-12.   5.5kg = ________ ℓ 13-12. ________________

13-13.   How many metric tons of water are in a swimming   13-13. ________________
          pool 50m x 25m that is filled to an average
          depth of 180cm?

          Converting from one metric unit to another score   ________________________

TEMPERATURE CONVERSION

13-14.   72° F = ________°C   13-14. ________________

13-15.   −10° C = ________°F   13-15. ________________

                          Temperature conversion score   ________________________

METRIC AND ENGLISH SYSTEM CONVERSIONS

13-16.   5' 7" = ________ cm   13-16. ________________

13-17.   121 lb = ________ kg   13-17. ________________

13-18.   Find the cost of carpeting a room that is   13-18. ________________
          15' x 18' if the cost is $21.98/m3.

13-19.   In returnable bottles, Pepsi sells at 99¢ for   13-19. ________________
          2 ℓ. What is the cost per ounce to the nearest
          tenth of a cent?

13-20.   A compact car manufacturer claims his car will   13-20. ________________
          get 14.03km/ℓ. This rate is equivalent to how
          many miles per gallon?

          Metric and English system conversion score   ________________________

After you have checked your answers on the Self-Test,
transfer your scores to the Chapter 13 Objectives.

Chapter 13

The Metric System

Upon successful completion of this chapter, you will
know the basic units and properties of the metric
system, including length, area, volume, weight, and
temperature. You will also be able to convert from
the metric system to the English system and from the
English system to the metric system.

Objectives

	Self-Test Scores	If you did poorly on the Self-Test, turn to:
· Basic units of the metric system	__________	Section 14-2
· Converting from one metric unit to another	__________	Section 14-3 through 14-6
· Temperature conversion	__________	Section 14-7
· Metric and English system conversions	__________	Section 14-8

13–1   Introduction to the Metric System

The metric system is much easier to understand than
the English system. The metric system is a common
language of measurement for the world. In the United
States, the movement toward the metric system is
increasing. Certain U.S. industries now use the metric
system, and many more will do so as international trade
and exchange of equipment and parts become more common.

The official name for the international metric
system is the International System of units (SI). The
system was proposed by Gabriel Mouton in France in 1670.
The metric system is based on multiples of 10, just like
our number system, and is very similar to our money
system. The metric system does not use commas in
numbers or periods in abbreviations.

13–2   Basic Units of the Metric System

There are seven basic units of measure in the SI
metric system: meter (length), kilogram (mass or
weight), Kelvin (temperature), second (time), ampere
(electricity), candela (luminous intensity), and mole
(amount of substance). Other commonly used metric units
are liter (capacity or volume) and Celsius (temperature).
We will study the four most commonly used units: meter,
liter, kilogram, and Celsius degree.

The metric system is based on powers of 10. Each
unit may be multiplied or divided by 10 to change a
base unit. Study the chart in Figure 13-1, which indicates metric units and the equivalent decimal units.

Prefixes (beginnings of words) are added to the
base unit as a way to identify the unit. Greek prefixes
are used for units larger than the basic metric unit:
kilo = 1000, _hecto_ = 100, and _deka_ = 10. Latin prefixes
are used for units smaller than the basic metric unit:
deci = 0.1, _centi_ = 0.01, _milli_ = 0.001, and _micro_ =
0.0001. Therefore, a kilometer equals 1000 meters and
100000 centimeters.

METRIC UNITS ←→	DECIMAL UNITS
MICRO + BASE UNIT ←→	MILLIONS
MILLI + BASE UNIT ←→	THOUSANDS
CENTI + BASE UNIT ←→	HUNDREDS
DECI + BASE UNIT ←→	TENS
BASE UNIT ←→	ONES
DEKA + BASE UNIT ←→	TENTHS
HECTO + BASE UNIT ←→	HUNDREDTHS
KILO + BASE UNIT ←→	THOUSANDTHS
MEGA + BASE UNIT ←→	MILLIONTHS

Figure 13-1

Procedure for Converting from One Metric Unit to Another

• •

1. To convert from a smaller unit to a larger unit,
   move the decimal point to the left. The number of
   places moved should be the same as the number of
   zeros in the larger unit.
2. To convert from a larger unit to a smaller unit,
   move the decimal point to the right. The number of
   places moved should be the same as the number of
   zeros in the larger unit.
3. Another way to convert from one unit to another is
   to multiply by the "1 unit."

• •

13–3   Metric Length

The meter is the standard unit of measure for length.
Originally, the length of a meter was defined as one
ten-millionth of the distance from the North Pole to the
equator measured along a line running through Paris.
Today the length of a meter is defined as exactly
1,650,763.73 times the wavelength of the orange light
emitted by krypton-86 gas.

The approximate length of a meter is the distance
from the tip of the fingers of the extended arm to the
tip of the nose of an average adult when turning his or
her head away from the extended arm. See the sketch in
Figure 13-2.

The approximate length of a millimeter is the
thickness of a dime. There are 1000 millimeters in a
meter.

The metric system, built on powers of 10, makes it
easy to convert from one unit to another by multiplying
or dividing by 10. Even easier conversions may be made
by moving the decimal point to the right or to the left.

Table 13-1 illustrates the various meter units,
their abbreviations, and their equivalents in meters.
Abbreviations are generally used in the metric system;
the complete words are seldom stated.

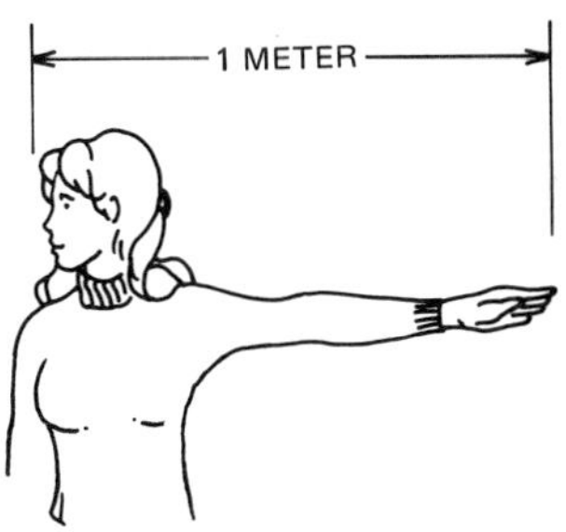

Figure 13-2

Table 13-1. Meter Table of Equivalents

Meter Unit	Abbreviation		Equivalent in Meters	
1 kilometer	km	=	1000	meters
1 hectometer	hm	=	100	meters
1 dekameter	dam	=	10	meters
1 meter	m	=	1	meter
1 decimeter	dm	=	0.1	meter
1 centimeter	cm	=	0.01	meter
1 millimeter	mm	=	0.001	meter

The most frequently used units of the meter are
kilometer, centimeter, and millimeter. One unit can be
changed to another by merely moving the decimal point
to the right or to the left.

Example 13–1

PROBLEM

Convert 6500m (meters) to km (kilometers).

SOLUTION

1. Note from the table of equivalents that 1000m = 1km.

2. To convert m to km, move the decimal point three
    places to the left:

    6500m = 6.500km = 6.5km

3. Another method of conversion is to multiply by the
    "1 unit":

    1000m = 1km   or $1 = \dfrac{1km}{1000m}$

    $$\left(\dfrac{6500m}{1}\right)\left(\dfrac{1km}{1000m}\right) = 6.5km$$

ANSWER

    6500m = 6.5km

Exercises: Converting from one Unit of Metric Length
to Another

Convert the following as indicated:

13-1.   8000m = _________ km

13-2.   5.6cm = _________ mm

13-3.   1467mm = _________ m

13-4.   567000cm = _________ km

13-5.   8.54km = _________ cm

13-6.   81cm = _________ mm

13-7.   35km = _________ m

13-8.   0.526m = _________ dm

13-9.   1.83km = _________ m

13-10.   17760000mm = _________ km

13–4   Metric Area

Area is the number of square units an object contains.
Area is always expressed in square units. Examples of
area in the metric system include square centimeters,
written cm^2; square meters, written m^2; and so on. In
the metric system, units of area are abbreviated with an
exponent of 2 (above and to the right of the unit).

    Land is measured by hectares (ha) in the metric
system. A hectare is equal to $10000m^2$ or to a plot of
land 100m x 100m:

    $1ha = 10{,}000m^2$

To find the hectares in a plot of land, find the area in m^2. Then convert to hectares by dividing by 10000, since 1ha = 10000m^2.

Example 13–2

PROBLEM

Convert 156700m^2 to ha (hectares).

SOLUTION

1. Note that 1ha = 10000m^2.

2. To convert m^2 to ha, move the decimal point 4 places to the left:

    156700m^2 = 15.67ha

3. Another method of conversion is to multiply by the "1 unit":

    $$\left(\frac{156700m^2}{1}\right)\left(\frac{1ha}{10000m^2}\right) = 15.67ha$$

ANSWER

   156700m^2 = 15.67ha

Exercises: Converting from One Unit of Metric Area to Another

Convert the following as indicated:

13-11.   1540000m^2 = ________ ha

13-12.   76.4ha = ________ m^2

13-13.   6643m^2 = ________ ha

13-14.   How many hectares are there in a plot of land 44m x 40m?

13-15.   A tract of land is 0.903km x 3.09km. How many hectares does it contain?

13–5   Metric Volume and Capacity

The volume or capacity of an object is determined by the number of cubic units in the object. It is found by multiplying the length by the width by the height: $V = lwh$.
   The basic unit for determining volume in the metric system is cubic meter (m^3). The exponent 3 indicates cubic measure.

Volume is the amount of space within a container.
Capacity is the amount of substance a container will
hold. Because their meanings are so similar, the words
are used interchangeably.

The most common unit of volume in the metric
system, although non-SI, is the liter (ℓ). The most
common unit of capacity in the metric system is the
cubic meter, or m^3. A liter occupies one cubic
decimeter ($1dm^3$)--that is, a cube 10cm long, 10cm wide,
and 10cm high, or $1000cm^3$.

Figure 13-3 shows the relationship between volume
and capacity. In general, volume refers to a solid or
nonliquid measure, and capacity to a liquid measure.

In the medical field, a common cubic measurement
is cc, the medical abbreviation for cubic centimeter.
(However, the unit abbreviation should be cm^3 according
to the metric system.)

Table 13-2 shows the relationship of 1 liter in
liquid to other units of liquid volume in the metric
system.

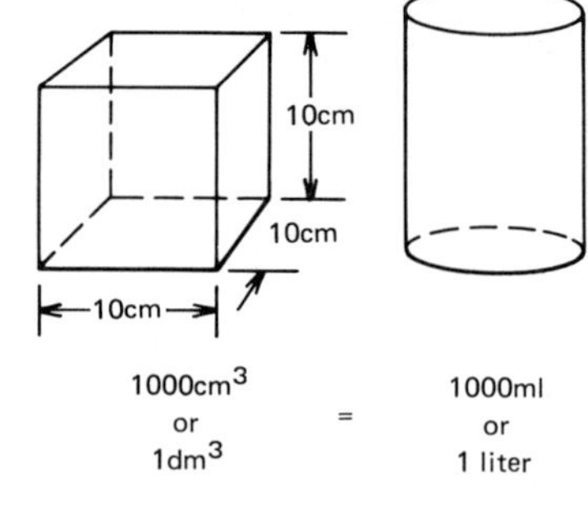

Figure 13-3

Table 13-2. Liter Table of Equivalents

Liquid Capacity		Volume		
1 kiloliter (kℓ)	= 1000	liters	= $1000dm^3$	= $1.0m^3$
1 hectoliter (hℓ)	= 100	liters	= $100dm^3$	= $0.1m^3$
1 dekaliter (daℓ)	= 10	liters	= $10dm^3$	= $0.01m^3$
1 liter (ℓ)	= 1	liter	= $1dm^3$	= $0.001m^3$
1 deciliter (dℓ)	= 0.1	liter	= $100cm^3$	= $0.0001m^3$
1 centiliter (cℓ)	= 0.01	liter	= $10cm^3$	= $0.00001m^3$
1 milliliter (mℓ)	= 0.001	liter	= $1cm^3$	= $0.000001m^3$

Changing from one unit of metric volume or
capacity may be done by merely moving the decimal point
to the right or to the left.

Example 13–3

PROBLEM

Convert 5.4 ℓ (liters) to mℓ (milliliters).

SOLUTION

1. Note that 1 ℓ = 1000mℓ.

2. To convert ℓ to mℓ, move the decimal point 3
    places to the right:

    5.4 ℓ = 5400mℓ

3. Another method of conversion is to multiply by the "1 unit":

$$\left(\frac{5.4\ \ell}{1}\right)\left(\frac{1000m\ell}{1\ell}\right) = 5400m\ell$$

ANSWER

5.4 ℓ = 5400mℓ

Exercises: Converting from One Unit of Metric Volume to Another

Convert the following as indicated:

13-16.   9.3 ℓ = _________ mℓ 13-17.   5670mℓ = _________ ℓ

13-18.   7640 ℓ = _________ kℓ 13-19.   1.4kℓ = _________ ℓ

13-20.   7500cc = _________ ℓ 13-21.   3.6 ℓ = _________ cm^3

13-22.   5807cℓ = _________ ℓ 13-23.   94.3 ℓ = _________ cℓ

13-24.   5.86dm^3 = _________ ℓ

13-25.   A 24 ℓ cooling system contains how many cubic decimeters?

13-26.   How many cubic meters must be excavated for a pit 32.4m long x 12.6m wide x 3.8m deep?

13-6   Metric Weight or Mass

Weight is the measure of the force of gravity upon an object. Mass is the quantity of material contained in an object. These definitions of mass and weight are technical ones. For everyday application, the words are used interchangeably. Most of the time, we refer to the weight of an object.

The kilogram is the basic SI metric unit of measurement for mass or weight. An average-sized paper clip weighs 1 gram (g). Thus, 1000 paper clips equal 1 kilogram (kg).

The beauty of the metric system is in the relationship between volume, liquid capacity, and weight (mass). Figure 13-4 and Table 13-3 show this relationship. Thus, 1cm^3 = 1mℓ = 1g, and a metric ton = 1m^3 = 1kℓ = 1000kg.

Example 13-4

PROBLEM

Convert 29g (grams) to mg (milligrams).

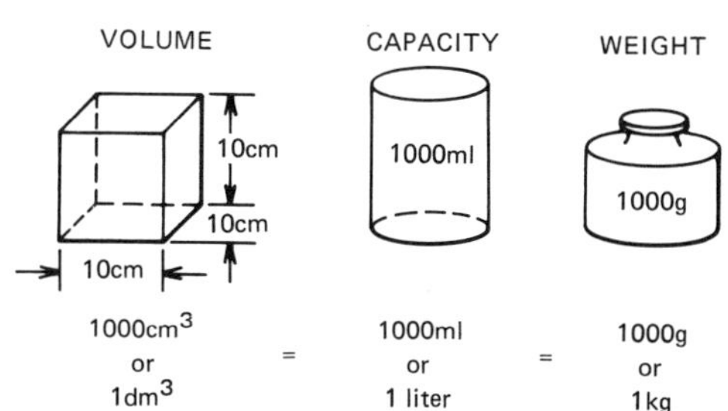

Figure 13-4

SOLUTION

1. Note that 1mg = 0.001g, or 1000mg = 1g.

2. To convert g to mg, move the decimal point 3 places
    to the right:

    29g = 29000mg

3. Another method of conversion is to multiply by the
    "1 unit":

    $$\left(\frac{29\cancel{g}}{1}\right)\left(\frac{1000mg}{1\cancel{g}}\right) = 29000mg$$

ANSWER

    29g = 29000mg

Table 13-3. Kilogram Table of Equivalents

Volume	Liquid Capacity	Weight	
$1m^3$	$= 1k\ell$	$= 1000$ kilograms (kg)	$= 1$ metric ton (t)
$1dm^3$	$= 1\ell$	$= 1$ kilogram (kg)	$= 1000$ grams (g)
$100cm^3$	$= 0.1\ell$	$= 1$ hectogram (hg)	$= 100$g
$10cm^3$	$= 0.01\ell$	$= 1$ dekagram (dag)	$= 10$g
$1cm^3$	$= 1m\ell$	$= 1$ gram (g)	$= 1$g
$1mm^3$	$= 0.1m\ell$	$= 1$ decigram (dg)	$= 0.1$g
$0.1mm^3$	$= 0.01m\ell$	$= 1$ centigram (cg)	$= 0.01$g
$0.01mm^3$	$= 0.001m\ell$	$= 1$ milligram (mg)	$= 0.001$g

Exercises: Converting from One Unit of Metric Weight to Another

13-27. 5.2kg = _______ g   13-28. 4050g = _______ kg

13-29. 46900mg = _______ g   13-30. 867cg = _______ g

13-31. 5280g = _______ ℓ 13-32. 12 ℓ = _______ g

13-33. 5.7kℓ = _______ g   13-34. 12000mℓ = _______ kg

13-35. 1980mg = _______ cg   13-36. 59.8dm^3 = _______ ℓ

13-37. 43g = _______ mℓ = _______ cm^3

13-38. 57.8dm^3 = _______ kg = _______ t

13-39.   What weight (in kg) does a bricklayer lift when
          he lifts 8 bricks? The bricks are 6.1cm x
          9.5cm x 20.2cm, and the bricks weigh 2.4g per
          cm^3.

13-40.   The conversion on torque wrenches is from foot
          pounds (English) to newton meters (metric).
          Facts: 1 ft lb = 1.356 N·m, and 1 N·m = 0.738
          ft lb. Find the torque for a Cadillac V-8 in
          foot pounds if the engine develops 550N·m.

13-41.   Find the weight of a 12.5cm cube of brass if
          brass weighs 8.2kg per dm^3.

13–7   Metric Temperature

For measuring temperature, scientists use a basic SI
unit called the <u>Kelvin</u> (K). On the Kelvin scale, water
freezes at 273.15° K and boils at 373.15° K (a range
of 100°). The Kelvin scale is not practical for everyday use. Thus, the Celsius (C) scale was developed
within the metric system.

The <u>Celsius</u> (C) thermometer divides the range of
temperature between the freezing and boiling points of
water into 100 units or degrees. Thus, one degree
Celsius (C) is exactly equal to one degree Kelvin (K).

In the United States, the Fahrenheit scale is
most commonly used to determine temperature. On the
Fahrenheit scale, water freezes at 32° and boils at
212°.

All three temperature scales are shown in Figure
13-5.

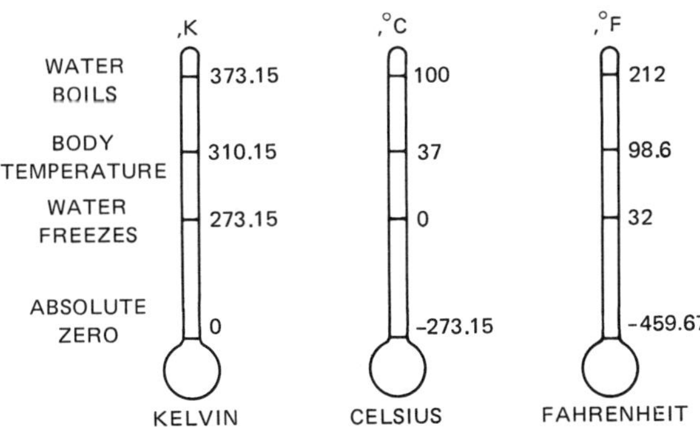

Figure 13-5

Eventually all the world will use the Celsius
temperature scale. During the transition period, the
following formulas make conversions easy, especially
with the use of a calculator: To convert from Celsius
(C) to Fahrenheit (F), use $F = 1.8(C) + 32°$. To
convert from Fahrenheit (F) to Celsius (C), use
$C = 0.6(F - 32°)$.

Procedure for Making Temperature Conversions

●●

1. To convert from Celsius to Fahrenheit, use the following formula:

$$F = 1.8(C) + 32°$$

2. To convert from Fahrenheit to Celsius, use the following formula:

$$C = 0.6(F - 32°)$$

●●

Example 13–5

PROBLEM

Convert 50° F (Fahrenheit) to C (Celsius).

SOLUTION

Use the formula $C = 0.6(F - 32°)$:

$$C = 0.6(F - 32°)$$
$$= 0.6(50° - 32°) = 0.6(18°) = 10.8°$$

ANSWER

$$50° F = 10.8° C$$

Exercises: Making Temperature Conversions

Convert the following as indicated:

13-42.   65° F = _______ ° C   13-43.   32° F = _______ ° C

13-44.   22° C = _______ ° F   13-45.   100° F = _______ ° C

13-46.   98.6° F = _______ ° C   13-47.   -40° F = _______ ° C

13-48.   -10° F = _______ ° C   13-49.   -10° C = _______ ° F

13-50.   The baking temperature for frozen cookies is 375° F. What is this temperature on a Celsius thermometer?

13-51.   The melting point of aluminum is 660.1° C. At what Fahrenheit temperature does aluminum melt?

13–8   Metric and English Systems Conversions

During the transition period from the English to the metric system, converting from one system to the other

will be necessary. Electronic calculators will be of
great help for conversions. Some industries will also
have conversion charts for frequently used quantities.
However, charts will not always be available for every
conversion. Thus, we will sometimes have to convert
from metric to English and from English to metric. In
some industries, both systems may be used for a while.
Tables 13-4 through 13-7 give some of the common
metric and English equivalents.

Table 13-4. Length Equivalents

Metric to English	English to Metric
1mm = 0.03937 in	1 in = 25.4mm
1cm = 0.3937 in	= 2.54cm
= 0.0328 ft	1 ft = 30.48cm
1m = 39.37 in	= 0.3048m
= 3.2808 ft	1 yd = 91.44cm
1km = 3280.8 ft	= 0.9144m
= 0.62136 mi	1 mi = 1609.344m
	= 1.609km

Table 13-5. Area Equivalents

Metric to English	English to Metric
$1cm^2$ = 0.155 sq in	1 sq in = $6.4516cm^2$
$1m^2$ = 10.7639 sq ft	1 sq ft = $0.09290m^2$
1ha = 2.471 acres	1 sq yd ‒ $0.83613m^2$
	1 acre = 0.4047ha

Table 13-6. Capacity and Volume Equivalents

Metric to English	English to Metric
1mℓ = 0.0338 oz	1 oz (liquid) = 29.75mℓ
1ℓ = 1.0568 qt	1 qt (liquid) = 0.946 ℓ
= 0.2642 gal	1 gal (liquid) = 3.785 ℓ
$1cm^3$ = 0.061 cu in	1 cu in = $16.387cm^3$
$1dm^3$ = 61.0237 cu in	1 cu ft = $28.317dm^3$
= 0.0353 cu ft	= $0.0283m^3$
$1m^3$ = 35.3147 cu ft	1 cu yd = $0.7646m^3$
= 1.3079 cu yd	

Table 13-7. Weight and Mass Equivalents

Metric to English	English to Metric
1mg = 0.0000352 oz	1 oz (dry) = 28.349g
1g = 0.0352 oz	1 lb (dry) = 0.4536kg
1kg = 2.2046 lb	1 lb = 0.4536kg
1 metric ton = 2204.6 lb	1 short ton = 907.2kg
	1 long ton = 1016kg

Procedure for Making Metric and English System Conversions

●●●

1. Refer to the tables of metric and English system equivalents.
2. Multiply by the "1 unit" to convert from one system to the other.

●●●

Example 13–6

PROBLEM

Convert 6 ft to cm.

SOLUTION

1. Note from Table 13-4 that 1 ft = 30.48cm.

2. Use the "1 unit" to convert ft to cm:

$$\left(\frac{6 \text{ ft}}{1}\right)\left(\frac{30.48\text{cm}}{1 \text{ ft}}\right) = 182.88\text{cm}$$

ANSWER

6 ft = 182.88cm

Exercises: Making Metric and English Conversions

Convert the following as indicated:

13-52.   6 ft = _______ cm   13-53.   220cm = _______ ft

13-54.   12 in = _______ mm   13-55.   500mm = _______ in

13-56.   2 qt = _______ ℓ   13-57.   10 ℓ = _______ qt

13-58.   500mℓ = _______ pt   13-59.   1/2 pt = _______ mℓ

13-60.   3 lb = _______ kg   13-61.   20kg = _______ lb

13-62.   5 long tons = _________ kg

13-63.   50t = _________ lb

13-64.   60 mi/hr = _________ km/hr

13-65.   120km/hr = _________ mi/hr

13-66.   A 6 oz can of frozen orange juice is equal to
         how many milliliters?

13-67.   A 747 jet liner travels at 620 mi/hr. How long
         will it take to go 580km?

13-68.   A prime rib at Victoria Station weighs at least
         230g. How many ounces does it weigh?

13-69.   A 6-pack of 2 ℓ bottles of 7-Up costs $3.98.
         What is the cost per ounce?

13-70.   What will it cost to carpet a room 11.5m x
         8.25m if carpet costs $18.95 per square yard?

13-71.   A 4' x 8' sheet of Formica is purchased to
         cover a counter 120cm x 200m. What percent is
         waste?

13-72.   The spark plug gap for a 350 V-8 Chevy engine
         is 0.035". What is this dimension in
         millimeters?

13-73.   Find the cost of filling a 50m x 20m swimming
         pool 3m deep if the water cost is $0.0048/cu ft.

13-74.   On a Canadian hunting trip, John Emmer bought
         $28.54 of gas for his Suburban. The stated
         price was 22.3¢/ℓ. Find the cost per gallon.

13-75.   At the cost of 22.3¢/ℓ, what would gasoline
         cost if you traveled 2450 mi and got 11.5
         mi/gal?

13-76.   John's Suburban makes 11.5 mi/gal. How many
         kilometers per liter would this rate equal?

Chapter 13

The Metric System

···

Solving Applied Welding Problems

Now that you have mastered the fundamental operations
with the metric system, you are ready to apply these
math skills to typical problems that welders encounter
on the job. After studying the examples that follow,
complete the applied welding problems.

Example 13–7

PROBLEM

A length of round stock 4dm long is welded to a length
of square stock 6dm long. What is the total length in
meters?

SOLUTION

1. Note that 10dm = 1m.

2. Add the two lengths and convert to meters:

    4dm + 6dm = 10dm = 1m

ANSWER

The total length of the round and square stock is 1m.

Example 13–8

PROBLEM

A 1m length is cut from a length of square tubing 175cm
long. What is the remaining length in centimeters?

SOLUTION

1. Note that 100cm = 1m.

2. Convert the 1m length to cm and subtract from the
   175cm length:

   175cm - 100cm = 75cm

ANSWER

The remaining length is 75cm.

Example 13–9

PROBLEM

What is the length in centimeters of a 24" length of
angle iron?

SOLUTION

1. Note that 1" = 2.540cm.

2. To convert 24" to cm, multiply 24 by 2.540:

   24 x 2.540 = 60.96

ANSWER

A 24" length of angle iron is 60.96cm long.

Example 13–10

PROBLEM

The amount of weld needed to fabricate a shovel bucket
is 75m. What is the length of the weld in feet?

SOLUTION

1. Note that 1' = 0.3048m.

2. To convert 75m to feet, divide 75 by 0.3048:

   75/0.3048 = 246

ANSWER

A 75m length of weld is 246' long.

Applied Welding Problems

Use this chart as a reference for answering questions
that require conversions:

```
    10dm = 1m
   100cm = 1m
 39.37" = 1m
     1" = 2.54cm
     1" = 25.4mm
```

13-77.   A length of pipe is 1m long. What is its length in decimeters?

13-78.   A length of pipe is 50dm long. What is its length in meters?

13-79.   A pipestand is 50cm high. What is its height in meters?

13-80.   Refer to Figure 13-6.
a.   Change the length of part A to decimeters.
b.   Change the length of part B to decimeters.
c.   Change the length of part C to decimeters.
d.   What is the total length in decimeters?

13-81.   Round stock is cut into lengths 15cm long. What will be left from a length 1m long?

13-82.   Refer to Figure 13-7.
a.   How long is part D in meters?
b.   How long is part E in meters?
c.   What is the total length in meters?

13-83.   A length of square stock is 2m long. How many inches long is it?

13-84.   A length of round stock is 14" long. How long is it in centimeters?

13-85.   Figure 13-8 shows a collector box.
a.   How many inches long is part F?
b.   How many inches long is dimension H?
c.   How many inches long is dimension G?
d.   What is the overall length of the collector box in centimeters?

13-86.   Figure 13-9 shows a spacer plate.
a.   How long is part I?
b.   How long is part J?
c.   What is the length in inches of parts I and J?
d.   What is the total length of the plate in inches?

13-87.   A length of steel strap iron is 3m long. What length would remain if 20dm were cut from it?

13-88.   How many inches of steel are required for the sides of the box shown in Figure 13-10?

13-89.   What is the width in meters of the channel shown in Figure 13-11?

13-90.   Refer to Figure 13-12.
a.   What is the height in centimeters of block L?
b.   What is the height in centimeters of block M?

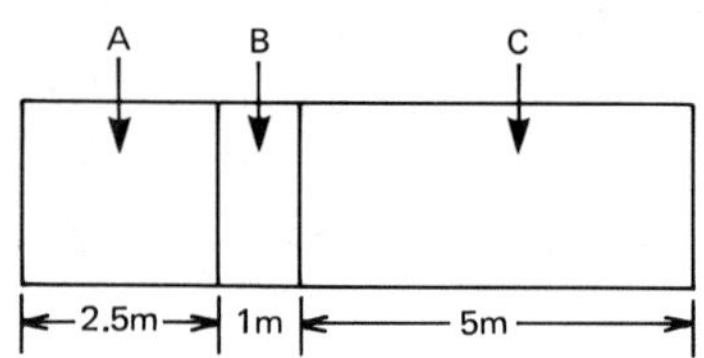

Figure 13-6

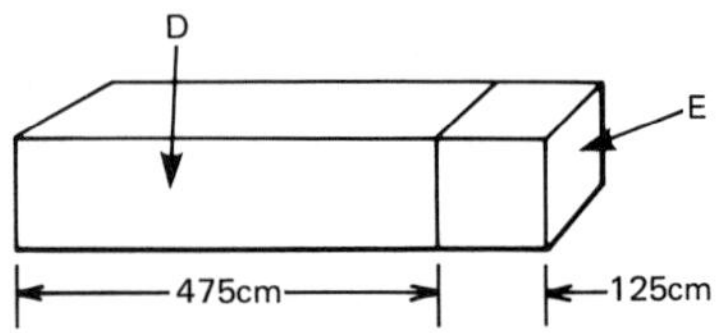

Figure 13-7

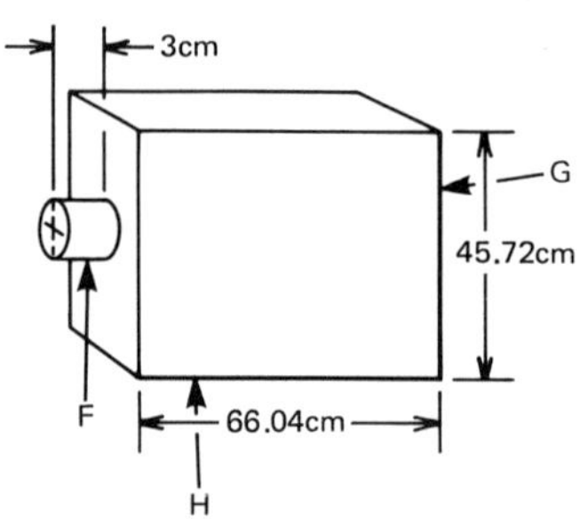

Figure 13-8

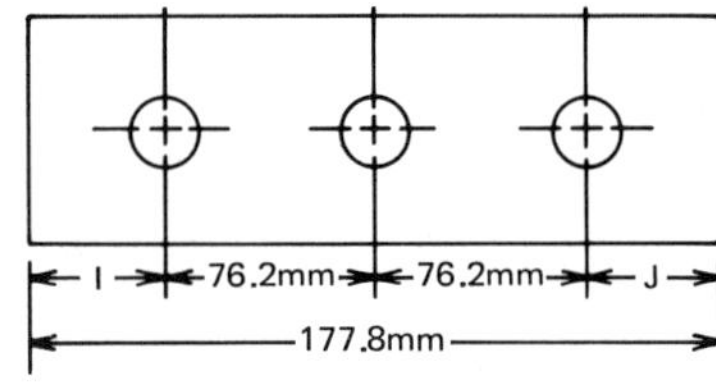

Figure 13-9

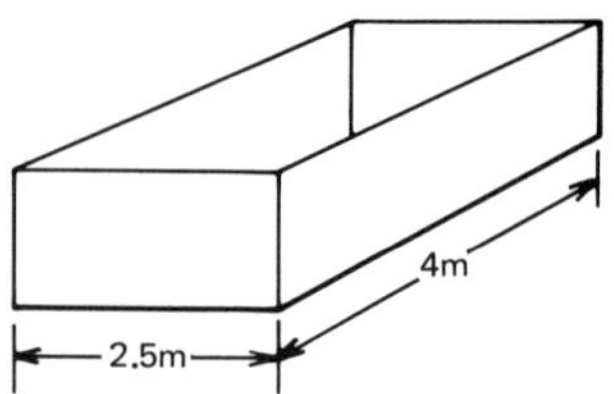

Figure 13-10

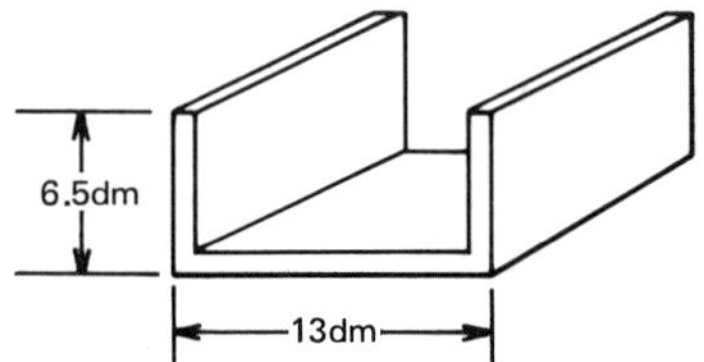

Figure 13-11

    c.   What is the height in centimeters of
        block N?

    d.   What is the total height in centimeters?

13-91. How many sections 6dm long can be cut from a
length of bar stock 3m long?

13-92. A length of channel iron is 2m long. How many
inches long is it?

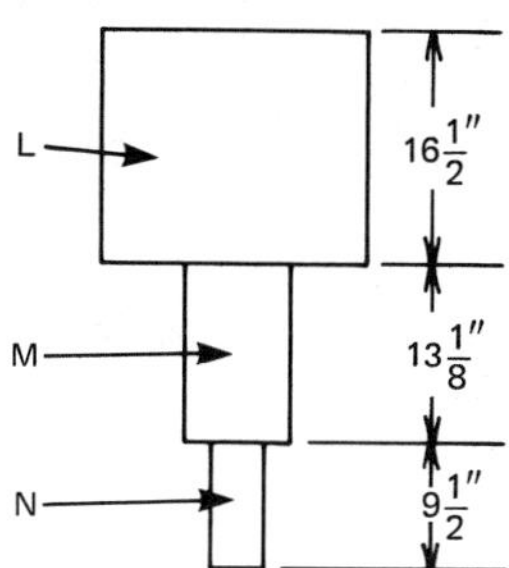

Figure 13-12

Chapter 13: The Metric System
Applied Welding Problems
Survey Test

Name ____

Course/Sec. _______________ Date ___________________

Use this chart as a reference for answering the
following questions:

```
1000mm = 1m
 100cm = 1m
39.37" = 1m
    1" = 2.54cm
    1" = 25.4mm
```

13-1. A length of angle iron is 2.5m long.

    a. How long is it in millimeters?   13-1a. ________________

    b. How long is it in inches?   13-1b. ________________

13-2. A section of bar stock is 12" long.

    a. How long is it in millimeters?   13-2a. ________________

    b. How long is it in centimeters?   13-2b. ________________

13-3. Figure 13-13 shows a length of round stock.

    a. What is its diameter in inches?   13-3a. ________________

    b. What is its length in inches?   13-3b. ________________

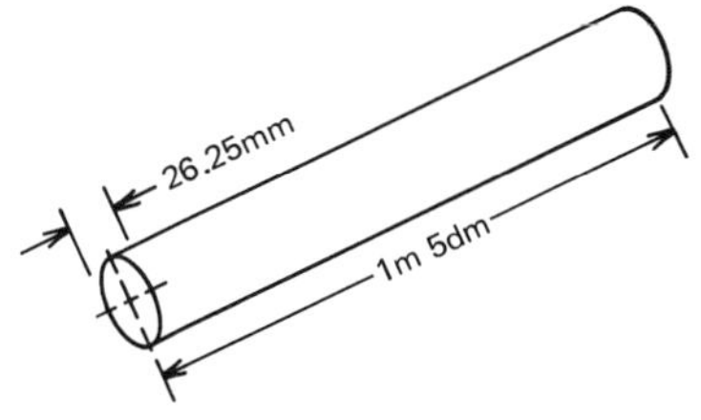

Figure 13-13

"#

13-4.   Figure 13-14 shows a jack stand.

    a.   What is the length of part A in centimeters?

    b.   What is the length of part B in centimeters?

    c.   What is the overall length in millimeters?

13-4a. _________________

13-4b. _________________

13-4c. _________________

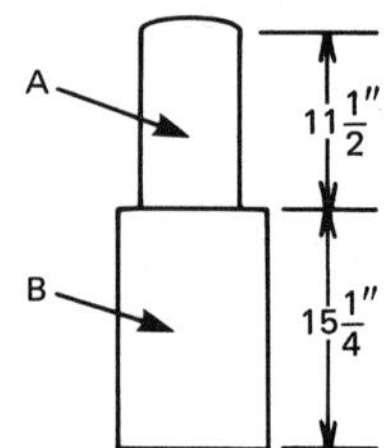

Figure 13–14

13-5.   What is the length in meters of a bar 393.7" long?

13-5. _________________

Answers to
Preview and Self-Tests

Answers to Preview and Self-Tests

CHAPTER 1

1-1.	99	1-8.	82,540	1-15.	121 gal
1-2.	sixty-six	1-9.	6111	1-16.	36,960'
1-3.	42,000,042	1-10.	89,760	1-17.	32 mpg
1-4.	491	1-11.	308	1-18.	87 bottles
1-5.	7161	1-12.	76	1-19.	58'
1-6.	5588	1-13.	89	1-20.	$11
1-7.	52,580	1-14.	37 bushels		

CHAPTER 2

2-1.	24/64	2-10.	1/3	2-19.	2-23/48
2-2.	12/17	2-11.	1-3/16	2-20.	3-1/32"
2-3.	1	2-12.	24-13/15	2-21.	1-25/32"
2-4.	1/6	2-13.	36-3/4	2-22.	51 gal
2-5.	2-1/2	2-14.	1/16	2-23.	5 shelves
2-6.	41-69/80	2-15.	39/40	2-24.	67-1/2"
2-7.	4/7	2-16.	24-33/64	2-25.	118-1/2"
2-8.	64	2-17.	18/41		
2-9.	2-4/7	2-18.	2-8/11		

CHAPTER 3

3-1.	0.052	3-10.	68,476.44	3-18.	$1.439 or
3-2.	4-4/100	3-11.	0.438		$1.43-9/10
3-3.	0.375	3-12.	4.47"	3-19.	43
3-4.	0.3125	3-13.	0.6875"	3-20.	$0.048
3-5.	14.21875	3-14.	4.13"	3-21.	1-7/8"
3-6.	1/8	3-15.	649.648	3-22.	1/16"
3-7.	525/16		cu ft	3-23.	62.89 cu ft
3-8.	8-3/32	3-16.	184.375'	3-24.	12 rev
3-9.	0.825	3-17.	5.3 oz	3-25.	3.5 lb

CHAPTER 4

4-1.	0.45	4-9.	7/40	4-18.	3374.4 lb
4-2.	.085	4-10.	277/200	4-19.	$316.25
4-3.	750%	4-11.	619.11	4-20.	368 sq yd
4-4.	8.75%	4-12.	7.496775	4-21.	105 hours
4-5.	12.5%	4-13.	102.5641	4-22.	1.381875" –
4-6.	46.$\overline{6}$%	4-14.	0.5632		1.368125"
4-7.	515.625%	4-15.	$22,500	4-23.	$239.64
4-8.	37/200	4-16.	$254.11	4-24.	$104,652.28
		4-17.	$814.63	4-25.	$36,505.17

CHAPTER 5

5-1.	like	5-10.	-7670	5-19.	$34x - 7$
5-2.	literal	5-11.	-173	5-20.	2
5-3.	algebraic	5-12.	36	5-21.	x^7
5-4.	smaller	5-13.	-4704		
5-5.	numerical	5-14.	0	5-22.	$s^8 j^6$
5-6.	11	5-15.	12	5-23.	105
5-7.	1098	5-16.	-99	5-24.	$8yz^2$
5-8.	2390	5-17.	-7		
5-9.	90	5-18.	$-40a$	5-25.	$81x^6 y$

CHAPTER 6

6-1.	4	6-6.	450	6-11.	108.97
6-2.	11	6-7.	1385.4456	6-12.	352
6-3.	720	6-8.	233.05	6-13.	53.3
6-4.	12	6-9.	97	6-14.	-0.67763
6-5.	17	6-10.	1.143	6-15.	261

CHAPTER 7

7-1.	32	7-8.	48.8	7-15.	12
7-2.	-1	7-9.	8	7-16.	-10
7-3.	-12	7-10.	0.56	7-17.	6
7-4.	-36	7-11.	5	7-18.	-3
7-5.	5	7-12.	24.5	7-19.	1
7-6.	34	7-13.	2.5	7-20.	6
7-7.	-184.2	7-14.	49.758916		

CHAPTER 8

8-1. $d = y - c$
8-2. $c = P - a - b$
8-3. $B = P - 2s - b$
8-4. $r = I/pt$
8-5. $s = P/3$
8-6. $b = 2A/h$
8-7. $a = 4A/\pi b$
8-8. $s = \sqrt{A}$
8-9. $d = \sqrt{\dfrac{A}{0.7854}}$
8-10. $r = \sqrt{\dfrac{V}{\pi h}}$
8-11. $b = P/\pi - a$
8-12. $r = \sqrt{\dfrac{360A}{\Theta\pi}}$

8-13. $h = \dfrac{\sqrt{3}s}{2}$
8-14. $s = \sqrt{\dfrac{2A}{3\sqrt{3}}}$
8-15. $h = (T - aA)/p$
8-16. $N = \dfrac{1.372}{(D - D_1)}$
8-17. $d = \sqrt[3]{6V/\pi}$
8-18. $r = A/4\pi^2 R$
8-19. $a = \sqrt{\dfrac{P^2}{s\pi^2} - b^2}$
8-20. $M = \dfrac{3V}{2h} - \dfrac{B_1}{4} - \dfrac{B_2}{4}$

CHAPTER 9

9-1.   29/13
9-2.   16/1
9-3.   3
9-4.   5.3
1-5.   5
9-6.   1.645
9-7.   220

9-8.   24
9-9.   2
9-10. 0.0014
9-11. 28
9-12. 6
9-13. 12.6
9-14. -2

9-15. 5760 lb
9-16. $20,625.00
9-17. 183 mi
9-18. 5-1/7 hours
9-19. 531.878 lb
9-20. 2596 rev

CHAPTER 10

10-1.   3-5/8
10-2.   4-7/8
10-3.   6-3/16
10-4.   11.5
10-5.   14.4
10-6.   18.9
10-7.   straight
10-8.   parallel

10-9.   polygon
10-10. triangle
10-11. square
10-12. trapezoid
10-13. octagon
10-14. rhombus
10-15. vertex

10-16. counter-clockwise
10-17. complementary
10-18. 76°
10-19. 136°
10-20. See text

CHAPTER 11

11-1.   35.6 ft
11-2.   121 ft
11-3.   170 ft
11-4.   131.964 in
11-5.   103.686 in
11-6.   71.988 in
11-7.   54 ft
11-8.   20.423 ft
11-9.   11,100 ft
11-10. 141.390 ft

11-11. 79.21 sq ft
11-12. 730 sq ft
11-13. 1337 sq ft
11-14. 1385.622 sq in
11-15. 855.301 sq in
11-16. 307.877 sq in
11-17. $809.10
11-18. 7.68 acres
11-19. 143 sq yd
11-20. $1254.57

CHAPTER 12

12-1.	prism	12-11.	24,127.488 cu units
12-2.	right prisms	12-12.	4224 cu units
12-3.	lateral	12-13.	2144.939 cu units
12-4.	32,193 cu units	12-14.	12.216 cu units
12-5.	98.56 cu units	12-15.	381.753 cu ft
12-6.	6-3/4 cu in	12-16.	9800 cu ft
12-7.	21,166.530 cu units	12-17.	.233 pounds
12-8.	1.352 cu units	12-18.	125.338 gallons
12-9.	1507.968 cu in	12-19.	8.182 cu in
12-10.	8042.496 cu units	12-20.	66.268 cu ft

CHAPTER 13

13-1.	kilo	13-11.	4.86
13-2.	centi	13-12.	5.5
13-3.	56.7	13-13.	2250 metric tons
13-4.	1.84	13-14.	22.2°
13-5.	4300000	13-15.	14°
13-6.	5.05ha	13-16.	170.18
13-7.	32.5ha	13-17.	54.89
13-8.	2.84	13-18.	$549.50
13-9.	8700	13-19.	$0.015/oz
13-10.	2718.76m^3	13-20.	33 mpg